BIOORGANIC CHEMISTRY

Volume III   Macro- and Multimolecular Systems

CONTRIBUTORS

HANS BROCKERHOFF

SIDNEY W. FOX

E. J. GABBAY

MURRAY GOODMAN

G. MARC LOUDON

J. W. LOWN

PAUL MELIUS

F. M. MENGER

MARVIN J. MILLER

FRED NAIDER

KOJI NAKANISHI

ALEXANDER L. NUSSBAUM

M. E. PARHAM

MORRIS J. ROBINS

ROBERT SHAPIRO

CLAUDIO TONIOLO

BIOORGANIC CHEMISTRY

Edited by
E. E. van Tamelen
Department of Chemistry
Stanford University
Stanford, California

Volume III
MACRO- AND MULTIMOLECULAR SYSTEMS

A treatise to supplement Bioorganic Chemistry:
An International Journal

Edited by
E. E. van Tamelen

ACADEMIC PRESS   New York   San Francisco   London   1977

A Subsidiary of Harcourt Brace Jovanovich, Publishers

CHEMISTRY

√ 6159- 7983

ACADEMIC PRESS, INC.
111 Fifth Avenue, New York, New York 10003

United Kingdom Edition published by
ACADEMIC PRESS, INC. (LONDON) LTD.
24/28 Oval Road, London NW1

Library of Congress Cataloging in Publication Data

Main entry under title:

Bioorganic chemistry.

    Includes bibliographies and index.
    CONTENTS: v. 1.   Enzyme action.–v. 2.   Substrate
behavior.–v. 3.   Macro- and multimolecular systems.
    1.   Biological chemistry.   2.   Chemistry, Organic.
I.   Van Tamelen, Eugene E.,   Date   [DNLM: 1.   Bio-
chemistry.   2.   Chemistry, Organic.   QU4 B61597]
QP514.2.B58   574.1'92   76-45994
ISBN 0–12–714303–3

Contents

List of Contributors   ix

Foreword   xi

Preface   xiii

Chapter 1   Molecular Designs of Membrane Lipids
HANS BROCKERHOFF
Introduction   1
Membrane Stratification   3
The Hydrophobic Core   5
The Hydrogen Belt   11
The Polar Zones   16
Conclusion   18
References   18

Chapter 2   Bioorganic Chemistry and the Emergence of the First
Cell
SIDNEY W. FOX
Introduction   21
Validity and Plausibility of Results   22
The Small Molecules   25
The First Informational Macromolecules   25
The Protocell   26
Photoreactive Proteinoid   29
Step (4) in the Flowsheet   30
Epilogue on Constructionism   30
References   31

Chapter 3   Interaction Specificities of Deoxyribonucleic Acid and
                the "Octopus" Model for Lac Repressor—Lac Operator
                Recognition and Chromatin Structure

                    E. J. GABBAY
                Introduction   33
                The Reporter Molecules   34
                Protein–DNA Interactions   50
                References   67

Chapter 4   Approaches to Carboxyl-Terminal Sequencing and
                End-Group Determinations in Peptides

                    G. MARC LOUDON, M. E. PARHAM, AND MARVIN J.
                    MILLER
                Introduction   71
                Enzymatic Degradations   73
                "Back Door" Methods   73
                Cyclization and Cleavage of a Carboxyl-Terminal Acyl Derivative 74
                Exclusive Labeling of the Carboxyl-Terminal Residue   75
                One-Electron Methods   79
                Migration to Electron-Deficient Nitrogen   80
                A New Carboxyl-Terminal End-Group Procedure   81
                Determination of β-Aspartyl and γ-Glutamyl Linkages in Proteins 87
                Degradation of Peptides on Solid Supports via a C-Terminal Azide 89
                Extension of the Solid-Phase Procedure to a Sequential Method 90
                Concluding Remarks   91
                References   92

Chapter 5   Interactions of Selected Antitumor Antibiotics with
                Nucleic Acids

                    J. W. LOWN
                Introduction   95
                Aminoquinone-Containing Antitumor Antibiotics   99
                Conclusions   118
                References   119

Chapter 6   Composition and Structure of Thermal Condensation
                Polymers of Amino Acids

                    PAUL MELIUS
                Introduction   123
                Effect of Amino Acid Composition of Reactants on Amino Acid Content
                    of Proteinoids   125
                Molecular Size of Proteinoids and Related Amino Acid Composition 130
                N- and C-Terminal Group Composition of Thermal Polymers   131
                Conclusion   134
                References   135

Chapter 7   The Bioorganic Chemistry of Aggregated Molecules
            F. M. MENGER
            Introduction   137
            Physical Properties of Micelles   139
            Examples of Micellar Chemistry   145
            References   151

Chapter 8   Recent Studies on Bioactive Compounds
            KOJI NAKANISHI
            Q*-Nucleosides   153
            An *in Vivo* Reaction Product of Benz[*a*]Pyrene with Bovine Bronchial
              Mucosa   157
            The β-Carbolines Isolated from Hydrolysis of Human Cataractous Lens
              Proteins   164
            An Antisickling and Desickling Agent: DBA   166
            Gonyautoxins-II and -III from the East Coast Toxic Dinoflagellate   169
            References   174

Chapter 9   Conformational Analysis of Oligopeptides by Spectral
            Techniques
            FRED NAIDER AND MURRAY GOODMAN
            Abbreviations   177
            Introduction   178
            Synthesis of Oligopeptides   179
            Optical Activity Measurements   181
            Ultraviolet Spectroscopy   183
            Circular Dichroism Investigations   185
            Nuclear Magnetic Resonance   190
            Studies of Oligopeptides Containing More than One Amino Acid Residue   194
            Conclusions and Plans for Future Studies   195
            References   199

Chapter 10   Chemical Synthesis of DNA Fragments: Some Recent
             Developments
             ALEXANDER L. NUSSBAUM
             Introduction   203
             Requirements for Repressor Binding   204
             Recognition of Foreign DNA   206
             Artificial Primers for DNA Sequencing   208
             Transfer RNA Genes   210
             Genes for Polypeptides   214
             Addendum   218
             References   219

Chapter 11   Synthetic Transformations of Naturally Occurring
               Nucleosides and Nucleoside Antibiotics
                    MORRIS J. ROBINS
               Text   221
               References   242

Chapter 12   Chemical Reactions of Nucleic Acids
                    ROBERT SHAPIRO
               Introduction   245
               Chemical Reactions of Nucleic Acids as Models for Events in Living Cells   246
               Chemical Reactions of Nucleic Acids in Studies of Their Structure and
                 Function   247
               Examples of Chemical Modification   249
               Conclusion   262
               References   262

Chapter 13   Bioorganic Stereochemistry: Unusual Peptide
               Structures
                    CLAUDIO TONIOLO
               Introduction   265
               The Intermolecularly H-Bonded Parallel-Chain Pleated-Sheet β Structure   267
               The $2 \rightarrow 2$ and $2 \rightarrow 3$ Intramolecularly H-Bonded Peptide Conformations   270
               The $3 \rightarrow 1$ Intramolecularly H-Bonded Peptide Conformations   274
               The $4 \rightarrow 1$ Intramolecularly H-Bonded Nonhelical *cis*-Peptide
                 Conformation   276
               The Oxy Analogs of the $3 \rightarrow 1$ and $4 \rightarrow 1$ Intramolecularly H-Bonded
                 Peptide Conformations   278
               The Intramolecularly Bifurcated H-Bonded Peptide Conformations   282
               cis-Amide Groups in Peptide Structures   284
               Addendum   286
               References   287

Index   293

List of Contributors

Numbers in parentheses indicate the pages on which the authors' contributions begin.

HANS BROCKERHOFF (1), Neurochemistry Department, Institute for Basic Research in Mental Retardation, Staten Island, New York

SIDNEY W. FOX (21), Institute for Molecular and Cellular Evolution and Department of Chemistry, University of Miami, Coral Gables, Florida

E. J. GABBAY (33), Department of Chemistry, University of Florida, Gainesville, Florida

MURRAY GOODMAN (177), University of California, San Diego, La Jolla, California

G. MARC LOUDON (71), The Spencer Olin Laboratory of Chemistry, Cornell University, Ithaca, New York

J. W. LOWN (95), Department of Chemistry, University of Alberta, Edmonton, Alberta, Canada

PAUL MELIUS (123), Department of Chemistry, Auburn University, Auburn, Alabama

F. M. MENGER (137), Department of Chemistry, Emory University, Atlanta, Georgia

MARVIN J. MILLER (71), The Spencer Olin Laboratory of Chemistry, Cornell University, Ithaca, New York

FRED NAIDER (177), Richmond College, City University of New York, Staten Island, New York

KOJI NAKANISHI (153), Havemeyer Hall, Columbia University, New York, New York

ALEXANDER L. NUSSBAUM* (203), Chemical Research Department, Hoffman-La Roche Inc., Nutley, New Jersey

M. E. PARHAM (71), The Spencer Olin Laboratory of Chemistry, Cornell University, Ithaca, New York

MORRIS J. ROBINS (221), Department of Chemistry, University of Alberta, Edmonton, Alberta, Canada

ROBERT SHAPIRO (245), Department of Chemistry, New York University, New York, New York

CLAUDIO TONIOLO (265), Institute of Organic Chemistry, University of Padua, and Center for the Study of Biopolymers, C.N.R., Padua, Italy

*Present address: Boston Biomedical Research Institute, Boston, Massachusetts.

Foreword

What is bioorganic chemistry? It is the field of research in which organic chemists interested in natural product chemistry interact with biochemistry. For many decades the natural product chemist has been concerned with the way in which Nature makes organic molecules. In the absence of any information other than that provided by structure, conclusions had necessarily to be derived from structural analysis. Broad groups of natural products could be recognized, such as alkaloids, isoprenoids, and polyketides (acetogenins), which clearly had elements of structure indicating a common biosynthetic origin. Indeed, for alkaloids and terpenoids, structural work was greatly helped by such biogenetic hypothesis. Similarly, after A. J. Birch had made an extensive analysis of polyketides, the repeating structural element postulated also helped in the determination of structure.

The alternative, and complement, to the above analysis is to consider the chemical mechanisms whereby the units of structure are assembled into the final natural product. For example, alkaloid structure can often be analyzed in terms of anion–carbonium ion combination. Also, the later stages of biosynthesis of many alkaloids can be analyzed by the concept of phenolate radical coupling. In polyisoprenoids the critical mechanism for carbon–carbon bond formation is the carbonium ion–olefin interaction to give a carbon–carbon bond and regenerate a further carbonium ion.

The analysis of natural product structures in terms of either structural units or mechanisms of bond formation has been subjected to rigorous tests since radioactively labeled compounds became generally available. It is gratifying that, on the whole, the theories developed from structural and mechanistic analysis have been fully confirmed by *in vivo* experiments.

Organic chemists have always been fascinated by the possibility of imitating in the laboratory, but without the use of enzymes, the precise steps of a biosynthetic pathway. Such work may be called biogenetic-type, or biomimetic, synthesis. This type of synthesis is a proper activity for the bioorganic chemist and undoubtedly deserves much attention. Nearly all such efforts are, however, much less successful than Nature's synthetic activities using enzymes. It is well appreciated that Nature has solved the outstanding problem of synthetic chemistry, viz., how to obtain 100% yield and complete stereospecificity in a chemical synthesis. It, therefore, remains a major task for bioorganic chemists to understand the mechanism of enzyme action and the precise reason why an enzyme is so efficient. We are still far from the day when we can construct an organic molecule which will be as efficient a catalyst as an enzyme but which will not be based on the conventional polypeptide chain.

Much of contemporary bioorganic chemistry is presented in these volumes. It will be seen that much progress has been made, especially in the last two decades, but that there are still many fundamental problems left of great intellectual challenge and practical importance.

The world community of natural product chemists and biochemists will be grateful to the editor and to all the authors for the effort that they have expended to make this work an outstanding success.

DEREK BARTON
Chemistry Department
Imperial College of Science and Technology
London, England

Preface

Although natural scientists have always been concerned with the development and behavior of living systems, only in the twentieth century have investigators been in a position to study on a molecular level the intimate behavior of organic entities in biological environments. By midcentury, the form and function of various natural products were being defined, and complex biosynthetic reactions were even being simulated in the nonenzymatic laboratory. As the cinematographic focus on biomolecules sharpened, one heard increasingly the adjective *bioorganic* applied to the interdisciplinary area into which such activity falls.

In 1971, publication of a new journal, *Bioorganic Chemistry*, was begun. As a follow-up, what could be more timely and useful than a well-planned, multivolume collection of bioorganic review articles, solicited from carefully chosen professionals, surveying the entire field from all possible vantage points? This four-volume work contains a collection, but it did not originate in this manner.

As the journal *Bioorganic Chemistry* developed, the number and quality of regular, original research articles were maintained at an acceptable level. However, comprehensive review articles appeared only sporadically, despite their intrinsic value at a time when general interest in bioorganic chemistry was burgeoning. In order to enhance this function of the journal, as well as to mark the fifth anniversary of its birth, we originally planned to publish in 1976 a special issue comprised entirely of reviews by active practitioners. After contact with a handful of stalwart bioorganic chemists, about two hundred written invitations for reviews were mailed during late 1975 to appropriate, diverse scientists throughout

the world. The response was overwhelming! More than seventy preliminary acceptances were received within a few months, and it soon became
evident that the volume could not be handled adequately through publication by journal means. After consultation with representatives of Academic
Press, we agreed to publish the manuscripts in book form.

Although the stringency of journal deadlines disappeared, the weightier
matter of editorial treatment had to be reconsidered. Should contributions
be published in the same, piecemeal, random fashion as received? Such
practice would be acceptable for journal dissemination, but for book
purposes, broader, more orderly, and inclusive treatment might be desirable and also expected. Partly because of editorial indolence, but mostly
because of a predilection for maintaining the candor and spontaneity
which might be lost with increased editorial control, we decided not to
attempt coverage of all identifiable areas of bioorganic chemistry, not to
seek out preferentially the recognized leaders in particular areas, and
even not to utilize outside referees. Consequently, we present reviews
composed by scientists who were not coerced or pressured, but who
wrote freely on subjects they wanted to write about and treated them as
they wanted to, at the cost perhaps of a certain amount of objectivity and
restraint as well as proper coverage of some important bioorganic areas.

We turn now to the results of this publication project. Because of the
inevitable attrition for the usual reasons, fewer than the promised number
of reviews materialized: fifty-seven manuscripts were received in good
time and accepted by this office. Eight countries are represented by the
entire collection, which emanates almost entirely from academia, as
would be expected. A great variety of topics congregated—greater than
we had foreseen. Inclusion of all papers in one volume was impractical,
and thus the problem arose of logically dividing the heterogeneous material into several unified subsections, each suitable for one volume, a
problem compounded by the fact that an occasional author elected to
treat, in one manuscript, several unconnected topics happening to fall in
his purview. Therefore, perfect classification without discarding or dis-
secting bodies of material as received was simply not possible.

After some reflection and a few misconceptions, we evolved a plan for
division into four more or less scientifically integral sections; these, hap-
pily, also constitute approximately equal volumes of written material, an
aspect of some importance to the publisher. The enzyme–substrate interaction was expected to be a well-represented subject, and, in fact, too
many manuscripts on this subject for one proportionally sized volume
were received. Although the separation of enzyme action and substrate
behavior is contrived and not basically justifiable, it turned out that, for
the most part, a group of authors heavily emphasized the former, while
another concentrated on the latter. Accordingly, Volume I was entitled

"Enzyme Action," and Volume II "Substrate Behavior." Admittedly, in a few cases, articles could be considered appropriate for either volume.

A gratifyingly significant number of contributions dealt with the behavior of biologically important polymers and related matters, sent in by authors having quite different investigational approaches. In addition, several discourses were concerned with molecular aggregates, e.g., micelles. All of these were incorporated into Volume III, "Macro- and Multimolecular Systems."

Whatever papers did not belong in Volumes I–III were combined and constitute Volume IV. Fortunately, in these remaining papers some elements of unity could be discerned; in fact, their entire content falls into the following categories: "Electron Transfer and Energy Conversion (photosynthesis, porphyrins, NAD^+, cytochromes); Cofactors (coenzymes, NAD^+, metal ions); Probes (cytokinin behavior, steroid hormone action, peptidyl transferase reactivity)."

Finally, early in this enterprise, we asked Derek Barton to compose a Foreword. Sir Derek complied graciously, and in every volume his personalized view on the nature of bioorganic chemistry appears.

E. E. VAN TAMELEN

Contents of Other Volumes

VOLUME I Enzyme Action

1. Direct Observation of Transient ES Complexes: Implications to Enzyme Mechanisms
   D. S. AULD

2. Models of Hydrolytic Enzymes
   MYRON L. BENDER AND MAKOTO KOMIYAMA

3. Aldol-Type Reactions and 2-Keto-4-hydroxyglutarate Aldolase
   EUGENE E. DEKKER

4. The Hydroxylation of Alkanes
   N. C. DENO, ELIZABETH J. JEDZINIAK, LAUREN A. MESSER, AND EDWARD S. TOMEZSKO

5. Intramolecular Nucleophilic Attack on Esters and Amides
   THOMAS H. FIFE

6. Chemical Aspects of Lipoxygenase Reactions
   MORTON J. GIBIAN AND RONALD A. GALAWAY

7. The Multiplicity of the Catalytic Groups in the Active Sites of Some Hydrolytic Enzymes
   E. T. KAISER AND Y. NAKAGAWA

8. Multifunctionality and Microenvironments in the Catalytic Hydrolysis of Phenyl Esters
    TOYOKI KUNITAKE

9. The Interplay of Theory and Experiment in Bioorganic Chemistry: Three Case Histories
    G. M. MAGGIORA AND R. L. SCHOWEN

10. Enzymatic Olefin Alkylation Reactions
    ROBERT M. MCGRATH

11. Enzymatic Catalysis of Decarboxylation
    MARION H. O'LEARY

12. The Catalytic Mechanism of Thymidylate Synthetase
    ALFONSO L. POGOLOTTI, JR., AND DANIEL V. SANTI

13. Studies of Amine Catalysis via Iminium Ion Formation
    THOMAS A. SPENCER

14. Fatty Acid Synthetase Complexes
    JAMES K. STOOPS, MICHAEL J. ARSLANIAN, JOHN H. CHALMERS, JR., V. C. JOSHI, AND SALIH J. WAKIL

15. Sulfane-Transfer Catalysis by Enzymes
    JOHN WESTLEY

VOLUME II   Substrate Behavior

1. Studies in Sesquiterpene Biogenesis: Implications of Absolute Configuration, New Structural Types, and Efficient Chemical Simulation of Pathways
    NIELS H. ANDERSEN, YOSHIMOTO OHTA, AND DANIEL D. SYRDAL

2. Mechanisms for Proton Transfer in Carbonyl and Acyl Group Reactions
    RONALD E. BARNETT

3. The Withanolides—A Group of Natural Steroids
    ERWIN GLOTTER, ISAAC KIRSON, DAVID LAVIE, AND ARIEH ABRAHAM

4. Novel Piperidine Alkaloids from the Fungus *Rhizoctonia leguminocola*: Characterization, Biosynthesis, Bioactivation, and Related Studies
    F. PETER GUENGERICH AND HARRY P. BROQUIST

5. Enzymatic Stereospecificity at Prochiral Centers of Amino Acids
    RICHARD K. HILL

6. Bioformation and Biotransformation of Isoquinoline Alkaloids and
    Related Compounds
    TETSUJI KAMETANI, KEIICHIRO FUKUMOTO, AND MASATAKA
    IHARA

7. Carbanions as Substrates in Biological Oxidation Reactions
    DANIEL J. KOSMAN

8. Synthetic Studies in Indole Alkaloids: Biogenetic Considerations
    JAMES P. KUTNEY

9. Mechanisms of *cis-trans* Isomerization of Unsaturated Fatty Acids
    WALTER G. NIEHAUS, JR.

10. The Synthesis and Metabolism of Chirally Labeled α-Amino Acids
    RONALD J. PARRY

11. Reactions of Sulfur Nucleophiles with Halogenated Pyrimidines
    EUGENE G. SANDER

12. Vitamin D: Chemistry and Biochemistry of a New Hormonal System
    H. K. SCHNOES AND H. F. DELUCA

13. Recent Structural Investigations of Diterpenes
    JAMES D. WHITE AND PERCY S. MANCHAND

    Index

**VOLUME IV   Electron Transfer and Energy Conversion; Cofactors;
Probes**

1. Photosynthetic Phosphorylation: Conversion of Sunlight into
    Biochemical Energy
    DANIEL I. ARNON

2. Oxidation and Oxygen Activation by Heme Proteins
    C. K. CHANG AND D. DOLPHIN

3. Affinity Labeling Studies on *Escherichia coli* Ribosomes
    BARRY S. COOPERMAN

4. Mechanisms of Electron Transfer by High Potential *c*-Type Cyto-
    chromes
    MICHAEL A. CUSANOVICH

5. Structural and Mechanistic Aspects of Catalysis by Thiamin
   ANTHONY A. GALLO, JOHN J. MIEYAL, AND HENRY Z. SABLE

6. Cytokinin Antagonists: Regulation of the Growth of Plant and Animal Cells
   SIDNEY M. HECHT

7. Specific Chemical Probes for Elucidating the Mechanism of Steroid Hormone Action: Progress Using Estrogen Photoaffinity Labeling Agents
   JOHN A. KATZENELLENBOGEN, HOWARD J. JOHNSON, JR., HARVEY N. MYERS, KATHRYN E. CARLSON, AND ROBERT J. KEMPTON

8. The Redox Chemistry of 1,4-Dihydronicotinic Acid Derivatives
   D. A. WIDDOWSON AND R. J. KILL

9. Models for the Role of Magnesium Ion in Enzymatic Catalysis, Phosphate Transfer, and Enolate Formation
   RONALD KLUGER

10. Some Problems in Biophysical Organic Chemistry
    EDWARD M. KOSOWER

11. Photoredox Reactions of Porphyrins and the Origins of Photosynthesis
    D. MAUZERALL

12. Mechanisms of Enzymelike Reactions Involving Human Hemoglobin
    JOHN J. MIEYAL

13. Interactions of Transition-Metal Ions with Amino Acids, Oligopeptides, and Related Compounds
    AKITSUGU NAKAHARA, OSAMU YAMAUCHI, AND YASUO NAKAO

14. Nonenzymatic Dihydronicotinamide Reductions as Probes for the Mechanism of NAD^+-Dependent Dehydrogenases
    DAVID S. SIGMAN, JOSEPH HAJDU, AND DONALD J. CREIGHTON

15. The Use of Puromycin Analogs and Related Compounds to Probe the Active Center of Peptidyl Transferase on *Escherichia coli* Ribosomes
    ROBERT H. SYMONS, RAYMOND J. HARRIS, PHILIP GREENWELL, DAVID J. ECKERMANN, AND ELIO F. VANIN

16. Hemoprotein Oxygen Transport: Models and Mechanisms
T. G. TAYLOR

Index

CHAPTER

1

Molecular Designs of Membrane Lipids

Hans Brockerhoff

INTRODUCTION

The lipid bilayer model of biological membranes (Fig. 1a), first proposed in 1925 [17], has now found general acceptance, and the question as to how proteins participate in membrane structure appears to be settled in favor of a liquid-mosaic model (Fig. 1b): Single or aggregated protein molecules float on one or both surfaces of the bilayer, with hydrophobic keels reaching into or through the lipid core [32]. The most common structure of the peptide keels is probably that of the α helix [31]. The lipid molecules are polar, with long aliphatic tails providing cohesion by "hydrophobic bonding" and hydrophilic heads for stabilization at the lipid–water interface. They must be in the liquid-crystalline mesophase, i.e., the aliphatic chain region must be molten, in order to permit lateral diffusion of membrane constituents and to give the membrane plasticity. The lipid matrix must function, first, as a wall between the cell and the outside or between organelles and cytoplasm, a wall more or less impenetrable to ions and dissolved molecules, and, second, as a frame for the membrane proteins that control the traffic of metabolites through the cell and its compartments. Lipid membranes probably also serve as a support for many enzymes of intermediary metabolism. The molecular structures of membrane lipids must be attuned to these functions.

Unfortunately, the abstract appearance of the model of Fig. 1 is the result not of pedagogic intention but of our ignorance. The polar lipids of the membrane matrix consist of an immense variety of molecular structures whose biogenesis is reasonably well understood but whose role in the mem-

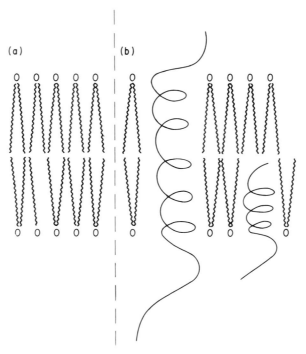

Fig. 1 Lipid bilayer membrane (a) with inserted membrane proteins (b).

brane remains mysterious. For example, no satisfactory explanation has yet been given for the need for two distinct major classes of polar lipids, glycero-lipids and sphingolipids, not to mention their numerous subclasses and species. Questions regarding such problems are usually answered by a description of the biosynthetic origins of the compounds. In this review, a fundamentally different approach will be taken. I shall try to interpret the structural features of membrane lipids under the assumption that each such feature contributes to making the lipid the best possible building block for the best of all possible membranes. This method can be justified by the argument that evolution would long ago have discarded any lipid that were not perfectly suited for its membrane, just as the membrane must be perfectly suited for its cell or organelle. The value of the teleological method lies, however, in its heuristic nature. In the course of this review we shall encounter many open questions which, if answered, would not only help to complete our theoretical under-standing but would tell pharmacochemists how to tailor drugs so that they will enter into specific membranes, stay in them, or penetrate through them.

MEMBRANE STRATIFICATION

The phospholipids and glycolipids of biological membranes have a tripartite structure: a hydrophobic aliphatic double chain; a hydrophilic head group of carbohydrate or phosphate esters, and the region where these two moieties are linked (Fig. 2). The lipid bilayers of membranes are accordingly stratified as shown in Fig. 3. The properties and function of the hydrophobic core, which also contains the bulk of the cholesterol molecule, have been extensively studied during the past several years, with the result that the concept of membrane fluidity has emerged as all-important. About the function of the polar elements very little is known, although some information on their distribution has become available: in erythrocyte membranes [40], endoplasmic reticulum membranes [11], and perhaps membranes in general, uncharged groups (the phosphorylcholine zwitterion or carbohydrate) occupy preferentially the outer or noncytosol side of the bilayer. The intermediate membrane region, the interface between the anhydrous and hydrated membrane layers, has as yet received very little attention. I call it the hydrogen

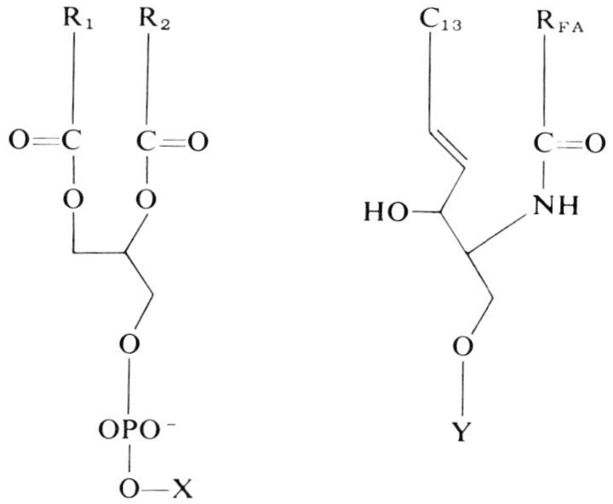

Phosphatidyl—X   N-Acylsphingosyl—Y

Fig. 2   Major phosphoglycerides and sphingolipids of mammals.

X,   Choline, ethanolamine, serine, inositol, phosphatidylglycerol

Y,   Phosphorylcholine, galactose, galactose 3-sulfate, polyglycosides

R_1,   Fatty acid chain, usually saturated; in plasmalogens, R_1—C—O is replaced by

R—C=C—O

R_2,   Fatty acid chain, most often unsaturated

R_{FA}, Fatty acid chain; in hydroxycerebroside, R—C—C=O

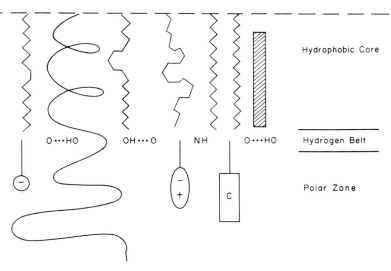

Fig. 3 Stratification of a biological membrane, one-half of the membrane shown. Zigzag lines represent aliphatic chains; the larger kinks show *cis* unsaturation. The shaded rectangle denotes cholesterol, and the looped line an intrinsic protein; C is carbohydrate.

belt [5] because it is composed of hydrogen bond acceptors (ester and amide groups) and hydrogen bond donors (hydroxyl groups of cholesterol, sphingosine, and α-hydroxy acids, NH groups, water, and perhaps also hydroxy groups of proteins).

In the following presentation, each stratum with its components is treated separately. The precedent for this procedure is set by nature. The various elements of the three regions of the polar lipids occur in almost all combinations, as if supplied independently in response to the higher demands of the membrane. Admittedly, there are preferred combinations; for example, polyunsaturation in the hydrophobic core tends to be linked to ester rather than amide groups in the hydrogen belt, and amide groups, at least in mammalian membranes, may link to phosphorylcholine but not to other phosphate esters. In general, however, the fabric of each stratum of a membrane seems to have been tailored independently. This becomes especially apparent when the strata of different membranes are compared. For example, the outer half-membranes of the erythrocytes of cow and pig differ drastically in their hydrogen belts (which contain no ester groups in the cow), whereas the hydrophobic core and polar zone are very similar in composition [25]. As another example, the hydrophobic cores of membranes respond to dietary manipulations; the other two regions do not [36].

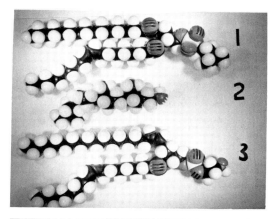

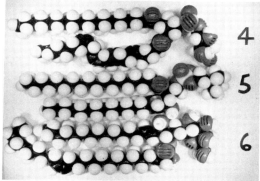

Fig. 4   Molecular models of membrane lipids. (1) phosphatidylcholine; (2) cholesterol; (3) plasmalogen (phosphatidalethanolamine); (4) phosphatidylinositol; (5) sphingomyelin; (6) hydroxycerebroside [galactosyl-N-(α-hydroxyacyl)sphingosine].

The molecular models of Fig. 4 represent some fairly typical, though not necessarily the most common, lipid species out of a legion, and although the molecules are lined up as they might be in a half-membrane they may perhaps never present themselves in these conformations and in this combination. They are intended as a visual reference for the following discussion. In the membrane, they may be as closely packed as in the lower half of Fig. 4.

THE HYDROPHOBIC CORE

Double Chains

Even the simplest bilayer model (Fig. 1a) poses some questions: Why are there no double-headed double chains with hydrophilic groups at both ends, and why are there double rather than single chains? An obvious answer to the first question is that there are no biosynthetic systems capable of performing the awesome chemistry necessary for the task, although it is also true that the two-headed lipids would not permit an independent diffusion of molecules in the membrane halves or their independent turnover. The second question, why double chains, can be answered in terms of design. The obligatory strongly hydrophilic groups, because they must contain several oxygens, always have cross areas larger than the 20 $Å^2$ occupied by a single hydrocarbon chain. Therefore, the hydrophobic part of the molecule must contain at least two chains to make packing in a planar sheet possible. Predictably, monochain lipids such as lysophosphatidylcholine (1-acylglycerol-3-phosphorylcholine) perturb and disintegrate membranes, and it has been proposed that such lysolipids are operative in membrane fusion [22]. The suggestion has also been made that the free hydroxyl group in position 2 of the glycerol of lysolipids may be the primary perturbing agent [5]. The question could be decided by experiments, which would also inform us of the role of lysolipids in membranes, if indeed they have a role apart from that of being intermediates of lipid metabolism.

Effective Hydrophobic and Geometric Chain Lengths

The most common carbon number of fatty acid chains is 18, which in a straight *all-trans* conformation would give the C_3–C_{18} hydrophobic region (see the next paragraph) a length of approximately 20 Å, with a total span of the hydrophobic membrane core of 40 Å. This distance is in fact approached in the myelin membrane, which contains much saturated *all-trans* fatty acid of an average chain length approaching C_{20}; X-ray diffraction studies suggest a hydrocarbon depth of around 40 Å [14]. Normally, hydrophobic cores are shallower, perhaps 30 Å or less, because not all CH_2—CH_2 conformations are *trans* [35] and because the *cis* configurations of the double bonds of unsaturated acids shorten the effective chain lengths. For example, the C_{20} tetraenoic acid of the phosphatidylinositol phosphate in Fig. 4, shown in an arbitrary but plausible conformation, appears to be markedly shorter than the neighboring C_{18} saturated acid.

The hydrophobic core consisting of aliphatic chains (together with cholesterol) is, of course, the heart and essence of the membrane, a liquid hydrocarbon barrier that not only blocks the uncontrolled entering and leaving of the compartment by ions and metabolites but also holds the enzymes or carrier proteins that govern the controlled transport of these solutes. The cohesion of the hydrocarbon residues with each other and the intrinsic peptides is traceable to the change of the intermolecular organization that water undergoes when in contact with hydrocarbon. This change appears thermodynamically as a loss in free energy of 0.8 Cal per mole CH_2 group when a hydrocarbon is transferred from water to a nonpolar environment [33]. The expression "hydrophobic bonding" is commonly used for this effect. Measurements show that a carbon which carries oxygen, and also its nearest neighbor, do not contribute to the effective hydrophobic chain length and that a double bond reduces this length by the equivalent of one CH_2 group [33]. Experience with chromatographic technology which relies on the partition of fatty acids between polar and nonpolar phases suggests that the antihydrophobic effect of a double bond may be larger, approximately equaling the loss of a CH_2–CH_2 segment. A C_{20} tetraenoic acid, therefore, has an effective hydrophobic length of C_{10}–C_{14}; the common C_{18} dienoic (linoleic) acid, C_{12}–C_{14}; the saturated palmitic acid, C_{16}, length C_{14}; stearic acid, C_{18}, length C_{16}. It appears that effective hydrophobic chain length and effective geometric chain length are roughly equal. The effective hydrophobic core span, then, is perhaps C_{24}–C_{30}, or 31–39 Å. This thickness must be sufficient to prevent any appreciable diffusion of cations or small metabolites. Permeation studies with artificial lipid membranes show that this is in fact true, especially if the membrane contains cholesterol besides phospholipids. Thinner membranes, such as the bilayer formed by dilauryl (C_{12}) phosphatidylcholine (hydrophobic span C_{20}), do not retain cations [4]; at even shorter hydrophobic chain length, lipid micelles rather than bilayers are formed.

The hydrophobic membrane core must also be ideally suited to accept the hydrophobic keels of intrinsic proteins (although, of course, proteins may have evolved to fit the lipid matrix rather than vice versa). Since the raise of a peptide α helix is about 1.5 Å per residue [28], such keels might contain around 20 hydrophobic amino acids. Glycophorin, a membrane-spanning protein of erythrocytes, has a hydrophobic middle section of about this length [30].

Double Bonds

DEEP-CORE UNSATURATION

The most common unsaturated acids have a double bond in their middle, most often at position 9–10. Other double bonds may occur toward the end of the chain, for example in 12–13, and also, especially in plants and poikilothermic animals, in 15–16. Such unsaturation resides in the interior of the hydrophobic core of a membrane. In animals, additional double bonds can be introduced into the di- and trienoic acids in the direction toward the carboxylic head; this results in unsaturation in the outer, shallower regions of the core. A good case can, I think, be made for the proposal that these two types of double bonds have different functions.

Deep-core unsaturation clearly is needed to give membranes the fluidity that makes lateral diffusion of both lipids and proteins possible. Saturated lipids of the required composition and chain length are crystalline at physiological conditions, even in their hydrated form. For example, the transition temperature of hydrated distearoyl phosphatidylcholine is 58°C; that of the dioleoyl analog, however, is only $-22°C$ [7]. An average of two double bonds per molecule can thus be expected to keep any membrane from crystallizing under the harshest conditions. Double bonds of the *cis* configuration are required because only these, not *trans* bonds, force the kinks on the chains that hinder close van der Waals packing and thus produce fluidity. Methylene (CH_2) groups are usually interposed between the *cis* double bonds; conjugated *cis* double bonds would impose too much rigidity on the chain since the rotational freedom in the six-carbon $C—C=C—C=C—C$ segment is limited to the middle single bond.

Deep-core unsaturation (and shallow-core unsaturation as well) is almost always confined to the fatty acid in position 2 of the glycerol of phospholipids. (Exceptions are found, for example, in the galactoglycerides of plants and the diphosphatidylglycerol of mitochondria; both contain linoleic acid in positions 1 and 2 [23].) The reason for the nonrandom distribution is not readily obvious. Numerous studies have been undertaken to find the cause of the phenomenon, all of them revolving around the specificity of the enzymes at this or that level of biosynthesis of the lipids [34]; here, however, we are asking for the purpose, not the cause. Perhaps the membrane, by insisting on microheterogeneity on the intramolecular level, secures the intermolecular disorder necessary for fluidity. As an illustration, a membrane consisting of a 1-palmitoyl-2-linoleoyl phosphoglyceride would be homogeneous and liquid; if, however, the fatty acids were randomly distributed, 25% of the lipid would be the saturated dipalmitoyl species which might aggregate and establish crystalline domains in the membrane. Phase separation effects have, in fact, been demonstrated in monolayer and in freeze-etching experiments [9,39]. A phospholipid with two polyenoic acids, on the other hand, might be too obstructive or hydrophilic to be held in the membrane. One saturated leg to stand on may be obligatory; the phosphatidylinositol model of Fig. 4 (4) may serve as an illustration.

Why is it always position 2 that carries the unsaturation? There have as yet been no indications that this specific distribution pattern is important for the overall structure of biological membranes, but it cannot therefore be assumed that it is merely a biosynthetic accident, important in fatty acid metabolism but immaterial for the functioning of membranes. It should rather be assumed that the effects of this distribution on the membrane have so far eluded detection. It is probably significant that the fatty acid distribution in some neoplasms has become randomized [3].

SHALLOW-CORE UNSATURATION AND WATER HOLES

The C_{20} tetraenoic (arachidonic) acid of the phosphatidylinositol model [Fig. 4 (4)] has its first double bond in position 5–6; the next most common of highly unsaturated acids, C_{22} hexaenoic acid, has a double bond in 4–5.

Arachidonic acid is a precursor of prostaglandin, but its abundant store in membrane lipids far exceeds whatever demand could be put to it by the synthesis of these compounds. Undoubtedly, the additional double bonds of these acids contribute to membrane fluidity; however, as we have seen, about two double bonds per lipid molecule would suffice for this purpose. An effect quite different from that of deep-core unsaturation is suggested in the phosphatidylinositol model. Shallow double bonds open the hydrocarbon core and the hydrogen belt toward the polar zone; they create water holes in the half-membrane. The new *cis* kinks make the chain too obstructive for van der Waals bonding to other lipids in the carboxyl-proximal region; in thermodynamic terms, this region becomes too hydrophilic.

Water holes need not be imagined as cavities that reach deep into the hydrophobic core, although the model [Fig. 4 (4)] appears to show that they do. It is the hydrogen belt that is actually disrupted because neighboring components come apart; the hydrophobic core becomes exposed to the aqueous zone but is not necessarily deeply penetrated by water.

What is the function of water holes? Perhaps they trap ions, such as calcium; this proposal goes well with the fact that the polyenoic acids are especially concentrated in lipids with an available anionic phosphate function: phosphatidylethanolamine, phosphatidylserine, and phosphatidylinositol. Phosphatidylinositol [Fig. 4 (4)] can be reversibly phosphorylated to phosphatidylinositol 4-phosphate and 4,5-diphosphate. Through such conversions, the inositol might be worked like a hinged lid on its proper water hole.

The negative aspects of shallow *cis* unsaturation can be speculated on with somewhat more confidence. First, bonding of cholesterol to the fatty acid is probably prevented. Second, the hydrophobic bonding of protein keels will be affected, probably negatively: a water-filled water hole may prevent lipid–protein bonding. If so, the residual lipids that are often so difficult to remove during the last stages of membrane protein isolation should contain much less highly unsaturated fatty acid than the same lipid class in the free lipid fraction. It has, in fact, been reported that the more saturated phospholipids bind more easily to mitochondrial proteins [12]. To extrinsic membrane proteins, which lack a penetrating hydrophobic keel, water holes may offer shallow berths that allow closer protein–membrane coupling.

Double Bonds of Cholesterol, Sphingosine, and Plasmalogens

The 5–6 double bond of cholesterol is introduced at a late stage in the biosynthesis of this lipid. Since it is not in the vicinity of a functional group, its purpose must be steric. In the precursors of cholesterol, the A and B rings are already *trans* linked, as they are in one of the two natural hydrogenation products of cholesterol, cholestanol, and these molecules have the same general contours as cholesterol. However, they are somewhat flatter, and the slightly more cylindrical shape of cholesterol may be advantageous for its function as a filler in the membrane. A comparison between molecular bond models (Dreiding) of cholesterol and cholestanol shows that there is a slight difference in the orientation of the hydroxyl group. This group is equatorial in both lipids but in cholesterol somewhat more so, if its conformation is related to the equatorial plane of the whole molecule. The difference is quite small, certainly less than 10°. In fact, any influence that the 5–6 double bond, as compared to a single bond, might exert on hydrophobic or hydrogen bonding in other regions of the molecule must be quite subtle.

The 4–5 double bond of the C_{18} amine diol, sphingosine [the lower chain in the sphingomyelin of Fig. 4 (5); the upper chain of the hydroxycerebroside of Fig. 4(6)], is introduced into this long-chain base as the last biosynthetic step. (The precursor, dihydrosphingosine, is also found in membrane lipids as a minor component.) The double bond is *trans* and therefore hardly alters the *all-trans* conformation that a saturated chain also would be expected to display. Sterically, little advantage seems to have been gained. In this case, a change in molecular reactivity induced by the double bond could be more important. The free OH group of sphingosine is in allyl position to the double bond; the mobility of electron clouds is increased in this system. I suggest that this makes the hydroxy group especially suited to serve as a hydrogen bond donor but also as an acceptor; the possible purpose of this mobility is discussed in the section on the hydrogen belt.

Plasmalogens [Fig. 4 (3)] have a *cis*-vinyl ether group, with an alkyl tail, linked to position 1 of glycerol. The 1'–2' double bond is introduced as a biosynthetic afterthought, the precursor being the completely assembled alkyl ether lipid. There is no reactive hydrogen bond donor–acceptor in the vicinity, only the weakly accepting ether —O—. Moreover, the double bond is *cis*. Its purpose is to be looked for in its geometry rather than its reactivity. The model [Fig. 4 (3, upper chain)] shows that the vinyl ether grouping may be exposed to the aqueous region and thus may serve a function in the hydrogen belt.

Cholesterol in the Hydrophobic Core

The fused rings, A–D, of cholesterol, with the A ring in the stable chair conformation, form a rigid, elongated disk. Figure 4 (2) shows how the molecule (which, in reality, would touch the neighboring phospholipid molecules) must be oriented and aligned in the membrane. If it were farther out into the polar region (to the right) or tilted, hydrophobic C—H groups would have to be hydrated; further inside, the terminal OH would become dehydrated. Both processes are disfavored by thermodynamics. In the model, the flexible side chain has been slightly (and arbitrarily) bent from the plane in order to show that flexibility starts where deep-core unsaturation begins. The cholesterol end chain seems to be especially tailored to wag about in the environment of the kinked ends of unsaturated acids.

The action of cholesterol in membranes has been studied intensively in model membrane systems. Many of the results can be explained with the rigid, hydrophobic nature of the molecule. Mixed with saturated phospholipids and water, cholesterol suppresses and, in equimolar mixture, abolishes the energy jump at the crystalline–liquid-crystalline transition point; it thus fluidizes membranes that might crystallize without it [7]. This effect is achieved by the insertion of the molecule between phospholipid molecules and the breaking of their crystalline lattice. Unsaturated membranes, on the other hand, are differently affected by cholesterol; their fluidity decreases. Cholesterol is again inserted between phospholipids, but this time it fills pits that are available because the kinked fatty acid end chains take more space than the saturated front ends; in contrast, in cholesterol the head is bulkier than the tail. The stiffening of the fatty acid front ends by cholesterol has been demonstrated by nuclear magnetic resonance [7] and electron magnetic resonance [18] methods.

When cholesterol and phosphoglyceride are mixed in a monolayer, a condensing effect ensues: The area covered is smaller than the sum of the areas covered by the components at equal surface pressure [21]. Related sterols also condense phosphoglyceride films if they have an end chain and a planar structure [10]. When incorporated into spherical phospholipid bilayers (liposomes) or into the membranes of some microorganisms, cholesterol suppresses the permeation by cations [1], small metabolites [2], and water [16]; again, the effect is shown by other sterols with an end chain and a planar structure [10]. The function of cholesterol in membranes, then, seems to be well understood. It plasticizes as well as seals the lipid bilayer.

THE HYDROGEN BELT

Hydrogen Bond Acceptors

The hydrophobic core of the membrane is sandwiched between two planes characterized by the accumulation of functional groups that can donate or accept hydrogen bonds. These hydrogen belts [5] contain no ionic or ionizable groups. The hydrogen bond acceptors are aprotic oxygens: the carbonyl $=O$ of the phosphoglyceride esters [Fig. 4 (1, 3, and 4)] and the carbonyl $=O$ of the sphingolipid amide [Fig. 4 (5 and 6)]. In Fig. 4, the carbonyl oxygens are marked by having the slots in the plastic shells in horizontal orientations; the ether oxygens of the ester groups are hidden. In pure phosphoglyceride bilayers, the carbonyl oxygens must be hydrated [5]. The negative free energy of an $=O \cdots H$ hydrogen bond is around 6 Cal/mole [28]; this should be compared with the hydrophobic bonding energy, which does not exceed 12 Cal/mole per fatty acid chain. Hydrogen bonds must be roughly linear; i.e., the bonds $O—H \cdots O$ must form a straight line. The $H \cdots O=$ angle may vary considerably [13]; the bond probably can enter anywhere in the area marked by the slots in the oxygen spheres. It should be recognized that the hydrogen bonds are not static and permanent but are continuously being broken and formed.

Figure 5 defines the geometry of the hydrogen belt. Aliphatic chains and the long axis of cholesterol are longitudinally oriented. The direction parallel to the membrane surface I call latitudinal. A latitudinal angle gives the deviation from this direction, positive toward the hydrophobic core, negative toward the aqueous phase. The $C=O$ carbonyls have an ideal latitudinal orientation if arranged as in Fig. 4 but may in reality have more negative latitudinal angles. At any angle, they are accessible to bonding from water protons, and, unless the negative angles become improbably large, they are also approachable by hydrogen belt protons derived from lipids, the most obvious potential donor being cholesterol. (In Fig. 4 lipid hydrogen atoms that can participate in hydrogen bonding are shown as flattened spheres, and the direction of the hydrogen bond is indicated by the small pin sticking out of these.) The obligatory hydration of the lipid oxygens in the hydrogen belt implies that the water molecules adjoining the belt must be oriented with an $O—H \cdots$ bond in more or less longitudinal direction, especially when the membrane is packed tightly enough to prevent water from penetrating into the belt. Such close packing becomes likely with increasing saturation or cholesterol content.

The phosphoglycerides containing a vinyl ether bond [plasmalogen, Fig. 4 (3)] or an ether bond (ether phosphoglycerides) possess only one $C=O$ group. Less hydration is possible, but the *cis* $C—C=C—O$ grouping is perhaps a

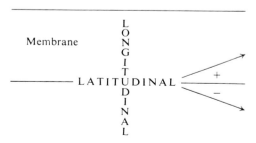

Fig. 5   Definition of latitudinal angles of bonds in the hydrogen belt.

reasonably strong hydrogen bond acceptor that secures the retention of the organization of the adjacent water layer.

Hydrogen Bond Donors

CHOLESTEROL

A striking difference in the composition of hydrogen belts is evident when membranes of mammalian cells are arranged in order of their biogenesis: endoplasmic reticulum (microsomal), Golgi apparatus, and plasma membrane. The hydrogen bond acceptors decrease in number, and hydrogen bond donating groups appear in their place. Microsomal (and also mitochondrial) membranes contain little sphingolipid or cholesterol; plasma membranes contain much. These hydrogen bond donor molecules, and the diminished acceptor plasmalogen, are found in highest concentrations in the specialized plasma of myelin. This membrane even has a unique lipid, hydroxycerebroside [Fig. 4 (6)], which carries an additional hydroxyl in the D-α position of its fatty acid. It appears that the biosynthesizing membranes need C=O groups unbalanced by lipid hydroxyls and that excess OH groups convert the membrane into a barrier or insulator. That cholesterol has membrane closing properties has already been discussed; sphingolipids may function similarly.

The rotational angular span of the cholesterol O—H vector is quite restricted if the molecule is assumed to retain its longitudinal orientation in the membrane. The limiting positive latitudinal direction, as it is shown in Fig. 4 (2) and 6, is around $+10°$; in the opposite conformation, the angle would be around $-50°$. In the $+10°$ conformation, cholesterol could donate a hydrogen bond to a phosphoglyceride ester carbonyl and at the same time attach closely to the fatty acid chain. Such lipid–lipid hydrogen bonding might be instrumental in the closure of the membrane [5]. Relatives of cholesterol that do not possess its β-hydroxyl group cannot, in fact, reduce the permeability of a phosphoglyceride membrane, although some of them have a condensing effect [10]. The reason why cholesterol does not possess an aromatic A ring, despite the seeming advantage that rigid planarity and an ideally equatorial OH group would confer, might be found here: The latitudinal O—H angle would be a constant $-18°$.

Cholesterol–phosphoglyceride bonding would necessitate the breaking of a water–carbonyl bond; I have argued [5] that the loss of this bond might be balanced energetically by hydrogen bonding from water toward the back of the cholesterol oxygen (Fig. 6). The generally longitudinal orientation of water O—H bonds would thus be left undisturbed. Cholesterol—OH · · · water bonding, on the other hand, would necessitate a reversal of water molecules. The fact that incorporation of cholesterol does not influence the hydration number (11–12 tightly bound water molecules) of phospholipids [19] may be

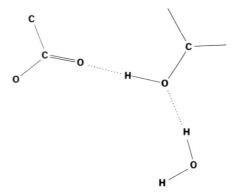

Fig. 6 Hypothetical alignment of phospholipid carbonyl, cholesterol hydroxyl, and water in the hydrogen belt (Brockerhoff [5]).

taken as an argument against this possibility. Incidentally, cholesterol cannot hydrogen bond to another cholesterol or accept a hydrogen bond from sphingolipids; Figure 4 (2) makes it clear that its oxygen can receive protons only from water. Lysophosphoglycerides, however (in which the fatty acid on the 2-hydroxyl of the glycerol is missing), could donate their secondary hydroxyl proton to cholesterol and at the same time accept the cholesterol proton at their ester carbonyl in position 1. Such concerted bonding might account for the existence of the cholesterol–lysophosphatidylcholine complex [29] and explain the protective action of cholesterol against membrane lysis by lysolipids.

The perfect steric fit that is apparent in Fig. 4, and even more so when the models are handled, is so powerful an argument for the existence of cholesterol–phosphoglyceride hydrogen bonding as to be nearly completely persuasive. However, the alternative, the existence of cholesterol—$H\cdots OH_2$ bonding next to phospholipid$=O\cdots HOH$ bonding, cannot be discarded on speculation alone, even if this alternative is less attractive; experimental evidence must be awaited. However, regardless of the final experimental outcome, it is clear that the donor–acceptor balance in the hydrogen belts has a fundamental influence in determining the character of the membrane.

SPHINGOLIPIDS

The C$=$O group of the fatty acid of sphingolipids makes the molecule a hydrogen bond acceptor; the 3—OH group of the sphingosine [Fig. 4 (5, upper chain) and Fig. 4 (6, lower chain)] and the α-OH of the fatty acid in hydroxy-cerebroside [Fig. 4 (6)] make it a donor. The location of all these functional groups in the hydrogen belt is secured by the *trans* configuration of the amide

TABLE 1

Role of the Components of Hydrogen Belts

Acceptor	Extender	Donor
Phosphoglyceride		Cholesterol
Plasmalogen	Hydroxysphingolipid	
	Sphingolipid	Protein (?)
Water (?)		Water

group connecting the chains. The amide N—H appears to be buried in our models but is potentially also a hydrogen bond donor, although weaker than the hydroxyls because of the smaller polarity of the N—H bond [28]. The sphingolipid O—H is sterically much less restricted than the cholesterol hydroxyl. It could more easily donate its proton to water. On the other hand, if it would bond to ester or amide C=O oxygens in the plane of the belt, the back of the OH oxygen would become approachable by other hydrogen bond donors, namely, cholesterol or sphingolipids. The allylic *trans* double bond may facilitate such a double role for the hydroxyl. Sphingolipids thus might be able to accept two molecules of cholesterol while donating one hydrogen bond to an acceptor. They have therefore been assigned, in Table 1, a position of hydrogen bond extenders leaning toward a net accepting function, except for the hydroxycerebroside, in which the donating functions predominate.

Stereochemically, phosphoglycerides (with a few odd exceptions) have the L-glycerol—3-phosphate structure; sphingosine has a D-*erythro* configuration. This stereospecificity cannot have much influence on the ligands situated in the hydrophobic core or polar zone. Its teleological significance, if there is any, must express itself in the hydrogen belt. Some evidence bearing on this question is perhaps offered by experiments which have shown that enantiomeric barbiturates have different anesthetic actions [37], since these drugs presumably act on the lipid matrix rather than on the protein of membranes.

PROTEINS

The intrinsic proteins are held in position by hydrophobic bonding in the membrane core. It is conceivable that they also engage in bonding in the hydrogen belts. They would probably do so as hydrogen bond donors, with residues such as serine or threonine participating, since the only potential acceptors, carboxylate groups, are charged and are more likely to be found in the polar zones, while the amide groups are donor–acceptors and are probably mutually bonded (as in the α helix) or hydrated. Membrane proteins have therefore been placed as potential donors in Table 1. There is some circumstantial evidence supporting the concept of phospholipid O···protein bonding. Isolated membrane proteins often cling tenaciously to phospholipids, in many cases to the acidic lipids, but not exclusively so. Cholesterol is never retained. Cholesterol has been found to hinder the penetration of phospho-glyceride monolayers by proteins [20], and the membrane with the belts richest in hydrogen bond donors, myelin, contains relatively little intrinsic protein. It appears, then, that cholesterol and protein compete for bonding to phosphoglycerides. This is puzzling if hydrophobic and ionic bonding are the only forces involved.

Membrane water holes could be expected to be as disruptive on the hydrogen belt bonding of proteins as on their hydrophobic bonding to the carboxyl-proximal fatty acid segment. As a consequence, lipids bound to the proteins would again be expected to contain mono- or dienoic rather than polenoic acids.

THE POLAR ZONES

Membrane Hydration

The foremost task of polar zone constituents is the stabilization of the membrane by anchoring both sides of it securely in the aqueous environment. This is accomplished by highly hydrophilic substituents. Thermodynamically, the effect achieved is the drastic reduction of the hydrocarbon–water interfacial tension which otherwise would tend to compress the lipid into spheric structures.

Among the lipid head groups, the phosphorylcholine residue of phospha-tidylcholine [Fig. 4 (1)] and sphingomyelin [Fig. 4 (5)] may be designed exclusively for membrane stabilization. This group binds 11 molecules of water firmly [15]. As a zwitterion of a strong base (choline) and a strong acid (a primary phosphate function) it is stable over a wide pH range and does not bind mono- or divalent cations [8], and it is therefore not likely to engage in any electrostatic bonding to membrane proteins. Thus, the function of the group may perhaps be described exhaustively as that of a hydrophilic stabilizer.

Phosphatidylethanolamine and phosphatidylserine also bind 10–11 water molecules and thus assist in membrane stabilization. The carbohydrates in the polar zone include phosphorylinositol and its 4- and 4,5-phosphates, and galactose in the cerebrosides of myelin and in the mono- and digalactosyl diglycerides of chloroplasts. Undoubtedly, these groups are strongly hydrated. The inositides have a very high turnover rate, which has generated a plethora of speculations [24]; possibly, inositides play a role in calcium metabolism. The fatty acid nearest to the inositol is usually highly unsaturated, creating a water hole that might permit the hydroxyls of inositol to participate in the hydrogen belt structure. Galactose might be a membrane stabilizer exclusively, just as is phosphorylcholine, from which it differs, however, by possessing numerous hydrogen bond donating as well as accepting sites. No suggestions have yet been made as to how this difference might benefit the myelin or chloroplast membranes in which galactolipids abound.

Membrane Surface Charge

Most phospholipids carry a net charge stemming from an unbalanced anionic phosphate on each polar head group. As a result, biological membranes have a negative surface charge. Since all net charges of the substituents have the same sign, the head groups in the polar zone are repellent toward each other, and the polar zone as a whole attracts counterions, that is, mono- and divalent cations together with their hydration shells. There have been many speculations but no solid evidence for intermolecular amine–phosphate bonding which would link the lipid molecules of a membrane together in an ionic lattice.

In phosphatidylethanolamine, the acidic phosphate function is not completely balanced by the more weakly basic alkylamino group; this means that a certain percentage of the amino groups in a membrane at physiological pH are unprotonated, and the membrane thus acquires a moderate negative charge. In phosphatidylserine, the amine is nearly balanced by a carboxyl group, and the polar end of the lipid carries a full negative charge. The two phosphoglycerides are interconvertible: The ethanolamine can be replaced by serine by base exchange, and the serine lipid can be converted to the ethanolamine compound by decarboxylation. The balance of these two lipids determines the surface potential of the membrane (assuming that other acidic lipids, especially phosphatidylinositol, are kept at constant concentrations). It is likely that the surface potential of a living membrane must be kept as constant as the intracellular pH, and the exchange–decarboxylation mechanism could possibly be used to establish such a homeostasis [6]. Although the exchange reaction is catalyzed by microsomal enzymes and the decarboxylation is catalyzed by mitochondrial enzymes, the lipids could be carried from one organelle to another with the help of transport proteins [38].

Cations and Proteins

Divalent ions, i.e., calcium and magnesium, are not merely bound ionically by the acidic lipids but are sequestered in strong complexes [26]. A role in calcium metabolism has often been proposed for the diphosphatidylglycerol of mitochondria and for the polyphosphoinositides [23]. Hard evidence is lacking, but it is reasonable to expect that these lipids do play such a role, if only as reservoirs for the cations. Whether phosphatidylserine has such a function is more questionable.

Many proteins can bind to the polar zone of lipid membranes [27]. Such bonding may hold some of the extrinsic proteins which can be dissociated from biological membranes by aqueous solutions of high ionic strength (although extrinsic proteins might be bound to intrinsic membrane proteins rather than to lipid). Since the lipids are negatively charged, the binding sites on the proteins must be positive. Protein–lipid bonding by ionic forces could be backed up by additional hydrophobic bonding, or, conversely, there is no reason why the intrinsic, hydrophobically bound proteins should not also engage in bonding in the polar zone. Evidence for such bonding is supplied by those enzymes, such as ATPases, that require acidic lipids, especially phosphatidylserine, for their activation [14].

Unless membrane proteins are synthesized directly into their lipid matrix, they must at some stage be dissolved in the cytosol, and we can ask by which mechanism they are inserted into their membrane. In solution, the outside of the molecule is covered by hydrophilic amino acid residues; any hydrophobic areas that cannot be accommodated in the interior of the molecule might be shielded by hydrophobic protein dimerization or polymerization. On approaching its membrane, the protein would orient itself with a positive head region toward the negatively charged polar zone. Ionic bonding would be followed by hydrogen bonding in the hydrogen belt, and these forces combined would open the protein molecule or aggregate sufficiently for hydrophobic residues to be exposed and bonding to the hydrophobic core to commence. Finally, the protein would be drawn inside-out into its membrane.

CONCLUSION

The stratification of biological membranes into hydrophobic core, hydrogen belts, and polar zones is preestablished in the tripartite structure of phospholipids and glycolipids. A complete teleological explanation of individual lipid structures in terms of architecture and functioning of the membrane is not yet possible; only a few structural designs are partially understood, in particular the double bonds which serve to fluidize the membrane core and the tailed hydrophobic disk of cholesterol which serves as a filler and plasticizer. This review is intended to present a survey of the structure–function problem of membrane lipids and to formulate some of the numerous unanswered questions.

REFERENCES

1. A. D. Bangham, M. M. Standish, and G. Weissmann, *J. Mol. Biol.* **13**, 253–259 (1965).
2. A. D. Bangham, J. De Gier, and G. D. Greville, *Chem. Phys. Lipids* **1**, 225–246 (1967).
3. L. D. Bergelson, *Prog. Chem. Fats Other Lipids* **13**, Part 1, 3–59 (1972).

4. M. C. Blok, E. C. S. Van Der Neut-Kok, L. L. M. Van Deenen, and J. De Gier, *Biochim. Biophys. Acta* **406**, 187–196 (1975).
5. H. Brockerhoff, *Lipids* **9**, 645–650 (1974).
6. H. Brockerhoff, *Bioorg. Chem.* **3**, 176–183 (1974).
7. D. Chapman, *in* "Form and Function of Phospholipids" (G. B. Ansell, J. N. Hawthorne, and R. M. C. Dawson, eds.), pp. 117–42. Elsevier, Amsterdam, 1973.
8. G. Colacicco, *Chem. Phys. Lipids* **10**, 66–72 (1973).
9. B. DeKruyff, R. A. Demel, and L. L. M. Van Deenen, *Biochim. Biophys. Acta* **255**, 331–347 (1972).
10. R. A. Demel, K. R. Bruckdorfer, and L. L. M. Van Deenen, *Biochim. Biophys. Acta* **255**, 311–320 (1972).
11. J. W. Depierre and G. Dallner, *Biochim. Biophys. Acta* **415**, 411–472 (1975).
12. G. G. De Pury and F. D. Collins, *Chem. Phys. Lipids* **1**, 1–19 (1966).
13. J. Donohue, *in* "Structural Chemistry and Molecular Biology" (A. Rich and N. Davidson, eds.), pp. 443–465. Freeman, San Francisco, California, 1968.
14. J. B. Finean, *in* "Form and Function of Phospholipids" (G. B. Ansell, J. N. Hawthorne, and R. M. C. Dawson, eds.), pp. 171–203. Elsevier, Amsterdam, 1973.
15. E. G. Finer and A. Darke, *Chem. Phys. Lipids* **12**, 1–16 (1974).
16. A. Finkelstein and A. Cass, *Nature (London)* **216**, 717–718 (1967).
17. E. Gorter and F. Grendel, *J. Exp. Med.* **41**, 439–443 (1925).
18. W. L. Hubbell and H. M. McConnell, *J. Am. Chem. Soc.* **93**, 314–326 (1971).
19. P. T. Inglefield, K. A. Lindblom, and A. M. Gottlieb, *Biochim. Biophys. Acta* **419**, 196–205 (1976).
20. H. K. Kimelberg and D. Papahadjopoulos, *Biochim. Biophys. Acta* **233**, 805–809 (1971).
21. J. B. Leathes, *Lancet* **1**, 853–856 (1925).
22. J. A. Lucy, *Biochem. Soc. Trans.* **3**, 611–613 (1975).
23. W. C. McMurray, *in* "Form and Function of Phospholipids" (G. B. Ansell, J. N. Hawthorne, and R. M. C. Dawson, eds.), pp. 205–252. Elsevier, Amsterdam, 1973.
24. R. H. Michell, *Biochim. Biophys. Acta* **415**, 81–147 (1975).
25. G. J. Nelson, *in* "Blood Lipids and Lipoproteins: Quantitation, Composition, and Metabolism" (G. J. Nelson, ed.), pp. 317–386. Wiley (Interscience), New York, 1972.
26. D. Papahadjopoulos, *Biochim. Biophys. Acta* **163**, 240–254 (1968).
27. D. Papahadjopoulos, *in* "Form and Function of Phospholipids" (G. B. Ansell, J. N. Hawthorne, and R. M. C. Dawson, eds.), pp. 143–169. Elsevier, Amsterdam, 1973.
28. L. Pauling, "The Nature of the Chemical Bond," 3rd ed. Cornell Univ. Press, Ithaca, New York, 1960.
29. R. P. Rand, W. A. Pangborn, A. D. Purdon, and D. O. Tinker, *Can. J. Biochem.* **53**, 189–195 (1975).
30. J. P. Segrest, R. J. Jackson, J. D. Morrisett, and A. M. Gotto, Jr., *FEBS Lett.* **38**, 247–253 (1974).
31. S. J. Singer, *in* "Structure and Function of Biological Membranes" (L. I. Rothfield, ed.), pp. 145–222. Academic Press, New York, 1971.
32. S. J. Singer and G. L. Nicolson, *Science* **175**, 720–731 (1973).
33. C. Tanford, "The Hydrophobic Effect: Formation of Micelles and Biological Membranes." Wiley, New York, 1973.
34. G. A. Thompson, *in* "Form and Function of Phospholipids" (G. B. Ansell, J. N. Hawthorne, and R. M. C. Dawson, eds.), pp. 67–96. Elsevier, Amsterdam, 1973.
35. H. Träuble, *J. Membr. Biol.* **4**, 193–208 (1971).

36. L. L. M. Van Deenen, J. De Gier, U. M. T. Houtsmuller, A. Montfoort, and E. Mulder, *in* "International Conference on the Biochemistry of Lipids" (A. C. Frazer, ed.), pp. 404–414. Am. Elsevier, New York, 1963.
37. H. Wahlström, H. Büch, and W. Buzello, *Acta Pharmacol. Toxicol.* **28**, 493–498 (1970).
38. K. W. A. Wirtz, *Biochim. Biophys. Acta* **344**, 95–117 (1974).
39. F. Wunderlich, A. Ronai, V. Speth, J. Seeling, and A. Blume, *Biochemistry* **14**, 3730–3735 (1975).
40. R. F. A. Zwaal, B. Roelofsen, and C. M. Colley, *Biochim. Biophys. Acta* **300**, 159–182 (1973).

2

Bioorganic Chemistry and the Emergence of the First Cell

Sidney W. Fox

INTRODUCTION

The strategy involved in identifying the first chemical steps leading to life is similar to the style of investigation and verification in bioorganic chemistry. What makes a salient difference, in the total study, are the facts that the biologist's unit of life is the cell and that the cell is a supramolecular system rather than a molecule. To study life comprehensively, a practitioner of any discipline must focus his attention beyond molecules and on to the cell. In one sense it is regrettable that we cannot simply analyze a cell and then "synthesize" one on the basis of the information obtained in the analysis. This inadequacy of the natural-products approach to the total problem of life's emergence is illustrated in Fig. 1. Distinctly different techniques, experimental philosophy, and modes of thinking are required for the steps beyond the first ones.

The primary task in the total problem is that of [step (a), Fig. 1] assembling a protocell, i.e., a primordial cell. In attempting to attain that goal, we are unable to use a protocell for comparison because we do not have, to our best knowledge, any protocell at hand. Even if we had one, its disassembly would not tell us how it had come into existence. Further difficulty arises from step (b) [Fig. 1]; the divagations of evolution have obscured the exact nature of the protocell. However, we must obtain our clues to the nature of the protocell from the contemporary cell, and we measure our progress by an increasing similarity of models relative to the contemporary cell. The

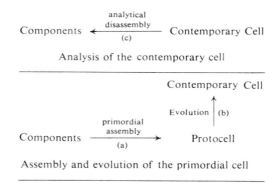

Fig. 1 Directions of flow of reductionistic research, (c), and of constructionistic evolution, (a) and (b).

rationale for use of the contemporary cell as a roster of properties for the protocell is the doctrine of the "unity of biochemistry" [1]. That principle permits us to believe that the first cells were not imponderably different from those of today.

However, we see no hope in backextrapolating to the primordial state by applying the "grind-and-find" technique to contemporary cells [step (c), Fig. 1]. Nor would the reverse of step (c), Fig. 1, serve our purpose. If successful, that exercise could only be *re*assembly of a contemporary cell, not assembly of a primordial cell.

We are thus left with the necessity of learning how steps (a) and (b), Fig. 1, occurred, in archaic time, by simulating them. This requires inductive reasoning. Although many scientists are skeptical of inductive reasoning, it is familiar to bioorganic synthesists and to a few other types of scientist. To extend his characteristic approach to include the protocell, the bioorganic synthesist must, however, develop an orientation in biochemistry and in cellular and molecular biology.

The "synthetic" approach applied to microsystems is thus the one that is necessary for solving this problem. Accuracy in finding clues for research and in judging results is aided by the fact that the problem is highly interdisciplinary. A meaningful result and view must comport simultaneously with chemical, biological, physical, and geological perspectives.

VALIDITY AND PLAUSIBILITY OF RESULTS

Like other sciences, that of the emergence of a protocell is developing its own criteria of validity and plausibility. Some of these are listed in Table 1.

TABLE 1

**Some Criteria of Validity and Plausibility of a Theory
for Emergence of the Protocell**

Experimental results obtained under geologically relevant conditions
Compatibility of geological conditions for reactions in sequence
Simplicity of reaction
Substantial efficiency of each reaction

The integrated overview of origins that has been derived from experiments suggests that a protocell arose very early in organic evolution. A main reason for inferring this is that the simulated protocell, i.e., the proteinoid microsphere [2], arises with such efficiency and ease—on mere contact of proteinoid with water. Once it had originated, a protocell would have had many advantages over other lines of evolution. These advantages are summarized in Table 2, which was compilable by 1974 [3].

The concept of an early protocell is implicit in Oparin's writing [4]. The same concept is explicit in the writings of van Niel [5] and of Ehrensvärd. Ehrensvärd reasoned that nucleoproteins would have appeared as precipitated by-products of metabolism in early cells [6]. Our experimental results are consistent with the views of those authors. Through the concept of an early protocell, we step in our discussion from bioorganic chemistry to supramolecular assemblies. Such study, with appropriate substances, is equivalent

TABLE 2

Benefits of a Cell (Protocell)

Benefits	Reference
Physical protection of organic material	
Organization of chemical reactions	4
Compartmentalization of functions	4, 33
As a site for emergence of nucleoprotein organelles	6
Maintenance of kinetically favorable concentrations	28
Promotion of dynamic interactions with the environment	28
Thermodynamically favorable hydrophobic zones	29
Reproduction at microsystemic level	30, 23
Adaptive selection (Darwinian) of individual variants at microsystemic level	31, 32
Screening of macromolecules from diffusible molecules	
Juxtaposition of enzymes, organelles, etc.	2
Systemic support and development of photosynthesis	
Enlargement of metabolic pathways through (re)combination	32
Favorable spatial relations for codonic interactions	34

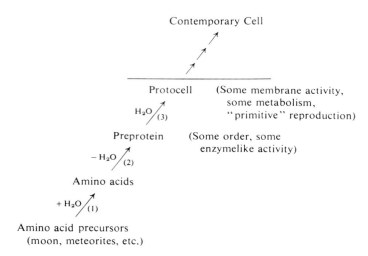

Fig. 2 Flowsheet of processes from primary reactants to first informational macro-molecule (preprotein) and to reproducing protocell. The model beyond that stage to the contemporary cell is incomplete.

to protobiology. In carrying out such investigation we cross the great divide, the conceptual barrier between a purely chemical world and one composed of micropackaged interactions which have been the province of the biological sciences. Only in recent years have biological phenomena been widely recognized as derivative of chemical functions.

The discernible benefits to evolution of a micropackage (Table 2) are so many and so powerful that it is difficult to see how any noncellular route to contemporary evolution could have competed with any development that included such advantages. A counterview is that the cell is so complex that it would have had to come later, but that idea is rooted in the premise that a primitive cell had to be much like a contemporary cell [cf. 7]. The experiments [7] demonstrate that a *minimal*, yet evolvable, cell would have arisen very easily; Lehninger [8] has explained this possibility most lucidly.

A simplified flowsheet that emerges from the research is found in Fig. 2. The evidence up to the stage represented by the horizontal line has been presented a number of times [2,7]. That this flowsheet is comprehensive has been noted in two textbooks of biochemistry that appeared in 1975 [8,9]. One of these volumes [9] has reassessed all of the experiments in the field and has spotlighted what the author, M. Florkin, regards as errors of interpretation and the reasoning supporting his views. The critical reappraisals are sum-marized in Table 3.

TABLE 3

Shibboleths in Early Experiments[a]

Reducing atmosphere
Thin primordial soup
Imputation to geological systems of reactions
  in walled flasks
General conditions for what may have been
  special events
Analytical inferences for synthetic occurrences

[a] Adapted from Florkin [9].

THE SMALL MOLECULES

Florkin's criticisms apply mainly to the first step in Fig. 2. Most of the early inferences were from laboratory experiments performed in flasks. As an alternative source of information, we can turn to studies of carbon compounds on the Moon, in meteorites, and in terrestrial lava. Indeed, the expectation of a "bonanza of organic molecules" [10] from the Moon had its roots in extrapolation from flasks to the open surface of the Moon. No bonanza was found [11].

What was identified were precursors of amino acids [11,12]. The indigenous nature of such materials to the lunar surface has been strongly supported by the evidence that the amino acids are not derived from human contamination [13–15] nor from the jet exhaust of the descent engine [16]. We thus have a tested view of the first step in the flowsheet of molecular evolution.

The availability of organic material from interstellar matter [17] has also provided an enlarged view of the possibilities for spontaneous organic synthesis. The techniques used for step (1) in Fig. 2 are those that most resemble the work of the bioorganic chemist. Here, however, the emphasis must be on thin-layer synthesis rather than on the usual flask-enclosed reactions. The goals of the experimental molecular evolutionist are most fully attained by a synthesis that yields a mixture of related compounds as in organisms, e.g., a set of amino acids rather than a single, pure compound.

THE FIRST INFORMATIONAL MACROMOLECULES

In the one experimentally based comprehensive flowsheet [2,8,9] that is available, the first informational macromolecule is a copolyamino acid arising by geological heating of amino acids [step (2), Fig. 2]. The reasons for assigning this role to a proteinoid (not a protein) have been given [2,18];

without that type of molecule serving that function, a comprehensive, experimentally based flowsheet was not visualized. The structural information on linkages in such molecules has been recently summarized [19,20]. Other structural data were summarized earlier [2].

While the catalogued properties are numerous [2], the salient aspects are the self-ordering of amino acids in thermal polycondensation [18,19], the numerous catalytic activities found [2], and the tendency of the polymers to aggregate into cell-like structures in the presence of water [2]. This takes us to step (3) of Fig. 2.

THE PROTOCELL

The simulation of the protocell is the proteinoid microsphere. For this entity, as for the proteinoids, the accumulation of information is extensive; only the most salient features are mentioned here.

In examining proteinoid microspheres for the first time, one is struck with the swarms of these units and their uniformity in size. One gram of thermal polymer typically yields 10^{10} microspheres. Each microsphere typically is composed of 10^{10} molecules of proteinoid.

The physical structure of an individual microsphere is difficult to distinguish, in micrographs, from those of some of the less evolved microorganisms (Fig. 3). The size, shape, and types of association resemble those of the prokaryotic [21] cocci. The boundaries have some of the quality of membranes; they allow small molecules to pass through, while retaining macromolecules (the inwardly directed diffusibility of small molecules must have been necessary until early cells developed their own ability to synthesize such biochemical intermediates).

Combined units also transfer informational particles from one to another [22] (Fig. 4). They have also been shown to be capable of proliferation (reproduction) in four ways, by budding (Fig. 5), binary fission, sporulation, and parturition [23], each in a manner suggestive of an evolutionary precursor of a modern process.

The associations of microspheres is undoubtedly an extension of some of the forces that caused aggregation of the proteinoid molecules to form microspheres. These forces complicate attempts to fractionate components during some of the chemical experiments, but strong binding forces might well have been critical in the early stages of evolution that yielded the first cells.

The catalogued properties of proteinoid microspheres have elicited a number of comments from biochemists, cell biologists, and others [see 8,9,18]. What has been mainly missing from this simulated protocell is identification of a cellular type of synthetic ability. Recent research, however, indicates

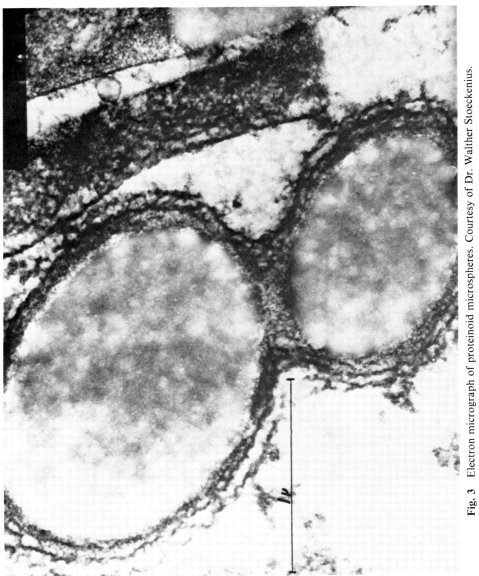

Fig. 3 Electron micrograph of proteinoid microspheres. Courtesy of Dr. Walther Stoeckenius.

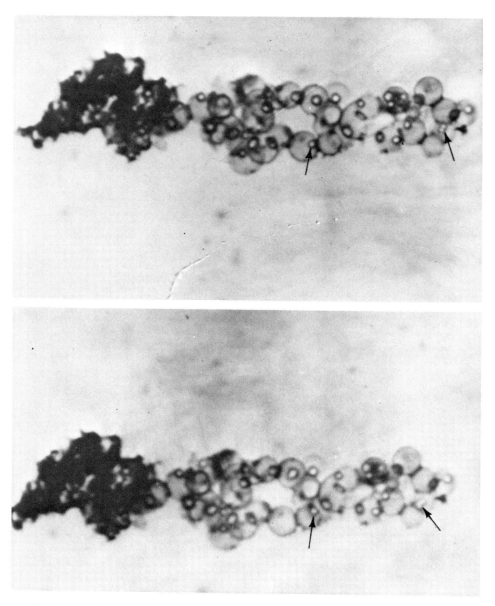

Fig. 4 Two frames from a cinematomicrographic sequence. Water has flowed through a cluster of proteinoid microspheres. They are joined by hollow connections. The endoparticles become small enough to pass through the junctions (arrows). The boundaries of the microspheres persist.

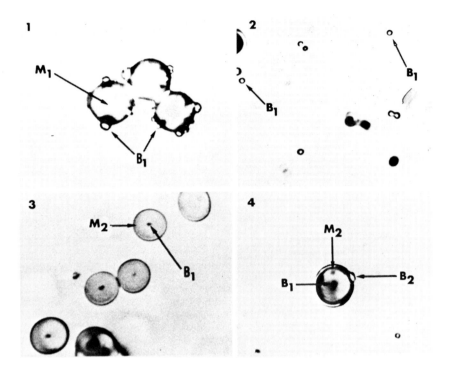

Fig. 5 Simulated protoreproduction of proteinoid microspheres through budding and growth of separated buds (M_1, first generation microsphere; B_1, first generation bud; M_2, second generation microsphere; B_2, second generation bud). Explanations and qualifications are presented in Hsu *et al.* [22].

that the proteinoid in microspheres is capable of using ultraviolet light, such as is believed to have been abundant on the primitive Earth, to activate chemical reactions.

PHOTOREACTIVE PROTEINOID

When condensation polymers of amino acids are made by heating the monomers, yellow pigment results. It is tenaciously bound to the polymers. The pigmented proteinoid has been found to have the ability to promote chemical reactions. This has been observed for the decarboxylation, by white light, of a variety of carboxylic acids: pyruvic acid, glyoxylic acid, and citric acid [24].

If such pigments will permit the synthesis of energy-rich phosphates, a means for understanding how a comprehensive metabolism could have originated will be provided. When photoactivity is combined with the demonstrated capacity for cellular replication and the inherently varied catalytic activities plus a membranous type of discrimination, the opportunities for evolution to a contemporary cell appear to have been rich at the stage represented by these experiments.

STEP (4) IN THE FLOWSHEET

The identification of a kind of primordial photoactivity covers part of the gap between a primordial cell and a contemporary cell [18]. In addition to the ability to transduce solar energy, the need for a molecular basis for coding was also crucial for evolution to the contemporary cell [25].

Some ability of appropriate proteinoid particles to synthesize internucleotide and peptide bonds has been demonstrated but so far only to a sharply limited extent [18]. None of the results to date, however, suggest that the fuller answers to these questions cannot be found.

EPILOGUE ON CONSTRUCTIONISM

As indicated in the Introduction, constructionistic techniques, as applied to the simulation of the first cells on Earth, are an extension to the systems level of typical bioorganic chemical approaches. The very word "bioorganic" recognizes organic chemistry in some of its relationships to biology. The missing link between a branch of pure organic (and some inorganic) chemistry and biology is the study of systems. This study to date, in the context of protobiogenesis, has had to deal with the construction ("synthesis") of systems. It has not enjoyed many benefits from analysis of contemporary systems except through background information.

The question has arisen as to what advances are attributable to the constructionistic approach to this problem that could not be attained by analysis. Examples of such new information begin with the self-ordering effects of diverse amino acids when those are thermally cocondensed. It is difficult to see how, or if, the principle of self-ordering could have been inferred from analysis of evolved amino acid sequences. While the principle is inferable from Lipmann's enzyme-controlled synthesis of antibiotic peptides [26] and from early enzymatic studies with peptide intermediates [2, p. 150], all of these studies were constructionistic.

A second example of the constructionistic possibilities is found in the production of the simulated protocell. This kind of advance was reported in 1959. By 1976, the much more difficult *re*construction of a contemporary cell from its macromolecular components had not been accomplished to more than a partial degree. The strength of binding forces in proteinoid molecules sufficient to yield a protocell had not been predicted, and probably could not have been.

Third, the tendency of proteinoid microspheres to participate in their own proliferation could not have been predicted. It was, rather, examination of microspheres by chemists with biological orientation that led to recognition of striking, but unforeseen, phenomena. These eye-catching phenomena included buds, simulation of binary fission, multisporelike inclusions, and parturitive processes [23]. The first two might have been anticipated in a general way from what is known of some other polymers, but they were not, to our knowledge, predicted in this context.

A fourth hint for construction of systems, from the experiments themselves, was the presence of pigment which, when finally tested, was found to have the ability to promote chemical reactions.

Finally, it has been possible to begin to analyze the sequence of the first few steps in the acquisition, or expression, of the earliest functions, i.e., a first-stage protocell, a second-stage protocell, etc. [27]. For this kind of inference, it seems impossible to imagine how a purely nonsynthetic study would serve.

It is now evident that it was necessary to identify unpredicted, or unpredictable, principles and phenomena by a constructionistic study only; waiting for a complete analysis would have been futile. The first step of the flowsheet (Fig. 2), however, has its roots in traditional bioorganic chemistry. Without that foundation, the later steps would not have been subject to study. The realistic view that one science merges into another is simply a reflection of the larger recognition that all of the sciences are increasingly being woven into a single huge tapestry.

ACKNOWLEDGMENT

Aided by Grant NGR 10-007-008 of the National Aeronautics and Space Administration. Contribution 297 of the Institute for Molecular and Cellular Evolution.

REFERENCES

1. M. Florkin and H. S. Mason, in "Comparative Biochemistry" (M. Florkin and H. S. Mason, eds.), Vol. 1, p. 1. Academic Press, New York, 1960.
2. S. W. Fox and K. Dose, "Molecular Evolution and the Origin of Life." Freeman, San Francisco, California, 1972.
3. S. W. Fox, Origins Life 7, 49 (1976).

4. A. I. Oparin, "The Origin of Life on Earth." Academic Press, New York, 1957.
5. C. B. van Niel, *in* "The Microbe's Contribution to Biology" (A. J. Kluyver and C. B. van Niel, eds.), p. 155. Harvard Univ. Press, Cambridge, Massachusetts, 1956.
6. G. Ehrensvärd, "Life: Origin and Development." Univ. of Chicago Press, Chicago, Illinois, 1962.
7. S. W. Fox, *Naturwissenschaften* **56**, 1 (1969).
8. A. L. Lehninger, "Biochemistry," 2nd ed. Worth Publ., New York, 1975.
9. M. Florkin, *in* "Comprehensive Biochemistry" (M. Florkin and E. H. Stotz, eds.), Vol. **29B**, p. 231. Elsevier, Amsterdam, 1975.
10. C. Ponnamperuma, *Space Life Sci.* **3**, 493 (1972).
11. S. W. Fox, *Bull. At. Sci.* **29** (10), 46 (1973).
12. S. W. Fox, K. Harada, and P. E. Hare, *Proc. Lunar Sci. Conf., 4th, 1973* Vol. 2, p. 2241 (1973).
13. K. Harada, P. E. Hare, C. R. Windsor, and S. W. Fox, *Science* **173**, 433 (1971).
14. K. Kvenvolden, *Annu. Rev. Earth Planet. Sci.* **3**, 183 (1975).
15. P. B. Hamilton and B. Nagy, *Anal. Chem.* **47**, 1719 (1975).
16. S. W. Fox, K. Harada, and P. E. Hare, *Geochim. Cosmochim. Acta* **40**, 1069 (1976).
17. G. Herbig, *Am. Sci.* **62**, 200 (1974).
18. S. W. Fox, *Int. J. Quantum Chem.* **QBS2**, 307 (1975).
19. P. Melius, this volume, Chapter 6.
20. S. W. Fox and F. Suzuki, *BioSystems* **8**, 40 (1976).
21. L. Margulis, *Am. Sci.* **59**, 230 (1971).
22. L. L. Hsu, S. Brooke, and S. W. Fox, *Curr. Mod. Biol.* **4**, 12 (1971).
23. S. W. Fox, *Pure Appl. Chem.* **34**, 641 (1973).
24. A. Wood and H. Hardebeck, *in* "Molecular Evolution: Prebiological and Biological" (D. L. Rohlfing and A. I. Oparin, eds.), p. 233. Plenum, New York, 1972.
25. S. W. Fox, *Mol. Cell. Biochem.* **3**, 129 (1974).
26. F. Lipmann, *Science* **173**, 875 (1971).
27. S. W. Fox, *in* "The Origin of Life and Evolutionary Biochemistry" (K. Dose *et al.*, eds.), p. 119. Plenum, New York, 1974.
28. A. I. Oparin, "The Origin and Initial Development of Life" (translated in NASA TT F-488), Washington, D. C., 1968.
29. S. W. Fox, *J. Sci. Ind. Res.* **27**, 267 (1968).
30. S. W. Fox, R. J. McCauley, and A. Wood, *Comp. Biochem. Physiol.* **20**, 773 (1967).
31. D. H. Kenyon, *in* "The Origin of Life and Evolutionary Biochemistry," p. 207 (K. Dose *et al.*, eds.), Plenum, New York, 1974.
32. L. L. Hsu and S. W. Fox, *BioSystems* **8**, 89 (1976).
33. S. Brooke and S. W. Fox *BioSystems*, **9**, in press, (1977).
34. J. C. Lacey, Jr., and D. W. Mullins, Jr., *in* "The Origin of Life and Evolutionary Biochemistry," p. 311 (K. Dose *et al.*, eds.), Plenum, New York, 1974.

3

Interaction Specificities of Deoxyribonucleic Acid and the "Octopus" Model for Lac Repressor– Lac Operator Recognition and Chromatin Structure

E. J. Gabbay

INTRODUCTION

A great deal of interest in the past decade has centered on the recognition process between nucleic acids and protein systems [1–10]. It is recognized that the interaction specificity between the two macromolecules is a problem of immense complexity and involves numerous types of forces operating at several sites along the nucleic acid and protein chains. In addition, recent studies on chromatin suggest that protein–DNA binding specificity not only is a dynamic process which is continually modified during the cell cycle [11–15], but may also involve specific protein aggregates–DNA recognition [16–20].

We have taken a simplified approach to the above problem by studying the interaction specificities of small molecules with DNA [21–34]. These studies have revealed that the nucleic acid helix may bind to small molecules via electrostatic, H bonding, and hydrophobic forces. Hydrophobic-type interactions are of particular interest since at least three have been noted: (1) intercalation between base pairs of DNA as exemplified by aromatic cations [23,25,27,31,35], (2) "partial" insertion (or intercalation) between base pairs exemplified by sterically restricted compounds containing an aromatic ring [28,29,31–34], and (3) external hydrophobic-type binding, which is noted in the nucleic acid–steroidal amine complexes [22,36]. In principle, therefore, nucleic acids may utilize the above forces, singly or in combination, to bind specifically to polypeptide chains.

This chapter is primarily a review of the work in our laboratories on the interaction specificities of small molecules (i.e., reporter molecules, steroidal amines, aromatic cations, etc.) and oligopeptides with DNA. It is shown that considerable information can be obtained concerning the dynamic and structural aspects of the Watson–Crick–Wilkins double helix.

THE REPORTER MOLECULES

In 1968, reporter molecules I, i.e., nitroaniline-labeled diammonium cations (Table 1), were synthesized and their interaction specificities with nucleic acids were examined [21]. Based on the electronic absorption and induced circular dichroism in the nitroaniline transition of I upon binding to different nucleic acids (native and denatured DNA and RNA) it was concluded that the complex is stabilized mainly by electrostatic-type interactions between the positively charged diammonium side chain of I and adjacent negatively charged phosphate anions of nucleic acids. The insertion of the nitroaniline ring of I between base pairs of nucleic acids was ruled out on the basis of (1) the ease of dissociation of the complex at high ionic strength, (2) the fact that removal of the positive charges from the side chain of I, e.g., compound 27, p-$NO_2C_6H_4NHCH_2CH_2OH$, abolishes the DNA binding, and (3) crude molecular framework models of the nucleic acid–I complex. Subsequent studies have shown that the early conclusion of Gabbay [21] concerning the mode of binding of the nitroaniline ring is erroneous and intercalation-type complexes are indeed obtained [23,27,28,37,38]. The latter conclusion is based on the results of proton magnetic resonance (^{1}H nmr), viscosity, and flow dichroism studies. A brief description of the methodologies employed in small-molecule–DNA studies is presented below together with specific application to the utilization of reporter molecules as probes of DNA structure.

Methods

ABSORPTION AND CIRCULAR DICHROISM STUDIES

The binding of 4-nitroaniline-labeled diammonium cations with DNA is accompanied by (1) a red shift, (2) hypochromism, and (3) induced circular dichroism (CD) in the 4-nitroaniline electronic transition. The origin of the red shift has been shown to be primarily due to changes in the electrostatic

TABLE 1

Reporter Molecules

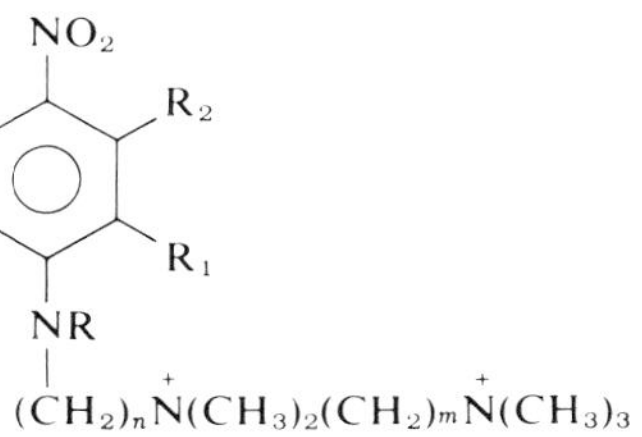

Structure I: benzene ring with NO_2, R_2, R_1, and NR substituents; NR chain: $(CH_2)_n \overset{+}{N}(CH_3)_2(CH_2)_m\overset{+}{N}(CH_3)_3$

I ($n = 2-3$; $m = 3-5$)

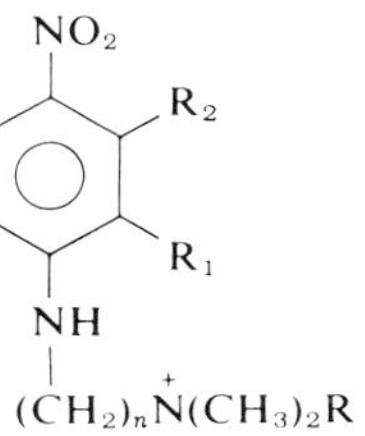

Structure II: benzene ring with NO_2, R_2, R_1, and NH substituents; NH chain: $(CH_2)_n\overset{+}{N}(CH_3)_2R$

II ($n = 2, 3$)

Structure III: benzene ring with R_1 substituent; chain: $(CH_2)_n\overset{+}{N}(CH_3)_2(CH_2)_3\overset{+}{N}(CH_3)_3$

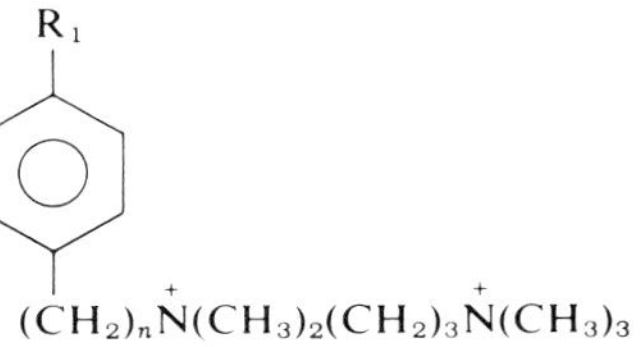

III ($n = 1-4$)

Reporter	Type	R	R_1	R_2	n	m
1	I	H	H	H	2	3
2	I	H	NO_2	H	2	3
3	I	H	CH_3	H	2	3
4	I	H	CF_3	H	2	3
5	I	H	CN	H	2	3
6	I	H	$CONH_2$	H	2	3
7	I	H	$CONHCH_3$	H	2	3
8	I	H	$CON(CH_3)_2$	H	2	3
9	I	H	$CON(C_2H_5)_2$	H	2	3
10	I	H	H	CH_3	2	3
11	I	CH_3	H	H	2	3
12	I	C_2H_5	H	H	2	3
13	I	$C_6H_5CH_2$	H	H	2	3
14	I	CH_3	NO_2	H	2	3
15	I	$C_6H_5CH_2$	NO_2	H	2	3
16	II	H	H	H	2	—
17	II	H	H	H	3	—
18	II	H	CH_3	H	2	—
19	II	CH_3	CH_3	H	2	—
20	II	C_5H_{11}	CH_3	H	2	—
21	III	—	H	—	1-4	—
22	III	—	CH_3	—	1-4	—
23	III	—	NO_2	—	1	—
24	III	—	NO_2	—	2	—
25	III	—	NO_2	—	3	—
26	III	—	NO_2	—	4	—

field as a result of binding of the diammonium side chain to vicinal phosphate groups of DNA (see Gabbay [21] for details). The hypochromic effect exhibited by the bound reporter molecule may be explained in terms of side-by-side coupling (card stack arrangements) of the low-energy nitroaniline transition moment with the higher-energy transition moments of the bases in nucleic acids. The theoretical treatment of hypochromism predicts that such an arrangement of transition moments leads to an intensity interchange whereby the lowest- and highest-energy transitions will be hypochromic and hyperchromic, respectively [39–42].

The result of the absorption studies is consistent with an intercalation model in which the nitroaniline ring is inserted between adjacent base pairs of DNA. Additional data consistent with this model have also been obtained by circular dichroism studies. For example, Gabbay [21] and Gabbay and Malin [43] have observed that ribose-containing double-stranded nucleic acids, i.e., rA-rU and rI-rC, induce a positive CD in the absorption band of the 4-nitroaniline transition of I irrespective of the nature of the substituents on the ring of the latter. Calf thymus and salmon sperm DNA exhibit a positive induced CD in the unsubstituted 4-nitroaniline reporter molecule (1) and a negative CD for the bound substituted 4-nitroaniline reporter molecules,

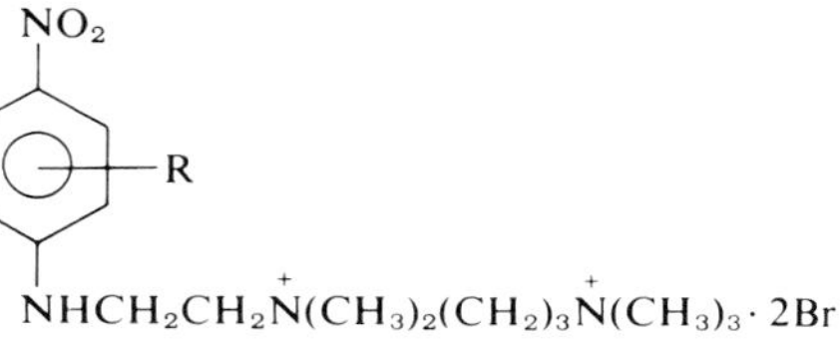

$$NO_2$$

R

$$NHCH_2CH_2\overset{+}{N}(CH_3)_2(CH_2)_3\overset{+}{N}(CH_3)_3 \cdot 2Br^-$$

I

i.e., 2,4-dinitroaniline (2), 2-CH$_3$-4-nitroaniline (3), 3-CH$_3$-4-nitroaniline (10), 2-cyano-4-nitroaniline (5), 2-CF$_3$-4-nitroaniline (4), and 2-carboxyamido-4-nitroaniline (6). Moreover, the interaction of RNA and DNA with reporter 28 leads to a positive and negative CD for the 450 nm 4-nitroaniline transition (a), respectively. However, only the DNA complex shows optical activity

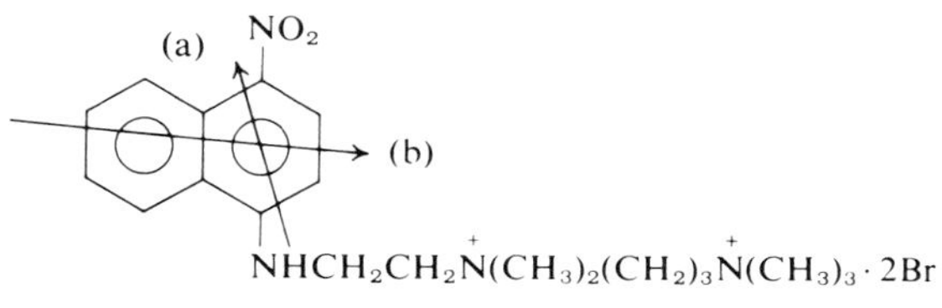

(a) $$NO_2$$

(b)

$$NHCH_2CH_2\overset{+}{N}(CH_3)_2(CH_2)_3\overset{+}{N}(CH_3)_3 \cdot 2Br$$

28

associated with the naphthylamine transition (b) of 28 at 340 nm. These results are best interpreted in terms of an "in" geometry of the substituent (in this case, a fused 2,3-phenyl ring) for the DNA complex (Fig. 1). It should be noted that, according to present theories, the induced molar ellipticity, [Θ], is given by Eq. (1) [44,45],

$$[\Theta] \alpha \sum_{j=1}^{n} f(\vec{R}_{ij} \cdot \vec{\mu}_i \times \vec{\mu}_j) \tag{1}$$

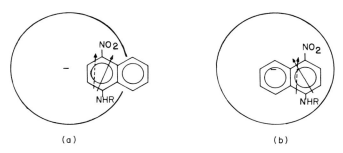

(a)   (b)

Fig. 1 Schematic illustration of a nucleic acid helix (top view) showing the "out" (a) and "in" (b) geometries. The direction of the transition moments of the 4-nitronaphthyla-mine (solid line) and of an adjacent base in the nucleic acid (dashed line) are also indicated.

where, for example, R_{ij} is the distance between the transition moment of the 4-nitroaniline and the transition moment of the bases (either purine or pyrimidine); μ_i and μ_j are the magnetic moment vectors associated with these transitions, respectively. An intercalation model in which the 2-substituents of the reporter molecules I may be pointing "in" and "out" of the helix could explain the oppositely induced CD observed for DNA and RNA complexes, since the cross-product term in Eq. (1) has a dependence on the sine of the angle between the magnetic moment vectors μ_i and μ_j. The approximate direction of the transition moments of the reporter molecule and the bases in nucleic acid is schematically shown in Fig. 1.

HYDRODYNAMIC STUDIES

Due to the elongation of the DNA molecule caused by intercalation of an aromatic ring, an increase in the viscosity of DNA solution is usually observed [35]. This phenomenon has been noted for DNA–I complexes [23,27,29,37]. For example, viscometric titration of salmon sperm DNA (at infinitely dilute solution, 0.2 mM P/liter i.e., 0.008%) with reporter molecule **2** leads to progressively increasing relative specific viscosity, $\eta_{sp}^{DNA2}/\eta_{sp}^{DNA}$, until a limiting value at phosphate–reporter (P/R) ratio of 8.0 $\pm$ 0.2 is obtained (Fig. 2a). Similarly, under identical ionic strength and buffer conditions, the reverse titration (using CD technique) of reporter molecule **2** (at 25 μM) with salmon sperm DNA leads to progressively increasing negative induced CD in the 4-nitroaniline transition until a limiting value at P/R ratio of 8.2 $\pm$ 0.3 is obtained (Fig. 2b). Moreover, Scatchard analysis [46] of the binding isotherms using equilibrium dialysis techniques showed that the value of the maximal binding, $\bar{n}_{max}$ (i.e., maximum number of reporter molecules per DNA phosphate), is approximately 0.12 $\pm$ 0.01 (Fig. 2c). The DNA binding site (i.e., the reciprocal of $\bar{n}_{max}$) is therefore found to be 8.3 $\pm$ 0.5 DNA phosphates per reporter, a value nearly identical with that obtained by viscometric

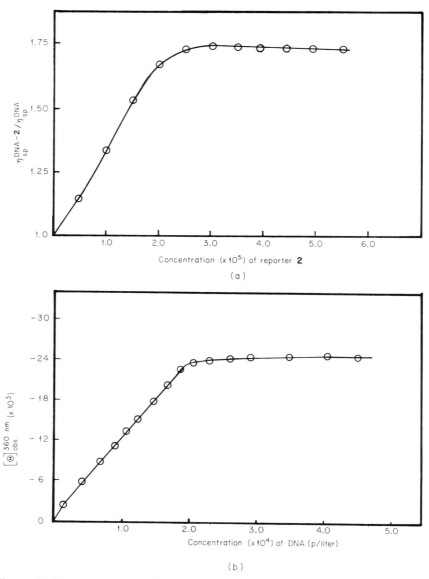

Fig. 2 (a) The effect of increasing concentration of reporter molecule **2** on the relative specific viscosity, $\eta_{sp}^{DNA2}/\eta_{sp}^{DNA}$ (where η_{sp}^{DNA} and η_{sp}^{DNA2} are the specific viscosities of complexed and free DNA, respectively). An almost infinitely dilute solution of DNA [0.2 mM p/liter, i.e., 0.008%, in 10 mM 2-(N-morpholino)ethanesulfonic acid (MES) buffer and 5 mM Na$^+$ at pH 6.2] was titrated with reporter molecule **2** at 25°C. (Taken from Destefano [47].) (b) The effect of increasing concentration of salmon sperm DNA on the observed ellipticity at 360 nm, $[\Theta]_{obs}^{360nm}$, of a solution containing 25 μM of reporter **2**. Ionic strength and buffer conditions used were identical to (a). (Taken from DeStefano [47].)

and CD titration techniques [38,47]. The above results strongly indicate that the strong binding of **2** to salmon sperm DNA, the induced CD in the absorption band of **2**, and the increase in viscosity of DNA–**2** solution arise from a common mode of binding, i.e., insertion (intercalation) of the aromatic ring between base pairs of DNA.

The results of flow dichroism studies are consistent with the above conclusion; e.g., it is found that the aromatic ring of **2** (in the DNA complex at $P/R > 8$) is in the same plane as the DNA base pairs. The dichroic ratios $A_\perp/A_\parallel$ (where $A_\perp$ and $A_\parallel$ are the absorbances of plane-polarized light perpendicular and parallel, respectively, to flow and hence helical axis) for free DNA (at 260 nm) and DNA–**2** complex (at 360 nm) were found to be nearly identical, i.e., 1.48 and 1.52, respectively [38,48].

PROTON MAGNETIC RESONANCE STUDIES

Proton magnetic resonance studies provide a unique method for studying the interaction of small molecules with macromolecules. The influence of a macromolecule on the line width and chemical shifts of the protons of a small molecule provides information concerning the degree to which motion of

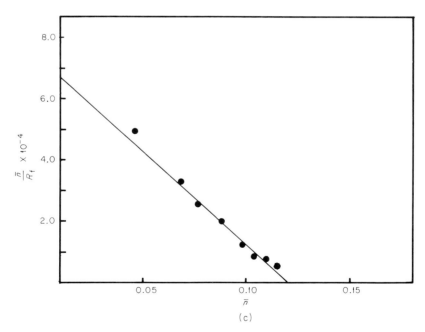

Fig. 2 (c) Scatchard plots of the binding data obtained by a spectrophotometric titration technique [47] for the interaction of **2** with salmon sperm DNA under conditions identical to (a).

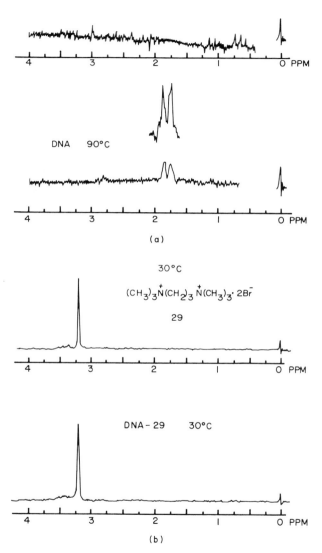

Fig. 3 (a) The partial ^{1}H nmr spectra of low molecular weight salmon sperm DNA (MW < 200,000) at 30° and 90°C at 160 mM phosphate/liter. (b) The ^{1}H nmr spectra of **29**, $(CH_3)_3N^+(CH_2)_3N^+(CH_3)_3 \cdot 2Br^-$ (10 mM), in the presence and absence of 160 mM P/liter of low molecular weight salmon sperm DNA. (c) The partial ^{1}H nmr spectra of reporter molecule **3** (10 mM) in the presence and absence of 160 mM P/liter of DNA at different temperatures. (Taken from Gabbay et al. [23].)

specific parts of the small molecule becomes restricted and concerning the environment of the small molecule when it is bound [49,50]. For example, it is well known that, if the rate of tumbling of molecules in solution is lower than the typical Larmor frequencies W_0 (of the order of 10^8–10^9 rads sec^{-1} for protons in the conventional magnetic field), then T_2, the transverse relaxation time, is considerably diminished, leading to substantial line broadening of the proton signal. This situation is noted if the proton is contained in a rigid macromolecule, e.g., DNA [51,52], or if the proton is contained in a slowly tumbling small molecule bound to a macromolecule [53–57].

Figure 3a shows the temperature-dependent partial ^{1}H nmr of free low molecular weight ($< 200,000$) salmon sperm DNA. No signal is observed at 30°C; however, at 90°C, where the DNA helix has melted to the random coils, two singlets are observed and are attributed to the resonance signals of the thymine methyl protons [51,52]. The results are simply interpreted in terms of restricted motions (tumbling or rotations) of the methyl protons of thymine in the DNA helix, which lead to incomplete averaging of the chemical shift environment and, hence, to considerable line broadening. The ^{1}H nmr spectra (Fig. 3b) of the simple diquaternary ammonium salt **29**, $(CH_3)_3N^+$ $(CH_2)_3N^+(CH_3)_3 \cdot 2Br^-$, are nearly identical for the free and the DNA-bound state at 30°C. This behavior is exhibited by a variety of alkyl diammonium salts, spermidine and spermine derivatives, and oligopeptides containing nonaromatic amino acids [24]. The results may be interpreted in terms of rapid tumbling of these molecules in the DNA complex and are indicative of

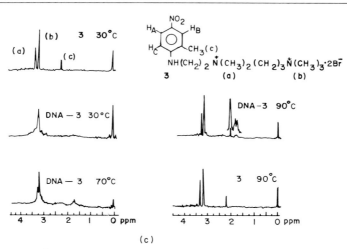

Fig. 3 *continued.*

an external electrostatic binding. It should be noted that under the conditions of ^{1}H nmr experiments (10 mM polyamine and 160 mM in DNA phosphate) the polyamines are fully bound to DNA as shown by binding studies [21,24,26, 38]. Figure 3c shows the temperature-dependent partial ^{1}H nmr spectra of the free and DNA-bound reporter molecule **3**. It is noted that the c-methyl proton signal is upfield shifted by 0.54 ppm and is extensively broadened, whereas the upfield chemical shift and line broadening are considerably less for the a-methyl protons and negligible for the b-methyl protons signals. The nmr signals of the aromatic protons H_A, H_B, and H_C of **3** are totally broadened and not observable in the presence of DNA at all temperatures. It is therefore reasonable to conclude that the ^{1}H nmr studies strongly indicate that the 4-nitroaniline ring system of the reporter molecule **3** is intercalated. Both the line broadening of the H_A, H_B, H_C, and c-methyl protons, indicative of restricted rotation, and upfield chemical shift of the c-methyl proton signal, indicative of anisotropic shielding by aromatic ring currents, support this conclusion.

Reporter Molecules as Probes of DNA Structure

STERIC REQUIREMENTS AND CRITERIA FOR INTERCALATION

In order to delineate the steric requirements for the intercalation process, the interaction of reporter molecules I (containing increasing size of R

$$
\begin{array}{ll}
R_1 = R_2 = H \\
R_1 = CH_3;\ R_2 = H \\
R_1 = R_2 = CH_3 \\
R_1 = R_2 = C_2H_5
\end{array}
$$

NO_2

H_B H_A

H_C $CONR_1R_2$

NH

$CH_2CH_2\overset{+}{N}(CH_3)_2(CH_2)_3\overset{+}{N}(CH_3)_3 \cdot 2Br^-$

I

substituents) with salmon sperm DNA was studied by absorption, circular dichroism, ^{1}H nmr, viscosity, flow dichroism, and equilibrium dialysis techniques [27,47,48]. The following observations were made. (1) The reporter molecules **6** and **7**, where $R_1 = R_2 = H$ and $R_1 = H$, $R_2 = CH_3$, respectively, show a red shift in λ_{max}, a strong hypochromic effect (47 and 41%), and a large induced circular dichroism ($[\Theta]^t = -6590$ and $-10,060$) in the 4-nitroaniline transition on binding to salmon sperm DNA; (2) total broadening of the ^{1}H nmr signals of the aromatic H_A, H_B, and H_C protons as well as the side-chain a- and b-methyl protons of **6** and **7** is observed in the DNA complex at 32°C; (3) low-shear viscometric studies show that the intrinsic viscosity of the DNA-bound reporters **6** and **7** relative to free DNA are increased by 0.26 and 0.30, respectively; and (4) flow dichroism studies indicate that the 4-nitroaniline ring of **6** and **7** are in the same plane as the DNA base pairs. This is in contrast to the behavior of reporter molecules **8** and **9** with N,N-dimethyl- and N,N-diethyl-substituted 2-carboxyamido groups, respectively. For example, (1) the latter reporters show little or no effect on the absorption and absence of an induced CD of the 4-nitroaniline transition on binding to DNA; (2) the ^{1}H nmr signals of the aromatic protons (H_A, H_B, and H_C) and the diammonium side-chain protons (a- and b-methyl protons) are clearly discernible in the presence of DNA at 32°C with only a slight broadening and upfield chemical shift; (3) a small decrease in the intrinsic viscosity of solution of DNA is observed upon binding **8** and **9**; and (4) the plane of the 4-nitroaniline ring of **8** and **9** is not oriented in the presence of DNA, i.e., the dichroic ratio $A_\perp/A_\parallel$ at 390 nm is found to be unity. The mode of binding of **8** and **9** to DNA seems to be external electrostatic binding to the phosphodiester chain with the 4-nitroaniline ring freely tumbling in solution. Intercalation between base pairs of DNA by **8** and **9** is inhibited, presumably due to steric hindrance by the bulky N,N-dialkylcarboxamido substituent. Thus, the four basic criteria, (1) hypochromism and induced CD in the absorption band of the DNA-bound reporter, (2) upfield chemical shift and broadening of ^{1}H nmr signals, (3) increase in the viscosity, and (4) an observed value of the dichroic ratio $A_\perp/A_\parallel$ greater than unity, are characteristic consequences of an intercalation process. In addition, it should be noted that the results of equilibrium dialysis studies of the binding of **6–9** to salmon sperm DNA show that the reporter molecules **6** and **7** are bound approximately 15 to 20-fold more strongly than **8** and **9**. This is in line with the proposed binding modes of these molecules as described above and would indicate that intercalation of the 4-nitroaniline ring of reporter molecules contributes approximately 1.5–2.0 kcal/mole to the binding process.

FORCES INVOLVED IN DNA–REPORTER COMPLEXES

The interaction of reporter molecules II with salmon sperm DNA has been studied using the above-described techniques [23]. In all cases, the criteria for

18, R = H
19, R = CH₃
20, R = C₅H₁₁

II

an intercalation mode of binding of the 4-nitroaniline ring of II to DNA are noted: (1) hypochromism and induced CD, (2) ^{1}H nmr broadening, (3) viscosity enhancement, and (4) flow orientation of the aromatic ring. Particularly interesting is the result of ^{1}H nmr studies on the a-methyl signal of II in the presence of salmon sperm DNA at P/R ratio of 10. For example, the nmr signal of a-methyl protons of II where R = H and C_5H_{11} is observed to be totally broadened and indistinguishable from baseline noise. On the other hand, the signal for the a-methyl protons of II where R = CH_3 in the DNA complex is clearly discernible as a slightly broadened singlet. The above results are consistent with a two-point contact between the reporter molecule II (R = H or C_5H_{11}) and DNA leading to restricted tumbling of the ammonium side chain. Intercalation of the 4-nitroaniline ring is the common mode of binding of II (where R = H, CH_3, and C_5H_{11}) to DNA. It is noted, however, that reporter **18** [containing $-N^+H(CH_3)_2$ side chain] is capable of electrostatic and H bonding with DNA, whereas reporter **20** [containing $-N^+(CH_3)_2C_5H_{11}$ side chain] is capable of electrostatic- and hydrophobic-type interactions with DNA. On the other hand, reporter **19** [containing $-N^+(CH_3)_3$ side chain] may bind to DNA principally by an electrostatic-type interaction, and therefore its motion is less restricted than the corresponding methyl protons of reporters **18** and **20** in the DNA complex. Thus, it is clear that DNA may bind small molecules via external electrostatic and H-bonding interactions. Moreover, two types of hydrophobic interactions are noted, i.e., internal (intercalation of planar aromatic ring between base pairs) and external hydrophobic binding of nonpolar substituent in the DNA groove.

NONEQUIVALENCE OF DNA INTERCALATION SITES

As a consequence of the right-handed Watson–Crick–Wilkins double helix of DNA with A–T and G–C base pairs, there are ten distinctly different intercalating sites [58]. Thus, it is possible that the reporter molecules I may have different affinities and may exhibit characteristic spectral properties for each of the ten possible sites. A systematic analysis of the induced CD in the absorption band of variously substituted reporter molecules I upon binding to salmon sperm DNA has been made. A large negative induced CD is observed in the 4-nitroaniline transition of I (**2**, R = NO_2; and **5**, R = CN) upon binding to DNA which is independent of the P/R ratio in the range of 6–70. On the other hand, the induced CD spectra of the DNA–I (**3**, R = CH_3; and **4**, R = CF_3) is found to be strongly dependent on the DNA phosphate to reporter ratio. For example, the molar ellipticity [Θ] at 395 nm (for R = CF_3) at P/R of 70 and 6 is found to be -4000 and 2400, respectively. In order to account for the above observations (as well as the viscometric, flow dichrois, and ^{1}H nmr studies, which show that the reporter molecules I form intercalated complexes at P/R ratio greater than 6 [47,58]) it has been suggested that the various possible intercalation sites in DNA may not necessarily induce the same sign and/or magnitude in the CD signal of the absorption band of the DNA-bound reporter molecule. In fact, it has been shown that (A–T)-rich DNA exhibits the strongest affinity for reporter molecules and induces a large negative CD in the absorption band of the bound reporter [29]. The significance of the presence of differing intercalating sites in DNA which are selectively recognized by small molecules has led to considerable specula- tion on the recognition process involved in protein–nucleic acid interactions (see below).

DISSYMMETRIC RECOGNITION OF THE HELICAL SENSE OF DNA

The concept of employing an optically active agent to separate racemic mixtures is well known. The technique relies on the formation of two dias- tereomers with different physical properties. Somewhat analogously, the optically active DNA molecule can be used as the agent for the formation of diastereomeric complexes. For example, in an attempt to determine the helical sense of DNA in solution, the dinitrophenyl (DNP) derivatives of L- and D-prolines, **30** and **31** have been employed [28,29].

$$R = CONHCH_2CH_2\overset{+}{N}(CH_3)_2(CH_2)_3\overset{+}{N}(CH_3)_3 \cdot 2Br^-$$
$$\text{(a)} \qquad \text{(b)}$$

Due to the helical twist of DNA, successive base pairs can be visualized as in Fig. 4. The asymmetric center of the prolyl systems is such that the direction of the carboxyamido side chain (—CONHR') is sterically unfavorable for intercalation of the L-enantiomer **30** if the DNA helix is right-handed, whereas intercalation of the 2,4-dinitroaniline ring of the D-enantiomer **31** is relatively unhindered. The results of equilibrium dialysis, ^{1}H nmr, absorption, circular dichroism, and viscometric studies show that the D-enantiomer interacts to a greater extent with DNA (as compared to the L-enantiomer) via intercalation of the DNP moiety and electrostatic binding of the diammonium side chain [28,29]. For example, the following results are obtained. (1) Visible absorption studies indicate that (a) the D-enantiomer **31** exhibits a large hypochromic effect (34%) and 3 nm red shift in the 4-nitroaniline transition on binding to salmon sperm DNA, whereas the L-enantiomer **30** elicits a 12% hypochromism

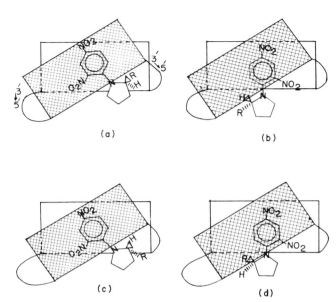

Fig. 4 Schematic illustration (top view) of the possible intercalated complexes of the D-prolyl derivative **31** (a, b) and L-prolyl derivative **30** (c, d) between base pairs of a right-handed DNA helix. The direction of the 3' → 5' sugar phosphate chain with respect to the base pairs at the top (shaded) and bottom is indicated. (Taken from Gabbay and Sanford [29].)

and no red shift; (b) denatured salmon sperm DNA (i.e., single stranded) shows no selectivity on the absorption spectra of the bound reporter molecules **30** and **31** (e.g., a 12 and 14% hypochromic effect and the absence of a red shift are noted, respectively); and (c) all double-stranded helical DNA studied [calf thymus, *Micrococcus luteus*, *Escherichia coli*, poly(dA-dT)-poly(dA–dT), and poly(dI–dC)-poly(dI-dC)] elicit a greater hypochromicity with the D- than with the L-enantiomer. (2) The circular dichroism spectra of **30** and **31** indicate that there are two optically active transitions, i.e., at 405 and 335 nm, which have been assigned to the 2- and 4-nitroaniline transitions [21]. It is noted that the D-enantiomer undergoes a larger change in the molar ellipticities, [Θ], of the peaks at 405 and 335 nm as compared to the [Θ] of the troughs of the L-enantiomer on binding to DNA. (3) The ^{1}H nmr studies are also consistent with a more intimate binding of **31** to DNA as compared to **30**; e.g., at 50° the H_A proton signal is upfield shifted by 64 and 50 Hz for the D- and L-enantiomers on binding to DNA. Moreover, the entire ^{1}H nmr spectrum of the DNA–D-**31** complex shows that the signals of **31** are more broadened than the corresponding ^{1}H nmr signals observed for the DNA–L-**30** complex. The results indicate that the dinitroaniline ring of **31** experiences greater shielding (i.e., via ring current anisotropy) and restricted tumbling than **30**. (4) Viscometric titration studies show that the specific viscosity, η_{sp}, of salmon sperm DNA solution increases in the presence of the D-enantiomer and significantly decreases in the presence of the L-enantiomer. The results are consistent with a model in which intercalation of the DNP ring of **31** between base pairs of DNA leads to lengthening of the helix, whereas the binding of **30** leads to a distortion of the helical rod (i.e., bending at the point of intercalation) which results in a decrease in the effective length of DNA (see also the section on "partial" intercalation). (5) The results of equilibrium dialysis binding studies show that the D-enantiomer is bound 2- to 5-fold more strongly than the L-enantiomer to salmon sperm DNA.

In summary, the dissymmetric recognition of the helical sense of native DNA *in solution* can be discerned via the interaction specificities of the DNP derivatives of L- and D-prolines. The results are consistent with a right-handed helical structure for native DNA in solution, in agreement with the Watson–Crick–Wilkins model.

BENDING OF HELIX VIA "PARTIAL" INTERCALATION

The mechanism(s) by which histone and nonhistone proteins influence the tertiary structure of DNA in condensed chromatin has been the subject of considerable interest in many laboratories [2,10,59–66]. Hanlon and co-workers [10] have suggested that supercoiling of DNA may occur in nucleo-histone via alternating B and C conformations of the DNA duplex, whereby the latter conformation is induced via histone binding. Bartley and Chalkley [2] have proposed that the histone proteins may act as a spring; i.e., an α-helical segment of the polypeptide chain may be involved in connecting the protein to two or more binding sites along the DNA helix, thus causing the latter to bend and assume a supercoil condensed form. Work from this laboratory [24,26] on the interactions of oligopeptides with DNA has shown that the peptides containing aromatic amino acids at the C-terminus cause a dramatic decrease in the specific viscosity, η_{sp}, of the DNA solution. The above data, together with the ^{1}H nmr studies of oligopeptide–DNA complexes led to the proposed nonclassical model of intercalation in which the aromatic residue of the oligopeptides is partially inserted between base pairs of DNA, thus leading to a bend of the helix at the point of intercalation.

In order to investigate the effect of partial and/or total insertion of an aromatic residue between base pairs on the tertiary structure of DNA, the interaction specificities of reporter molecules III with DNA have been examined [31]. Viscometric titration studies (at near infinite dilution of native salmon sperm DNA) showed that the effective length of DNA–III complexes is considerably diminished when $n = 1$ and enhanced when $n = 2, 3,$ and 4. Proton magnetic resonance spectra of the DNA–III complexes (where $n = 2, 3,$ and 4)

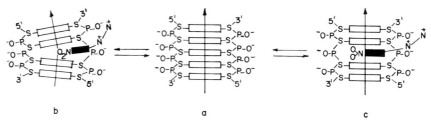

Fig. 5 Schematic illustration of a segment of DNA duplex (a) which can either partially intercalate a molecule to give structure b or fully intercalate a molecule to give structure c. These two processes either decrease or increase, respectively, the effective length of the DNA duplex [31].

show that the aromatic proton signals of III (H_A and H_B) are totally broadened and indistinguishable from baseline noise. On the other hand, the aromatic protons H_A and H_B (in the DNA–III complex with $n = 1$) undergo an upfield chemical shift of 10 and 6 Hz, respectively, upon binding to DNA. On the basis of the viscometric and ^1H nmr studies of these systems (as well as of

$$NO_2$$

$$H_A$$

$$H_B$$

$$(CH_2)_n \overset{+}{N}(CH_3)_2(CH_2)_3 \overset{+}{N}(CH_3)_3 \cdot 2Br^-$$
(b)   (c)

III (where $n = 1$–4)

other model compounds [33]) an intercalation model in which the *p*-nitro-phenyl ring of III (where $n = 1$) is "partially" inserted between base pairs of DNA (Fig. 5) has been proposed [31]. Such a model adequately accounts for (1) the observed decrease in the effective length of the DNA–III ($n = 1$) (i.e., via bending at the point of insertion) and (2) the differential upfield chemical shifts of the nmr signals of the aromatic protons H_A and H_B (i.e., the H_A proton experiences a larger ring current anisotropy than the H_B proton as a result of the "partial" insertion). It is clear that the single methylene group (CH_2), between the 4-nitrophenyl ring of III ($n = 1$) and the quaternary ammonium group of the side chain is not sufficient to allow for "full" insertion of the aromatic ring and hence lengthening of the helix as is the case for $n = 2$, 3, and 4.

EVIDENCE FOR A SEQUENCE-DEPENDENT DYNAMIC STRUCTURE OF
DNA IN SOLUTION

Through the use of hydrogen exchange techniques as well as the rates of reaction of formaldehyde with native and denatured DNA, von Hippel and collaborators [67,68] have shown that the conformation of DNA is "subject to continuous thermally-induced local fluctuations and distortions, i.e., 'breathing.'" We have reported on a novel kinetic technique of studying this aspect of DNA structure [30]. For example, the reporter molecules **32** and **33** were synthesized and their interaction specificities with nucleic acids were

$$--------(CH_2)_2\overset{+}{N}(CH_3)_2CH_2C_6H_5$$

11.2 Å

$$--------(CH_2)_2\overset{+}{N}(CH_3)_2CH_2C_6H_5$$

O N O

$(CH_2)_2$

$(CH_3)_2\overset{+}{N}(CH_3)_2(CH_2)_3N(CH_3)_3$

32

33

examined. It was shown by ^{1}H nmr, ultraviolet absorption, induced circular dichroism, flow dichroism, and viscometric studies that the aromatic ring of **32** and **33** intercalates between base pairs of nucleic acids. It is noted that the 1,8-naphthylimide ring of **32** may intercalate readily between base pairs of DNA without the necessity of breaking H bonds. However, in order to intercalate the 1,8,4,5-naphthyldiimide ring of **33** between base pairs of DNA, unstacking of adjacent base pairs and local melting (H bond breakage) of helix must occur. If the 1,8,4,5-naphthyldiimide ring of **33** intercalates, the N,N-dimethyl side chains must occupy opposite grooves in DNA (i.e., one side chain in the minor, the other in the major, groove) since the distance between side chains of **33** is approximately 11.2 Å. Moreover, it has been shown in this laboratory that a *tert*-butyl group, on an aromatic ring blocks the intercalation process [47]. Since the N-benzyl-N, N-dimethyl group of **33** is larger than a *tert*-butyl group, it is concluded that intercalation of the aromatic ring of **33** must be preceded by local opening, i.e., melting of the helix.

Kinetic studies of the binding process of **32** and **33** to DNA as well as the dissociation of the DNA–**32** and DNA–**33** complexes are consistent with the above interpretation. For example, the following results are obtained. (1) Intercalation of the 1,8-naphthylimide ring of **32** between base pairs of nucleic acids occurs rapidly (< 1 msec), whereas the corresponding process of intercalation of **33** with nucleic acids is significantly slower. (2) The binding of **33** to poly(dAT-dAT) and to poly(dG-dC) exhibits first-order kinetics for at least three half-lives; however, complex kinetics are obtained with salmon sperm DNA. (Complex kinetics of binding of reporter **33** to DNA may be due to the presence of ten different possible intercalation sites in the latter [58]). (3) The reaction of **33** with poly(dAT-dAT) and poly(dG-dC) is first order with respect to reporter concentration. For example, first-order kinetics are observed for the binding of **33** to poly(dAT-dAT) at DNA phosphate to reporter ratios that varied from 103/1 to 3.5/1. (4) A sequence-dependent dynamic structure of nucleic acid is indicated; i.e., the observed rate constant, k_{obs}^{assoc}, for intercalation of **33** to poly(dAT-dAT) is approximately 2.5 times as great as that noted for intercalation to poly(dG-dC). (5) The dissociation of **33** from poly(dAT-dAT)–**33** and poly(dG-dC)–**33** complexes exhibits first-order kinetics for at least three half-lives; however, complex kinetics are obtained with salmon sperm DNA [47]. (6) The dissociation of **33** from poly(dAT-dAT) and poly(dG-dC) complexes is first order with respect to the concentration of the complex [47]. (7) A sequence-dependent dynamic structure of nucleic acid is indicated; i.e., the observed rate constant for dissociation, k_{obs}^{dis}, of **33** from poly(dAT-dAT) is approximately three times as fast as that noted from the poly(dG-dC) complex [47].

The above kinetic data are consistent with a "dynamic" structure of DNA helix in solution and support the "breathing" model of DNA as postulated by von Hippel. More significantly, they demonstrate for the first time a *sequence-dependent* effect, i.e., a sequence-dependent "breathing" of DNA as noted by the relatively faster kinetics of binding to and dissociation from A–T relative to G–C base pair sites. Such effects are consistent with the structure of the Watson–Crick–Wilkins helix, since the G–C base pairs are less likely to melt (break H bond) than the A–T base pairs.

It should be noted that compounds which exhibit slow dissociation rates of the DNA complex are observed to possess template inhibitory and anti-metabolic activity [69,70]. Compound **33**, for example, is found to inhibit DNA-dependent *E. coli* RNA polymerase and to exhibit potent and selective bacteriostatic activity against gram-positive bacteria [71].

PROTEIN–DNA INTERACTIONS

The mechanism(s) by which proteins of defined amino acid sequence may recognize specific sequences of nucleic acid has been the subject of considerable interest in many laboratories [1–20,59–66,72–76]. As indicated earlier, our approach to this problem has been centered on model systems composed of small oligopeptide amides interacting with DNA of various AT/GC base pair compositions [24,26]. Although the relationship of these model studies to the overall problem of the recognition process between macromolecules (DNA and proteins) is not immediately obvious, nonetheless, the results obtained are valuable in elucidating the interaction specificities of nucleic acids with small oligopeptide systems.

Peptides Containing Aliphatic and Basic Amino Acids (IV)

Results of temperature-dependent ^{1}H nmr, ultraviolet, circular dichroism, viscometric, flow dichroism, equilibrium dialysis, and melting-temperature studies, which dealt with the interaction of over 50 peptides with nucleic acids, showed the following. Amino acid amides and di-, tri-, and tetrapeptide amides (IV) containing only aliphatic amino acids (e.g., glycine, alanine, valine, serine, and leucine) and/or basic amino acids (e.g., lysine, arginine, histidine, and ornithine) form "external"-type complexes with DNA. This conclusion is based on the following observations:

1. Proton magnetic resonance studies showed that the proton signals of the peptides in the free and DNA-bound state are nearly identical; e.g., sharp resonance signals accompanied by small upfield chemical shifts (<2–3 Hz) are observed in the complex. The results are indicative of an external binding process whereby the peptides containing aliphatic amino acids undergo rapid tumbling in the DNA complex, a behavior similar to that observed for simple polyammonium salts (see Fig. 3b). It should be noted that under the conditions of ^{1}H nmr experiments the oligopeptides were shown to be totally bound to DNA by equilibrium dialysis studies [26].

2. The interactions of the oligopeptide amides IV with salmon sperm DNA led to little or no change in the absorption and circular dichroism spectra of DNA; e.g., the intensity at 260 nm (absorption) and the CD spectrum of DNA varied by $\pm 3\%$ in the presence of IV. The results suggest that no gross alteration in the native structure of DNA is occurring [24,26].

3. In all cases, the T_m of the helix–coil transition of salmon sperm DNA was observed to increase in the presence of the oligopeptides IV. A primary sequence effect on ΔT_m ($\Delta T_m = T_m - T_{m0}$, where T_m and T_{m0} are the melting temperatures in the presence and absence of peptide) is observed; e.g., peptides containing lysine, arginine, and/or histidine exhibit a high value of ΔT_m. Moreover, a sequence-dependent effect on ΔT_m of isomeric dipeptide amide is observed; e.g., N-terminal lysyl-containing dipeptide amide, L-Lys-X-A, stabilizes the DNA helix to a greater extent than the C-terminal isomer, X-L-LysA [26,38]. Unfortunately, the interpretation of the T_m data is complicated by at least two competing processes; i.e., relative interaction of the peptide (P) with the helix (H) and with the random coils (C) and a knowledge of the value of K_H and K_C *at* or *near* the melting temperature are necessary before any valuable conclusions can be drawn.

$$H + P \underset{\longleftarrow}{\overset{K_H}{\longrightarrow}} H \cdot P \underset{\longleftarrow}{\overset{}{\longrightarrow}} C \cdot P \underset{\longleftarrow}{\overset{K_C}{\longrightarrow}} C + P$$

4. Direct binding studies of oligopeptides amide IV with DNA of various AT/GC base pair compositions have been carried out using equilibrium dialysis techniques [26,38]. The value of the apparent binding affinity, K_a, is

found to be dependent on (a) the number of positively charged ammonium groups in the peptide and (b) the ionic strength of the medium. In general, the values of K_a are found to increase by a factor of 10 per increase in the number of positive charges in the peptide amides [38]. For example, the values of K_a for the binding of L-leucylamide (monocation), L-lysylglycylamide (dication), and L-lysyl-L-lysylamide (trication) are found to be approximately 1.0–2.0 × 10^2, × 10^3, × 10^4, respectively (at pH 7.0 and 10 mM NaCl). Moreover, it is found that the peptide amides IV (especially those containing hydrophobic amino acids) exhibit a slight preference for A–T base pairs of DNA. For example, the values of K_a for the binding of L-Lys-L-LeuA to poly(dA-dT)-poly(dA-dT) (100% A–T), salmon sperm DNA (58% A–T), and *M. luteus* DNA (28% A–T) are found to be 2.0, 1.2, and 0.8 × 10^3, respectively [26]. These results are consistent with the observation that hydrophobic cations (e.g., tetramethylammonium) have a higher affinity to A–T than to G–C sites [77].

5. The oligopeptide amides IV are found to slightly decrease the specific viscosity, η_{sp}, of solution containing native salmon system DNA. Similar results are also obtained with increasing salt concentrations (e.g., NaCl) and in the presence of diammonium salts (e.g., 1,5-diaminopentane·2HCl) [24,26]. Presumably, the effect is due to shielding of neighboring negatively charged phosphate groups by the positively charged counterions, which would lead to electrostatic constriction of the DNA polymer [78].

The results (1–5) described above are consistent with a model in which the peptide amides IV (composed of aliphatic and/or basic amino acids) bind to DNA via external electrostatic, H bonding, and hydrophobic types of interaction similar to that observed for simple ammonium and/or polyammonium cations [24,26,31,38].

Peptides Containing Aromatic Amino Acids (V)

The interaction of aromatic amino acids (in the free acid form and in esters, amides, oligopeptides, etc.) has been extensively studied in a number of laboratories [24,26,32–34,79–85]. In 1970, Brown [79] was the first to propose a "bookmark" hypothesis according to which aromatic amino acids of proteins are anchored to DNA via an intercalation mechanism. Recent studies from this laboratory [24,26,32–34] on tryptophanyl-, tyrosyl-, and phenylalanyl-containing peptides utilizing ^{1}H nmr, flow dichroism, equilibrium dialysis, viscometric, and circular dichroism techniques have presented evidence that is consistent with a modified intercalation model, i.e., "partial" intercalation. The results (reviewed below) of such studies suggest that protein–DNA binding (the recognition process) may involve a "selective bookmark" as well as single-stranded β-chain–DNA interaction [26,34].

TABLE 2

Peptides That Exhibit External- and Internal-Type DNA Binding

L-Lys-L-PheA (**34**)	L-Lys-L-Phe-(Gly-L-Leu)$_2$A (**45**)
L-Lys-D-PheA (**35**)	L-Lys-L-Phe-D-Leu-L-Leu-D-LeuA (**46**)
L-Lys-L-TyrA (**36**)	L-Lys-L-Phe-Gly-L-LeuA (**47**)
L-Lys-L-TrpA (**37**)	L-Lys-L-Phe-L-AlaA (**48**)
L-Lys-L-Phe-GlyA (**38**)	L-Lys-L-Phe-D-AlaA (**49**)
L-Lys-Gly-L-PheA (**39**)	L-Lys-L-Phe-L-Ala-ValA (**50**)
L-Lys-L-Phe-diMeA (**40**)	L-Lys-L-Phe-D-Ala-L-ValA (**51**)
L-Lys-L-Phe-Gly-GlyA (**41**)	L-Lys-L-Lys-L-PheA (**52**)
L-Lys-L-Phe-L-Leu-GlyA (**42**)	L-Lys-L-Phe-L-LysA (**53**)
L-Lys-L-Phe-D-Leu-GlyA (**43**)	L-Lys-L-Phe-L-PheA (**54**)
L-Lys-L-Phe-D-Ala-L-LeuA (**44**)	

EXTERNAL- VS. INTERNAL-TYPE COMPLEXES AND THE "SELECTIVE BOOKMARK"
HYPOTHESIS

External and internal (i.e., "partial" intercalation) types of DNA binding
are observed for peptides that contain an aromatic amino acid. For example,
external-type binding is noted for the DNA complexes of L-Lys-D-PheA,
L-Lys-L-Phe-L-LeuA, and L-Lys-L-Leu-L-PheA (Table 2) upon binding to
salmon sperm DNA. Such binding is characterized by a sharp ^{1}H nmr
spectrum for all the protons of the peptide in the DNA-bound state and a
small decrease in the viscosity and reduced dichroism of DNA solution due
to electrostatic constriction of the polymer (Figs. 6–8). On the other hand, the

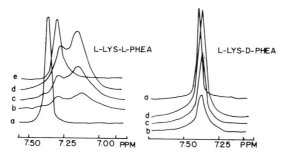

Fig. 6 The ^{1}H nmr signal of the aromatic protons of L-Lys-L-PheA and L-Lys-D-PheA
in the absence (a) and presence of DNA at a base pair to peptide ratio of 7.2 (b), 3.6 (c)
2.4 (d), and 0.5 (e). Low molecular weight salmon sperm DNA was used at 60–70 mM
phosphate/liter in the presence of 1 mM EDTA in D$_2$O (pD 7.0). The concentration of
34 and **35** was varied from 4 to 60 mM. Spectra were recorded at 34°C using a Varian
XL-100-15 spectrometer equipped with a Nicolet Technology FT accessory. Sodium
3-trimethysilylpropionate-2,-2,3,3,-d_4 was used as the internal standard [33]. (Reprinted
with permission from Gabbay *et al.*, *Biochemistry* **15**, 146 (1976). Copyright by the
American Chemical Society.)

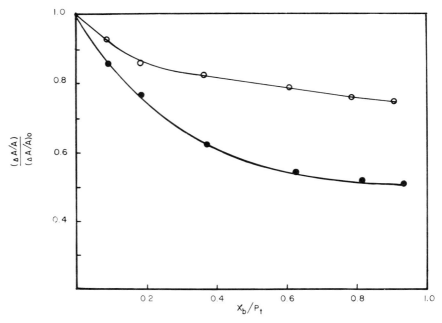

Fig. 7   The effect of bound L-Lys-L-PheA (●) and L-Lys-D-PheA (○) on the relative reduced dichroism of DNA (X_b represents the concentration of bound peptide and P_t the total DNA concentration in phosphates per liter). The relative reduced dichroism is defined as $(\Delta A/A)/(\Delta A/A)_0$, where $\Delta A = A_{\parallel} - A_{\perp}$ and A is the absorbance of a stationary DNA solution at 260 nm; $(\Delta A/A)$ and $(\Delta A/A)_0$ refer to the reduced dichroism of DNA–peptide complex and free DNA, respectively. Flow dichroism measurements were carried out at $25° \pm 1°C$ using a Cary 15 spectrometer with a Glan-Taylor calcite polarizing prism; DNA (3 mM P/liter) solution was flowed through a quartz capillary (0.415 mm radius) by means of a Sage syringe pump. The shear rate in all experiments was maintained constant at 2600 sec^{-1}. At the highest concentration used in these studies, the peptide contribution to the absorbance at 260 nm is found to be less than 1%. It should be noted that identical results are also obtained at lower DNA concentrations (0.5 mM P/liter), which indicates that the effects are caused by a molecular conformational change rather than aggregation [33]. (Reprinted with permission from Gabbay *et al.*, *Biochemistry* **15**, 146 (1976). Copyright by the American Chemical Society.)

internal-type binding exhibited by a number of oligopeptides (e.g., L-Lys-L-PheA, L-Lys-L-TyrA, L-Lys-L-TrpA; see Table 2) is characterized by large upfield chemical shifts and line broadening of the ^{1}H nmr spectrum of peptide in the DNA-bound state and a large decrease in the viscosity and reduced dichroism of DNA solution which cannot be solely attributed to electrostatic constriction of the polymer (Figs. 6–8). Presumably, for the above peptides, the single methylene group (CH_2), between the peptide backbone and the

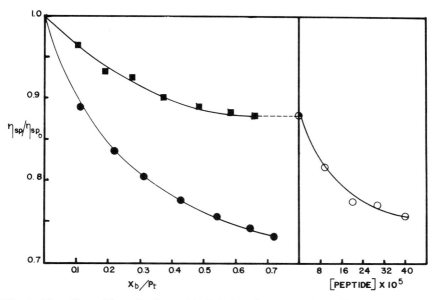

Fig. 8 The effect of bound L-Lys-L-PheA (●) and L-Lys-D-PheA (■) on the relative specific viscosity of rodlike sonicated low molecular weight salmon sperm DNA (MW < 2×10^5) at 17°C in 10 mM MES buffer (pH 6.2) (5 mM Na$^+$) using 0.2 mM DNA P/liter. Note that titration (dashed line) of the DNA–35 complex (at a bound peptide to phosphate ratio of 0.64) with L-Lys-L-PheA (**34**) (○) results in an additional lowering of η_{sp}/η_{sp_0} to a value approaching that of DNA–35 complex [33]. (Reprinted with permission from Gabbay *et al.*, *Biochemistry* **15**, 146 (1976). Copyright by the American Chemical Society.)

aromatic ring of tyrosine, tryptophan, and phenylalanine is not sufficient to allow for "full" insertion and lengthening of the helix. Identical results have been noted for the reporter molecules III ($n = 1$) and have been interpreted in terms of "partial" insertion of the aromatic ring between DNA base pairs leading to shortening of the helix by a local bending process (Fig. 5 and 9).

In addition, direct equilibrium dialysis binding studies of the interaction of dipeptide amides that form internal-type complexes (e.g., L-Lys-L-PheA, L-Lys-L-TyrA, L-Lys-L-TrpA) with DNA of various AT/GC base pair compositions show a 4-fold (or greater) preference for A–T sites [26,38]. It is noted, for example, that the affinity to A–T binding sites for the aromatic-containing lysyl dipeptide amides is tryptophan > phenylalanine > tyrosine. The biological significance of the intercalation specificity found for the aromatic-containing peptides is difficult to evaluate at present. It has been suggested that the ten distinctly different intercalation sites possible in DNA may serve as a specific means for recognition of DNA sequence by an intercalating molecule [58]. In a sense, the different intercalation sites may be considered as pages in a book with the intercalating molecule acting as a selective bookmark. Thus, the "bookmark" hypothesis proposed by Brown [79] takes on added significance now that it has been shown that the aromatic amino acid residues not only can (partially) intercalate, but do so with some degree of selectivity with respect to primary sequence of the peptide as well as base pair specificity.

STEREOSPECIFIC BINDING OF DIASTEREOMERIC PEPTIDES TO DNA

The ^{1}H nmr data (Fig. 6a and b) indicate that the protons of the aromatic ring of L-Lys-L-PheA (34) experience a large upfield chemical shift (23.5 Hz) and line broadening, whereas the aromatic protons of the diastereomeric peptide L-Lys-D-PheA (35) are relatively unaffected upon binding to DNA. A model which assumes that the aromatic ring of 34 points into the helix (i.e., partial insertion between base pairs of DNA) and the aromatic ring of 35 points outward toward the solvent can best explain the data. The larger upfield chemical shift experienced by the aromatic protons of 34 as compared to 35 is indicative of closer contact to the DNA base pairs and is due to ring current anisotropy [49,50]. On the other hand, the large ^{1}H nmr signal broadening of the aromatic protons of 34 ($\Delta\nu_{1/2} = 31$ Hz) as compared to 35 ($\Delta\nu_{1/2} = 6.5$ Hz) could be explained by several mechanisms: (1) slower tumbling rate of the aromatic ring of 34 in the DNA complex as compared to 35, (2) slow exchange between various DNA binding sites for DNA–34 as compared to DNA–35 complexes (3) the larger differences in the chemical shifts experienced by the ortho, meta, and para protons of the aromatic protons of 34 in the DNA complex as compared to 35, and/or (4) combination of all three mechanisms.

In order to discriminate between the above alternatives, spin-lattice relaxation time (T_1) measurements were performed on the DNA–34 and DNA–35 complexes under conditions of total binding [32,33]. Since the value of T_1 is determined (among other things) by the correlation time (τ_c) and the mean residence time (τ_m) [50,56,57], and because of the observation that the T_1 values of the aromatic protons of 34 and 35 in the DNA complex are nearly identical ($T_1 \simeq 0.65$ sec [32]), it is concluded that the tumbling rate ($1/\tau_c$) and the chemical exchange rate ($1/\tau_m$) of the aromatic protons of 34 and 35 in the DNA complex are very similar in magnitude. Therefore, the large ^{1}H nmr signal line broadening observed for the aromatic protons of 34 in the DNA complex ($\Delta\nu_{1/2} = 31$ Hz) can be due only to large differences in the chemical shifts experienced by the ortho, meta, and para protons. The observation of two ^{1}H nmr signals for the aromatic protons of 34 in the presence of excess DNA (Fig. 6a) would result from the greater upfield chemical shift experienced by the meta and para protons than by the ortho

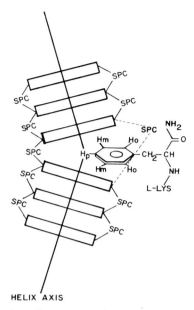

HELIX AXIS

Fig. 9 Schematic illustration of a segment of DNA duplex showing partial insertion of the aromatic ring of L-Lys-L-PheA (**34**). Under these conditions, the para and meta protons (H_p and H_m) are expected to undergo greater upfield chemical shifts than the ortho protons (H_o) due to ring current anisotropy of the neighboring bases of DNA (SPC-sugar phosphate chain).

protons as a consequence of ring current anisotropy of neighboring base pairs (Fig. 9). The observed relative areas of the two aromatic signals of DNA–**34** are consistent with this interpretation. The results of the flow dichroism (Fig.7) and viscometric (Fig. 8) studies provide added support for the "partial" insertion (or intercalation) model. The selective lowering of the relative specific viscosity and reduced dichroism of DNA solution upon binding the dipeptide amide **34** suggest that the effective length of the DNA helix is smaller in the DNA–**34** than in the DNA–**35** complex. Tilting (or bending) of the helix at the point of insertion of the aromatic ring of the dipeptide amide **34** between base pairs (schematically shown in Fig. 9) would adequately account for all the observed data. The aromatic ring of L-Lys-D-PheA (**35**), on the other hand, points outward toward the solvent and experiences small upfield chemical shifts and similar chemical shifts, δ, for the ortho, meta, and para protons ($\Delta\nu_{1/2} = 6.5$ Hz) in the DNA complex. The results of the relative specific viscosity and reduced dichroism of DNA solution upon binding the dipeptide amide **35** are consistent with the above interpretation. For example, the effects on the viscosity and dichroism of

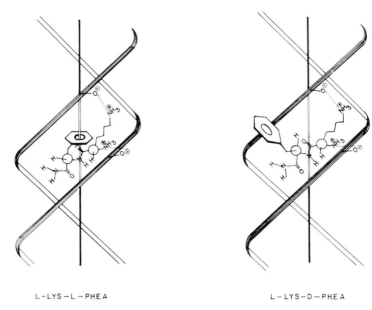

L–LYS–L–PHE A   L–LYS–D–PHE A

Fig. 10 Schematic illustration of a DNA segment showing a possible mechanism by which the α- and ε-amino groups of the N-terminal L-lysine residue are anchored stereospecifically, thus dictating the positioning of the C-terminal aromatic ring of phenylalanine in the DNA–**34** and DNA–**35** complexes.

DNA exhibited by L-Lys-D-PheA are similar to those observed for L-Lys-L-LeuA and 1,5-diaminopentane·2HCl [26,33]. Such effects are characteristic of an external-type binding and as mentioned previously are due to shielding of neighboring negatively charged counterions, which would lead to electrostatic constriction of the DNA polymer [78].

In order to account for the DNA binding specificity of **34** and **35** it is necessary to conclude that the α- and ε-amino groups of the N-terminal L-lysyl residue interact in a stereospecific manner with DNA which dictates the positioning of the aromatic ring of the C-terminal phenylalanine residue. Such an effect is schematically illustrated in Fig. 10. It should be noted that no significant differences in the binding of *L*- and *D*-phenylalanine amide to DNA have been found by ^{1}H nmr studies [24], and therefore the specificity observed with **34** and **35** cannot be attributed to the chirality of the phenylalanine residue itself.

EVIDENCE FOR PEPTIDE β-CHAIN–DNA BINDING

Wilkins [86] proposed a model based on X-ray studies in which basic proteins are wound in a helical fashion along the grooves of DNA. The polypeptide chain appears to assume a slightly modified β-chain conformation with an approximate helical increment angle of 20° per residue. The Wilkins model has received added support from the observation that β chains are not linear but helical (with dimensions similar to the DNA helix), as evidenced by the X-ray data of many proteins [87]. Carter and Kraut [88] argue, on the basis of the available protein X-ray data, for a model of RNA–protein complex in which the protein assumes a double-stranded helical β-chain configuration wrapped around the RNA helix. Their model, however, is specific for RNA and involves H bonding between the 2'-OH of ribose and the polypeptide double helix. More recently, Kim *et al.* [89] proposed a similar model for the recognition scheme between double-stranded DNA and proteins, i.e., a double-stranded antiparallel β ribbon bound to the minor groove of DNA.

In an attempt to provide experimental evidence for and/or against β-chain–DNA binding, the synthesis and interaction specificities of the oligopeptide amides **34–54** (see Table 2) were undertaken [34]. The ¹H nmr evidence strongly suggests that L-lysine in the dipeptide amides **34** and **35** binds stereospecifically to DNA and dictates the positioning of the aromatic ring of the C-terminal residue (Fig. 10). Thus, the latter technique can be used to differentiate between and/or evaluate the extent of the "in" or "out" geometry of the aromatic ring of phenylalanine residues in the DNA–peptide complexes.

Figure 11 schematically illustrates the Wilkins model for peptide–DNA binding in which the polypeptide chain assumes a helical β-chain structure that is wrapped around the nucleic acid helix. It is noted that, for polypeptides composed of L-amino acids, the side chains of the amino acid residues would alternately point "into" and "out" of the helix (Fig. 11a). Polypeptides composed of alternating L- and D-amino acids can form two types of complexes with DNA, i.e., all the side chains pointing into (Fig. 11b) or out of the helix (not shown). The rationale for the synthesis of the oligopeptide amides **34–54** now becomes a little clearer; i.e., the effect of peptide chain elongation on the ¹H nmr signal of the aromatic ring protons of the phenylalanine residue can be used to provide experimental evidence for (or against) the Wilkins model. The initial starting point, namely, the N-terminal L-lysine residue of **34–54**, is assumed to be anchored to the DNA (as evidenced from the extensive studies on the dipeptide amides **34** and **35**) in a stereospecific manner (see Fig. 10) which would allow the side chain of the neighboring amino acid (as well as amino acids at the even-numbered positions, i.e., fourth, sixth, etc.) of the L configuration to point into the helix. In addition, it is assumed that substitution of hydrophobic amino acids of the L and D configuration at the even- and odd-numbered positions of the peptide, respectively, would allow close contact of the aromatic ring "probe" of

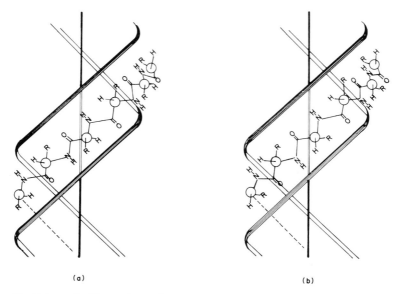

(a)   (b)

Fig. 11 Schematic illustrations of a model for peptide–DNA binding in which the polypeptide chain assumes a helical β-sheet structure that is wrapped around the nucleic acid helix. For polypeptides composed of L-amino acids, the side chains alternately point "into" and "out" of the helix (a). Polypeptides composed of alternating L- and D-amino acids can form two complexes with DNA, i.e., all the side chains pointing "into" the helix (b) or "out" of the helix (not shown).

L-phenylalanine with DNA base pairs if the former is present at an even-numbered position. On the other hand, it is reasoned that hydrophobic amino acids that are not "in register" with the aromatic ring "probe" would compete for the internal DNA site and thus weaken the interaction of the "probe" with the DNA base pairs.

If the above model is valid it should predict that the aromatic ring in L-Lys-L-TyrA (**36**), L-Lys-L-TrpA (**37**), L-Lys-L-PheA (**34**), L-Lys-L-Phe-GlyA (**38**), L-Lys-L-Phe-Gly-GlyA (**41**), L-Lys-L-Phe-D-Leu-GlyA (**43**), L-Lys-L-Phe-D-Ala-L-LeuA (**44**), L-Lys-L-Phe-(Gly-L-Leu)$_2$A (**45**), L-Lys-L-Phe-D-Leu-L-Leu-D-LeuA (**46**), L-Lys-L-Phe-D-AlaA (**49**), and L-Lys-L-Phe-D-Ala-L-ValA (**51**) is in close contact to the bases of DNA. Similarly, the model should predict a lesser contact between the aromatic ring "probe" and the DNA bases for the following peptides: L-Lys-D-PheA (**35**), L-Lys-Gly-L-Phe (**39**), L-Lys-L-Phe-L-Leu-GlyA (**46**), L-Lys-L-Phe-L-AlaA (**48**), and L-Lys-L-Phe-L-Ala-L-ValA (**50**). The ^{1}H nmr evidence is completely consistent with the above, except for two peptides [34]. For example, a large up-field chemical shift, Δδ (indicative of close contact to the nucleic acid bases), and large signal line broadening, $\Delta\nu_{1/2}$ (indicative of the "in" geometry), are observed for the peptide amides **34**, **36–38**, **40**, **41**, **44**, **45**, **49**, and **51** but not for **43** and **46**. The small upfield chemical shift and ^{1}H nmr signal line broadening observed for the peptide amides **35**, **39**, **42**, **48**, **50**, **51**, **53**, and **54** are accurately predicted [34].

The inaccurately predicted binding of the oligopeptide amides [i.e., L-Lys-L-Phe-D-Leu-GlyA (**43**) and L-Lys-L-Phe-D-Leu-L-Leu-D-LeuA (**46**)] to DNA by the model may be due to steric factors which prohibit the simultaneous "internal" binding of the L-Phe and D-Leu side chains at the 2 and 3 position of the peptides, respectively. In line with the above is the observation that substitution for the D-leucine at the 3 position by an amino acid containing a small side chain (glycine and/or D-alanine) results in a peptide in which the aromatic ring "probe" is in closer contact with the DNA base pairs, e.g., the peptide amides **38**, **49**, and **51** [34].

In summary, the ^{1}H nmr studies on the interaction specificities of 21 different oligopeptides amides are found to be consistent with the Wilkins model [86] in which the peptide chain assumes a single-stranded helical β-chain that is wrapped around the nucleic acid helix. It is reasonable to suggest that protein–DNA interactions are mediated via a single-chain helical peptide β-chain structure, especially since the primary structure of histones contains a statistically significant number of sequences in which basic and/or hydrophobic amino acids alternate with small hydrophilic amino acids, e.g., glycine and serine [90].

*Proposed Recognition Model for Protein–DNA Interaction—the
"Octopus" Model for Lac Repressor–Lac Operator Interaction*

The interaction of proteins with DNA is mediated via forces similar to those observed for DNA–small molecule complexes, i.e., electrostatic-, H-bonding-, and hydrophobic-type interactions. It is reasonable to suggest that the electrostatic interaction does not play an important role in the specific recognition between the two macromolecules. For example, an electrostatic-type interaction may occur at each of the DNA phosphate groups.

Hydrogen-bonding interactions are known to be highly specific in non-hydroxylic solvents and/or in aqueous solution under conditions of water molecule exclusion as in the case of Watson–Crick base pairs. Formation of *specific* H bonds between protein amino acid side chains and the externally exposed H bond donor and acceptor groups of the DNA base pairs has been postulated by many groups as the contributing element in the DNA–protein recognition process. It is our view that such interaction (although it may strengthen the binding) cannot contribute significantly to the recognition specificity for the following reasons: (1) the presence of H bond donor and acceptor at all DNA base pair sites and all along the protein chain; (2) the known flexibility of both macromolecules, which may accommodate the formation of numerous H-bonding schemes; and (3) the possible modification or exchange of such bonds by the intermediacy of water molecules. In line with the latter is the observed rapid diffusion of lactose repressor protein along the *E. coli* circular DNA [91] (see below).

It is reasonable to conclude that hydrophobic-type interaction (which shows a preference for A–T binding sites and is noted in the DNA binding of aromatic and hydrophobic amino acids) is primarily responsible for the recognition process. The observed "partial" intercalation of the aromatic ring of the side chain of Phe, Tyr, and Trp of small peptides (see above) and presence of A–T clusters in native DNA [92–95] are consistent with this interpretation. It should be noted that recent ^{1}H nmr studies by Bradbury *et al.* [96–98] on chromatin and histone–DNA complexes indicate that the aromatic and hydrophobic amino acids of histones are not involved in the binding to DNA but rather are involved in histone–histone intermolecular binding processes. This finding is not particularly disturbing since histones are a class of nonspecific proteins uniformly found with nearly identical sequences in all eukaryotic cells [90]. It is generally thought that the primary role of the histone proteins involves the cellular packaging of DNA [2,10–15,59–66]. Specific recognition between histones (and/or histone aggregates) and DNA is not essential for the formation of condensed DNA (see below). On the other hand, tissue-specific nonhistone proteins have been found to be involved in the control of gene expression, which is presumably mediated via a specific recognition process [11–15].

Protein–DNA recognition is best exemplified by the specific interaction of lactose repressor protein with lactose operator (lac op) DNA. For example, the *E. coli* circular genome is composed of 6×10^6 nucleotide sequences, yet the lac repressor protein specifically distinguishes a unique stretch of 35 nucleotides (i.e., the lac op) and binds to it with an affinity ($K_a \simeq 10^{13}$) that is seven orders of magnitude greater than that observed for nonoperator DNA ($K_a \simeq 10^6$). Considerable progress has been made in the past few years which makes it possible to postulate a reasonable model for this remarkable specificity. First, however, the important findings are briefly summarized.

The lac repressor protein is a tetramer (MW 150,000) with identical subunits [99]; each chain is composed of 347 amino acids of known sequence [100,101]. The N-terminal region of the lac repressor has been shown to be involved in the binding to lac operator (as well as nonoperator DNA) on the basis of genetic studies [102–105], enzymatic cleavage [106,107], and differential chemical reactivity studies [108]. The N-terminal sequence (Fig. 12a) is considerably more hydrophylic and basic than the internal and C-terminal sequences and, moreover, the protein is slightly acidic [101].

1	2	3	4	5	6	7	8	9	10
Met	Lys	Pro	Val	Thr	Leu	Tyr	Asp	Val	Ala
11	12	13	14	15	16	17	18	19	20
Glu	Tyr	Ala	Gly	Val	Ser	Tyr	Gln	Thr	Val
21	22	23	24	25	26	27	28	29	30
Ser	Arg	Val	Val	Asn	Gln	Ala	Ser	His	Val
31	32	33	34	35	36	37	38	39	40
Ser	Ala	Lys	Thr	Arg	Glu	Lys	Val	Glu	Ala
41	42	43	44	45	46	47	48	49	50
Ala	Met	Ala	Glu	Leu	Asn	Tyr	Ile	Pro	Asn

(a)

Fig. 12 (a) The amino acid sequence of the N-terminus (1–50) of the lac repressor. (Taken from Beyreuther *et al.* [101].) (b) The nucleotide sequence of the lac operator gene. Note the 2-fold symmetry of the outer (1–6 and 30–35) and inner (8–13 and 23–28) sequences. (Taken from Dickson *et al.* [109].)

The sequence of the lac op was first elucidated by Gilbert and Maxam [73] and further elaborated by Dickson *et al.* [109] and is shown in Fig. 12b. The sequence is double stranded and is comprised of 35 base pairs with inner (high A–T content) and outer regions of 2-fold symmetry.

In addition, the following important observations have been made. (1) The molecularity of the reaction involves one lac operator per lac tetramer [99]; (2) electron micrograph studies show that the free repressor as well as the repressor–operator complex contains an axis of 2-fold rotational symmetry [110]; (3) fluorescence [111] and chemical reactivity [108] studies suggest that tryptophan (at positions 190 and 209) and tyrosine (at positions 47, 126, 193, 260, and 269) residues are buried in the lac repressor tetramer; (4) the tyrosyl residues (at positions 7, 12, and 17 of the N-terminal region) react rapidly with KI_3, and the binding of the resulting iodinated repressor to lac operator and nonoperator DNA is abolished [108]; (5) in the presence of lac operator, the iodination of tyrosine (at positions 5, 12, and 17) is 50% inhibited [108]; (6) lac repressor protein obtained from mutant strains and modified at either position 16 (Ser to Pro) or 19 (Thr to Ala) is sufficient to eliminate repressor–operator binding [103].

Several models have been proposed for the lac repressor–operator complex [76,105,112–114]. Sobell [113], expanding on a model proposed earlier by Gierer [112], suggested that the operator whose sequence contains 2-fold symmetry (Fig. 12b) forms a cloverleaf structure (i.e., a hairpin loop) in which the exposed bases are presumably recognized by the repressor protein. The model possesses an approximate 4-fold symmetry relating the arms of the operator as well as the four subunits of the repressor. Such a cloverleaf structure is unlikely on the basis of thermodynamic unstability and recent studies [115] showing that closed circular λplac DNA containing a high density of negative supercoil twists (which favors the cloverleaf structure) binds repressor only slightly better than nonsupercoiled DNA. In addition, the sequence of the lac operator [73,109] has been shown to be quite different than that assumed by Gierer [112] and Sobell [113].

Adler *et al.* [105] proposed a model for the complex in which a helical segment (17–33) on the repressor interacts with the double-stranded operator in its major groove via a specific H-bonding scheme involving the amino acid side-chain residues and the exposed H bond donor and acceptors of the DNA base pairs. More recently, Patel [76] has suggested that the sequence 16–19 is involved in the binding process whereby the aromatic ring of Tyr-17 forms an intercalation complex with DNA and has proposed a model for the specific complexation of lac repressor to lac operator. The model, which is also based on the empirical calculations of the secondary structure of lac repressor protein (based on the rules of Chou and Fasman [116]), involves (1) electro-static interaction of sequence 33–37, (2) specific H-bonding interactions between Gln-18 and Ser-16 and lac op, and (3) antiparallel β-pleated sheet–DNA interactions of sequences 4–9 and 15–20. Unfortunately, both of the above models ignore the following features. (1) The lac repressor is a tetramer with identical subunits [99]; (2) the isolated sequence of the lac operator (shown in Fig. 12) is double stranded and contains 35 base pairs with inner and outer regions of 2-fold symmetry [73,109]; (3) the molecularity of the complex involves one lac repressor tetramer per lac operator; and (4) electron micrograph studies show that the complex has an axis of 2-fold rotational symmetry.

The "octopus" model shown in Fig. 13 attempts to incorporate many of the interactions that have been noted in oligopeptide–DNA binding studies as

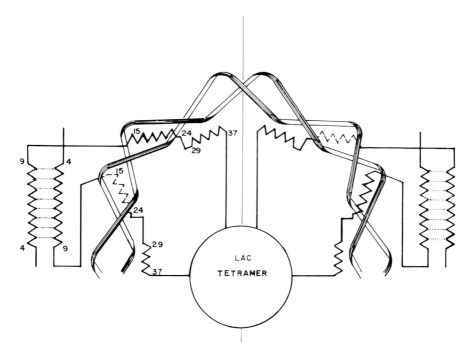

Fig. 13 Schematic illustration of the "octopus" model of DNA–protein recognition in lac repressor tetramer–lac operator complex. The model has the following features: (1) intermolecular antiparallel β sheet composed of the N-terminal sequence 4–9, (2) single-chain polypeptide β-chain–DNA interaction involving sequences 15–24 of the lac repressor with the outer (1–6 and 30–35) and inner (8–13 and 23–28) sequences of lac operator, (3) "partial" insertion of the aromatic ring of Tyr-17 between base pairs at positions 4, 5; 9, 10; 26, 27; and 31, 32 leading to bending of the helix, and (4) electrostatic interaction of His-29, Lys-33, Arg-35, and Lys-37 of each lac repressor chain with adjacent regions of lac operator. The model has 2-fold symmetry.

well as the specific criteria noted in the lac repressor–lac operator complex. It includes the following features: (1) C2 symmetry (i.e., a rotational axis of symmetry); (2) the symmetrical interaction of selected regions of sequence 15–37 of all four subunits of the lac tetramer, each of which is helically wound around one of the outer (1–6 or 30–35) or inner (8–13 or 23–28) sequences of the lac operator gene; and (3) "partial" intercalation of the aromatic ring of Tyr-17 at positions 4, 5; 9, 10; 26, 27; and 31, 32 of the lac operator (Fig. 12) with concomitant bending and unwinding of the helix. It should be noted that in the proposed model (Fig. 13) not all of the lac repressor sequence 15–37 is necessarily involved in the binding process. For example, molecular framework models suggest that the sequence 15–24 (Val*-15-Ser-16-Tyr*-17-Gln-18-Thr-19-Val*-20-Ser-21-Arg*-22-Val-23-Val*-24) may be helically wound about a cluster of five base pairs (i.e., the symmetrical outer and inner sequences) with the side chains of residues 15, 17, 20, 22, and 24 pointing "in." The remaining sequence, 25–37, may in fact be looped out and/or coiled back to allow for electrostatic interactions of His-29, Lys-33, Arg-35, and Lys-37 to adjacent regions of the lac operator. In addition, the four-armed "octopus" model includes the possible formation of an intermolecular antiparallel β sheet composed of the N-terminal sequence 4–9, the residues of which are strong β-sheet formers [76]. The latter type of binding leads to "catenane"-type structures and provides an additional locking mechanism which is in line with the observed slow dissociation rates of repressor–operator complex [116]. It is interesting that the unexpectedly high rates observed for the binding of repressor to operator have been interpreted in terms of an initial nonspecific binding to nonoperator DNA sequences followed by a rapid diffusion along the DNA molecule until the operator region is reached [91]. The proposed "catenane"-type structure, which is expected to be strongly dependent on the presence and correct spacings of symmetrical operator sequences, would provide a unique "capture" mechanism. This type of protein–DNA interaction has not been previously described and might well be of great importance in the recognition process.

The "octopus" model is admittedly highly hypothetical at this stage; however, unlike the previous models it is consistent with all the available data. Moreover, it is based on a considerable body of evidence obtained from the interaction specificities of small oligopeptides with DNA including (1) "partial" intercalation of an aromatic ring [24,26,33,34], (2) A–T cluster recognition (e.g., the inner sequences of lac operator, 8–13 and 23–28), and (3) single-stranded β-chain–DNA binding [34].

It is noteworthy that recent studies on the structure of chromatin have revealed that histones are present in oligomeric forms, i.e., tetramers and octamers [16–20]. Chromatin structure may also be envisioned in terms of a series of octopi (histone oligomer) with appendages (polypeptide chain), helically wrapped about the DNA, causing extensive compaction and condensation of the polymer. Such a model is consistent with the bead-like structure of chromatin seen by electron microscopy [16–20]. Specific recognition of a DNA sequence by proteins may occur in a similar manner but with the added provision for formation of internal hydrophobic-type binding (e.g., "partial" intercalation of aromatic residues and A–T cluster recognition) and possibly sequence-dependent "catenane"-type structures.

REFERENCES

1. Bekhor, I., Kung, G. M., and Bonner, J. (1969). *J. Mol. Biol.* **39**, 351.
2. Bartley, J., and Chalkley, R. (1973). *Biochemistry* **12**, 468.
3. Clark, R. J., and Felsenfeld, G. (1971). *Nature (London), New Biol.* **229**, 101.
4. Olins, D. E., and Olins, A. L. (1971). *J. Mol. Biol.* **57**, 437.
5. Riggs, A. D., Lin, S., and Wells, R. D. (1972). *Proc. Natl. Acad. Sci. U.S.A.* **69**, 761.
6. Shih, T. Y., and Fasman, G. D. (1972). *Biochemistry* **11**, 398.
7. Simpson, R. T., and Sober, H. A. (1970). *Biochemistry* **9**, 3103.
8. Yarus, M. (1973). *Annu. Rev. Biochem.* **38**, 841.
9. Zimmerman, E. (1972). *Angew. Chem., Int. Ed. Engl.* **11**, 496.
10. Johnson, R. S., Chan, A., and Hanlon, S. (1972). *Biochemistry* **11**, 4347.
11. Kleinsmith, L. J., Heidema, J., and Carroll, A. (1970). *Nature (London)* **226**, 1025.
12. Gilmour, R. S., and Paul, J. (1970). *FEBS Lett.* **9**, 242.
13. Spelsberg, T. C., and Hnilica, L. S. (1970). *Biochem. J.* **120**, 435.
14. Stein, G. S., and Farber, J. (1972). *Proc. Natl. Acad. Sci. U.S.A.* **69**, 2918.
15. Stein, G. S., Spelsberg, T. C., and Kleinsmith, L. J. (1974). *Science* **183**, 817.
16. D'Anna, J. A., and Isenberg, I. (1974). *Biochemistry* **13**, 4992.
17. Kornberg, R. D. (1974). *Science* **184**, 868.
18. Kornberg, R. D., and Thomas, J. O. (1974). *Science* **184**, 865.
19. Olins, A. L., and Olins, D. E. (1974). *Science* **183**, 330.
20. van Holde, K. E., Sahasrabuddhe, C. G., and Shaw, B. R. (1974). *Biochem. Biophys. Res. Commun.* **70**, 1365.
21. Gabbay, E. J. (1969). *J. Am. Chem. Soc.* **91**, 5136.
22. Gabbay, E. J., and Glaser, R. (1971). *Biochemistry* **10**, 1665.
23. Gabbay, E. J., Glaser, R., and Gaffney, B. L. (1970). *Ann. N. Y. Acad. Sci.* **171**, 810.
24. Gabbay, E. J., Sanford, K., and Baxter, C. S. (1972). *Biochemistry* **11**, 3429.
25. Gabbay, E. J., Scofield, R., and Baxter, C. S. (1973). *J. Am. Chem. Soc.* **95**, 7850.
26. Gabbay, E. J., Sanford, K., Baxter, C. S., and Kapicak, L. (1973). *Biochemistry* **12**, 4021.
27. Gabbay, E. J., and DePalois, A. (1971). *J. Am. Chem. Soc.* **93**, 562.
28. Gabbay, E. J., Sanford, K., and Baxter, C. S. (1972). *J. Am. Chem. Soc.* **94**, 2876.
29. Gabbay, E. J., and Sanford, K. (1974). *Bioorg. Chem.* **3**, 91.
30. Gabbay, E. J., DeStefano, R., and Baxter, C. S. (1973). *Biochem. Biophys. Res. Commun.* **51**, 1083.
31. Kapicak. L., and Gabbay, E. J. (1975). *J. Am. Chem. Soc.* **97**, 403.
32. Adawadkar, P. D., Wilson, W. D., Brey, W., and Gabbay, E. J. (1975). *J. Am. Chem. Soc.* **97**, 1959.

33. Gabbay, Z. J., Adawadkar, P. D., and Wilson, W. D. (1976). *Biochemistry* **15**, 146.
34. Gabbay, E. J., Adawadkar, P. D., Kapicak, L., Pearce, S., and Wilson, W. D. (1976). *Biochemistry* **15**, 152.
35. Lerman, L. S. (1961). *J. Mol. Biol.* **3**, 18.
36. Mahler, H. R., Goutarel, R., Khuong-Huu, G., and Ho, M. T. (1966). *Biochemistry* **5**, 1966.
37. Passero, F., Gabbay, E. J., Gaffney, B., and Kuruscev, T. (1970). *Macromolecules* **3**, 158.
38. Sanford, K. (1972) Ph.D. Thesis, University of Florida, Gainesville (unpublished results).
39. Tinoco, I. (1960). *J. Am. Chem. Soc.* **82**, 4745.
40. Tinoco, I. (1961). *J. Chem. Phys.* **34**, 1067.
41. Devoe, H., and Tinoco, I. (1962). *J. Mol. Biol.* **4**, 518.
42. Rhodes, W. (1961). *J. Am. Chem. Soc.* **83**, 3609.
43. Gabbay, E. J., and Malin, M. (1968). Unpublished results.
44. Bush, C. A., and Tinoco, I. (1967). *J. Mol. Biol.* **23**, 601.
45. Schellman, J. A. (1968). *Acc. Chem. Res.* **1**, 144.
46. Scatchard, G. (1949). *Ann. N.Y. Acad. Sci.* **51**, 660.
47. DeStefano, R. (1973), Ph.D. Thesis, University of Florida, Gainesville (unpublished results).
48. Gabbay, E. J., and Denham, S. (1971). Unpublished results.
49. Jardetsky, O., and Jardetsky, C. D. (1962). *Methods Biochem. Anal.* **9**, 235.
50. Pople, J. A., Schneider, W. G., and Bernstein, H. J. (1959). "High Resolution Nuclear Magnetic Resonance." McGraw-Hill, New York.
51. McDonald, C. C., Phillips, W. D., and Penswick, J. (1965). *Biopolymers* **2**, 169.
52. McDonald, C. C., Phillips, W. D., and Lazar, J. (1967). *J. Am. Chem. Soc.* **89**, 4166.
53. Jardetsky, O., Wade, N. G., and Fisher, J. J. (1963). *Nature (London)* **197**, 183.
54. Schmidt, P. G., Stark, G. R., and Baldeschweiler, J. D. (1969). *J. Biol. Chem.* **244**, 1860.
55. Gerig, J. T. (1968). *J. Am. Chem. Soc.* **90**, 2681.
56. Dwek, R. A. (1973). "Nuclear Magnetic Resonance in Biochemistry." Oxford Univ. Press (Clarendon), London and New York.
57. Mildvan, A. S., and Cohn, M. (1970). *Adv. Enzymol. Relat. Areas Mol. Biol.* **33**, 1.
58. Gabbay, E. J. DeStefano, R., and Sanford, K. (1972). *Biochem. Biophys. Res. Commun.* **46**, 155.
59. Ziccardi, R., and Schumaker, V. (1973). *Biochemistry* **12**, 3231.
60. Pardon, J. F., and Wilkins, M. H. F. (1972). *J. Mol. Biol.* **68**, 115.
61. Sadgopal, A., and Bonner, J. (1970). *Biochim. Biophys. Acta* **207**, 227.
62. Simpson, R. T. (1972). *Biochemistry* **11**, 2003.
63. Shih, T. Y., and Fasman, G. D. (1972). *Biochemistry* **11**, 2003.
64. Bradbury, E. M., Carpenter, B. G., and Rattle, H. W. E. (1973). *Nature (London), New Biol.* **241**, 123.
65. Clark, R. J., and Felsenfeld, G. (1971). *Nature (London) New Biol.* **229**, 101.
66. Axel, R., Cedar, H., and Felsenfeld, G. (1975). *Biochemistry* **14**, 2489.
67. Printz, M. P., and von Hippel, P. H. (1965). *Proc. Natl. Acad. Sci. U.S.A.* **53**, 363.
68. McConnell, B., and von Hippel, P. H. (1971). *J. Mol. Biol.* **50**, 297 and 317.
69. Muller, W., and Crothers, D. M. (1968). *J. Mol. Biol.* **35**, 251.
70. Gabbay, E. J., Grier, D., Fingerle, R. E., Riemer, R., Levy, R., Pearce, S. W., and Wilson, W. D. (1976). *Biochemistry* (in press).
71. Gabbay, E. J., Mayhew, D., and Rosen, I. (1973). Unpublished results.

72. Ptashne, M. (1967). *Nature (London)* **214**, 232.
73. Gilbert, W., and Maxam, A. (1973). *Proc. Natl. Acad. Sci. U.S.A.* **70**, 3581.
74. Platt, T., Files, J. G., and Weber, K. (1973). *J. Biol. Chem.* **248**, 110.
75. Adler, A. J., Moran, E. C., and Fasman, G. D. (1975). *Biochemistry* **14**, 4179.
76. Patel, D. J. (1975). *Biochemistry* **14**, 1057.
77. Shapiro, J. T., Stannard, B. S., and Felsenfeld, G. (1969). *Biochemistry* **8**, 3219.
78. Cohen, G., and Eisenberg, H. (1969). *Biopolymers* **35**, 251.
79. Brown, P. E. (1970). *Biochim. Biophys. Acta* **213**, 282.
80. Helene, C., Dimicoli, J. L., and Brun, F. (1971). *Biochemistry* **10**, 3802.
81. Dimicoli, J. L., and Hélène, C. (1974). *Biochemistry* **13**, 724.
82. Brun, F., Toulme, J. J., and Hélène, C. (1975). *Biochemistry* **14**, 558.
83. Durand, M. Maurizot, J. C., Borazan, H. N., and Hélène, C. (1975). *Biochemistry* **14**, 563.
84. Novak, R. L., and Dohnal, J. (1973). *Nature (London), New Biol.* **243**, 110.
85. Sundaralingham, M., and Arora, S. K. (1972). *J. Mol. Biol.* **71**, 49.
86. Wilkins, M. H. F. (1956). *Cold Spring Harbor Symp. Quant. Biol.* **21**, 75.
87. Chothia, C. (1973). *J. Mol. Biol.* **75**, 295.
88. Carter, C. W., and Kraut, J. (1974). *Proc. Natl. Acad. Sci. U.S.A.* **71**, 283.
89. Kim, S. H., Sussman, J. L., and Church, G. M. (1975). *In* "Structure and Conformation of Nucleic Acids and Protein-Nucleic Acid Interactions" (M. Sundarlingham and S. T. Rao, ed.), p. 571. Univ. Park Press, Baltimore, Maryland.
90. Delange, R. J., and Smith, E. L. (1972). *Acc. Chem. Res.* **5**, 368.
91. Richter, P. H., and Eigen, M. (1974). *Biophys. Chem.* **2**, 255.
92. Jones, K. W. (1970). *Nature (London)* **225**, 912.
93. Yunis, J. J., and Yasmineh, W. G. (1970). *Science* **168**, 263.
94. Walker, P. M. B. (1968). *Nature (London)* **219**, 228.
95. Blumenfeld, M., Fox, A. S., and Forrest, H. S. (1973). *Proc. Natl. Acad. Sci. U.S.A.* **70**, 2772.
96. Bradbury, E. M., Cary, P. D., Crane-Robinson, C., and Rattle, H. W. E. (1973). *Ann. N.Y. Acad. Sci.* **222**, 266.
97. Baldwin, J. P., Boseley, P. G., Bradbury, E. M., and Ibel, K. (1975). *Nature (London)* **253**, 245.
98. Bradbury, E. M., and Rattle, H. W. E. (1972). *Eur. J. Biochem.* **27**, 270.
99. Gilbert, W., and Muller-Hill, B. (1966). *Proc. Natl. Acad. Sci. U.S.A.* **56**, 1891.
100. Beyreuther, K., Adler, K., Giester, N., and Klemm, A. (1973). *Proc. Natl. Acad. Sci. U.S.A.* **70**, 3576.
101. Beyreuther, K., Adler, K., Fanning, E., Murray, C., Klemm, A., and Giester, N. (1975). *Eur. J. Biochem.* **59**, 491.
102. Muller-Hill, B., Crapo, L., and Gilbert, W. (1968). *Proc. Natl. Acad. Sci. U.S.A.* **59**, 1259.
103. Weber, K., Platt, T., Ganem, D., and Miller, J. (1972). *Proc. Natl. Acad. Sci. U.S.A.* **69**, 3624.
104. Davies, J., and Jacob, F. (1968). *J. Mol. Biol.* **36**, 413.
105. Adler, K., Beyreuther, K., Fanning, E., Giester, N., Gronnenborn, B., Klemm, A., Muller-Hill, B., Pfahl, M., and Schmitz, A. (1972). *Nature (London)* **237**, 322.
106. Platt, T., Files, J. G., and Weber, K. (1973). *J. Biol. Chem.* **248**, 110.
107. Lin, S., and Riggs, A. D. (1975). *Biochem. Biophys. Res. Commun.* **62**, 704.
108. Fanning, T. G. (1975). *Biochemistry* **14**, 2512.
109. Dickson, R. C., Abelson, J., Barnes, W. M., and Reznikoff, W. S. (1975). *Science* **187**, 27.

110. Ohshima, Y., Horiuchi, T., and Yanagida, M. (1975). *J. Mol. Biol.* **91**, 515.
111. Maurizot, J. C., Charlier, M., and Hélène, C. (1974). *Biochem. Biophys. Res. Commun.* **60**, 951.
112. Gierer, A. (1966). *Nature (London)* **212**, 1480.
113. Sobell, H. M. (1972). *Proc. Natl. Acad. Sci. U.S.A.* **69**, 2483.
114. Steitz, T. A., Richmond, T. J., Wise, D., and Engelman, D. (1974). *Proc. Natl. Acad. Sci. U.S.A.* **71**, 593.
115. Wang, J. C., Barkley, M. D., and Bourgeois, S. (1974). *Nature (London)* **251**, 249.
116. Chou, P. Y., and Fasman, G. C. (1974). *Biochemistry* **13**, 211 and 222.

4

Approaches to Carboxyl-Terminal Sequencing and End-Group Determinations in Peptides

G. Marc Loudon, M. E. Parham, and Marvin J. Miller

INTRODUCTION

The determination of the amino acid sequence of insulin by Sanger and his co-workers [1] heralded the beginning of an age in which the determination of the primary structure of proteins would evolve from an arduous effort requiring many scientist-years of labor to a procedure much of which borders on the routine. The explosion of sequence information in recent years is important on several fronts. Crystallographers rely quite heavily in general on sequence information in the interpretation of electron density maps in terms of specific structures. Sequence information has provided in at least one well-known case (sickle-cell anemia) an understanding of the molecular basis of a genetically transmitted disease [2]. Sequence information lies at the heart of "chemical taxonomy," which presents especially intriguing insights into the relationships in the evolution of various species [3]. Unexpected relationships in sequence among proteins that are apparently unrelated functionally, such as α-lactalbumin and lysozyme [4], have been uncovered, and sequence information is central to modern theories of antibody diversity [5]. Sequencing techniques are essential for the elucidation of the site of action of protein modification reagents. There are undoubtedly many other examples of the utility of primary structural information, and few would question the importance of this area of scientific endeavor.

Among the techniques for the elucidation of primary sequences, the use of the Edman reagent [6], especially in modern mechanized versions [7], for the sequential degradation of polypeptide chains from the amino terminus is best known and most widespread. This marvelous procedure presently allows degradation of peptides at a rate of up to 15 residues per day. Closely allied to the determination of protein sequence are "end-group" methods, some of the best known of which employ fluorescamine [8], 5-*N,N*-dimethyl-aminonaphthalenesulfonyl chloride (dansyl chloride) [9], and 2,4-dinitro-fluorobenzene [10] and its water-soluble analogs [11] (Sanger reagents). End-group methods also provide, in the absence of more complete sequence information, data bearing on the identity of the polypeptide chains in multichain proteins, and these methods are often valuable adjuncts to the use of other sequencing tools in the determination of the relative order in a protein of degradation peptides and in assaying the results of sequential degradation procedures themselves.

Despite the routine use of degradation procedures at the amino terminus of proteins and peptides, the situation at the opposite end of the polypeptide chain is considerably less well developed. One might reasonably ask why one would wish to carry out degradations of polypeptides from the carboxyl terminus when in fact methods for degradation from the amino terminus are so well developed. There are several important answers to this question. The attitude that one must take toward a carboxyl-terminal degradation is not so much that it will be competitive with the Edman method but that it will be complementary to it. If, for example, we envisioned the day when a carboxyl-terminal method were as rapid and simple as the Edman procedure, we could imagine dividing a sample of a 70-residue peptide in half and sequencing 40 residues from each end (overlapping in the center), thus doubling the information available from a given peptide. Another, more immediate reason for continuing the development of these methods is that it would be desirable to have C-terminal end-residue analyses that are as reliable and general as N-terminal methods for aid in specifying the number of polypeptide chains in multichain proteins, as well as for assistance in the ordering of degradation peptides in sequences of larger polypeptides. Some methods of specific cleavage of polypeptide chains [12], including one that we are developing, result in peptides that have a blocked amino terminus; a carboxyl-terminal method opens these peptides to analysis. Finally, if one has covalently modified a protein which, upon degradation, yields a peptide with the modified residue near the carboxyl terminus, it would be desirable to identify the modification site with the economy of effort inherent in a satisfactory carboxyl-terminal procedure rather than to carry out multiple degradations from the amino terminus to yield the same information.

It is the purpose of this review to summarize the chemistry that has characterized the efforts to develop degradations of peptides from the carboxyl terminus, as well as carboxyl-terminal end-group methods, and to report in this context our research in this area.

ENZYMATIC DEGRADATIONS

Ever since Lens [13] in 1949 proposed the use of carboxypeptidase A in terminal-residue analysis, this enzyme and carboxypeptidase B have been important tools for this purpose. The use of these enzymes has been the subject of thorough reviews [14]. A recent development, of importance, however, has been the isolation from baker's yeast [15] of a new enzyme, carboxypeptidase Y, which appears to have considerably broader specificity [16] than carboxypeptidases A and B, including an unprecedented ability to digest the peptide bond of carboxyl-terminal proline. This enzyme is also of considerable mechanistic interest, since it appears to be a serine protease rather than a zinc metalloenzyme like the other carboxypeptidases [17].

Although the utility of these enzymes is undeniable, enzymatic methods of degradation by nature suffer from the twin problems of specificity and phase. Although carboxypeptidase Y is evidently of broad specificity, it would indeed be surprising if this enzyme did not exhibit preferences for certain amino acids or sequences; in fact, this appears to be the case [16]. This situation, when it occurs, can result in the release of two or more amino acids at the same *apparent* rate, thus complicating the analysis of sequences. It is perhaps obvious that the use of any enzyme method means that the degradations of individual peptide molecules will in general be out of phase with respect to each other; that is, one cannot expect to isolate at any time peptides in which the same number of residues have been lost on each molecule. Although this problem can be solved in principle, a practical solution is hard to envision, and no attempts in this direction have been reported.

"BACK DOOR" METHODS

There are methods available that enable one to identify which of several peptides resulting from a chemical or enzymatic cleavage of a protein lies at the carboxyl terminus. Having been identified, the peptide itself, if not too long, can be sequenced from its amino terminus (from the "back door"), thus permitting the identification of the carboxyl-terminal sequence of the original protein. Hargrave and Wold [18] presented a method whereby one allows glycinamide and protein to condense in the presence of a carbodiimide; after proteolytic digestion, only the carboxyl-terminal peptide is totally devoid of free carboxylic acid groups and can in principle be separated from the remaining peptides, each of which has one carboxylate at its C-terminus resulting from digestion, on the basis of charge characteristics. The same general idea was employed by Duggleby and Kaplan [19], who preferred ethanolamine to glycinamide as a protecting group. These methods, of course, rely on quantitative protection of all free carboxyl groups in the original protein and will be useful when subsequent digestion of the protected protein yields peptides of manageable size.

CYCLIZATION AND CLEAVAGE OF A CARBOXYL-TERMINAL ACYL DERIVATIVE

The methods described in this section bear closest resemblance to the Edman degradation of proteins from the amino terminus, and this resemblance represents one of their attractive features. An early attempt to cyclize and hydrolyze a carboxyl-terminal acyl derivative was especially elegant in that it attempted to turn an annoying side reaction in peptide synthesis to advantage. This method, due to Khorana [20], is shown in Scheme 1. In this procedure, $O- \rightarrow N$-acyl shift of a C-terminal O-acylisourea gives an N-acylurea, which can be internally attacked by the amide nitrogen to yield the carbamylated peptide (1), which can then be selectively hydrolyzed. Problems were encountered in the application of this idea, however, because of side reactions and contaminating materials resulting from simple hydrolysis.

An idea nearly 50 years old which has seen recent revitalization is that of Schlack and Kumpf [21]; it has been developed in ensuing years by Tibbs [22], Waley and Watson [23], Kenner, Khorana, and Stedman [24], Yamashita [25], and Stark [26]. The chemistry of this method is shown in Scheme 2 and is contrasted there with the Edman degradation. This scheme has immediate and obvious appeal for the laboratory already accustomed to Edman procedures in both the similarity of the methodology and the close relationship of the derivatives produced, the thiohydantoin (2) from the carboxyl-terminal procedure and the phenylthiohydantoin (3) from the Edman method. The acknowledged difficulties with the method are the failure to remove proline and the problems with aspartic acid due to anhydride formation. The incomplete hydrolysis of the acylthiohydantoin at each step has precluded lengthy sequence determinations by this method. Recently, this procedure has been utilized in a solid-phase approach [27,28] as suggested by

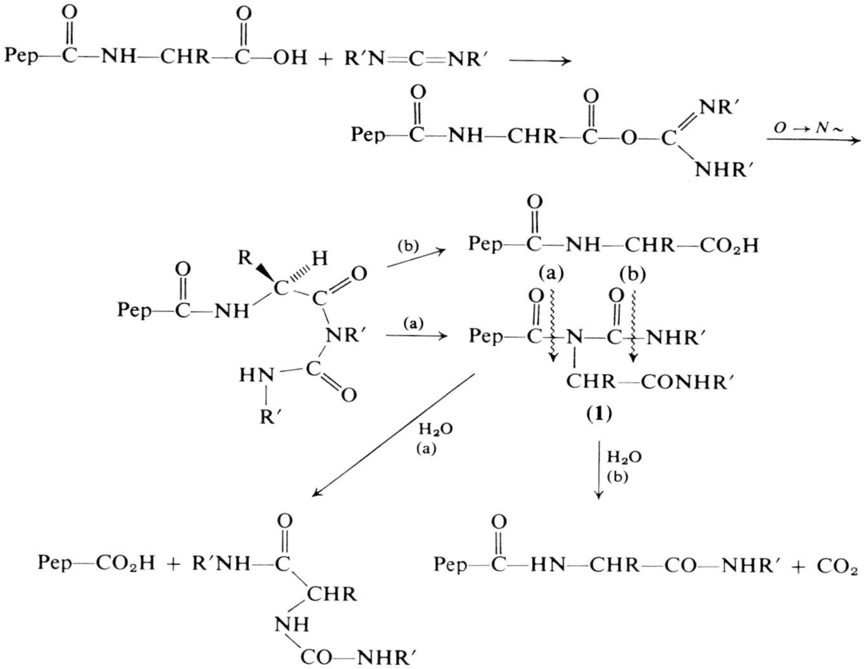

$$Pep-\overset{O}{\overset{\|}{C}}-NH-CHR-\overset{O}{\overset{\|}{C}}-OH + R'N=C=NR' \longrightarrow$$

$$Pep-\overset{O}{\overset{\|}{C}}-NH-CHR-\overset{O}{\overset{\|}{C}}-O-C\overset{NR'}{\underset{NHR'}{}} \xrightarrow{O \rightarrow N\sim}$$

Scheme 1 The Khorana carboxyl-terminal peptide degradation. (a) Productive paths, (b) nonproductive paths (Pep, Residual peptide).

Stark [29]. This modified solid-phase method is interesting from the point of view of the glass supports used and of the degradation itself.

An idea very similar to the one just presented is currently under development by Tarr [30] and involves the formation, cyclization, and subsequent hydrolysis of a C-terminal cyanamide (Scheme 3) to yield an "iminohydantoin" (4). The degradations of C-terminal proline and aspartic acid, not surprisingly, met the same fate as those in the Stark procedure.

EXCLUSIVE LABELING OF THE CARBOXYL-TERMINAL RESIDUE

Another philosophy of carboxyl-terminal residue identification in peptides which has been pursued in various ways is that of exclusively labeling the carboxyl-terminal amino acid in some way, hydrolyzing the labeled peptide, and discerning which amino acid or derivative carries the label. Hydride

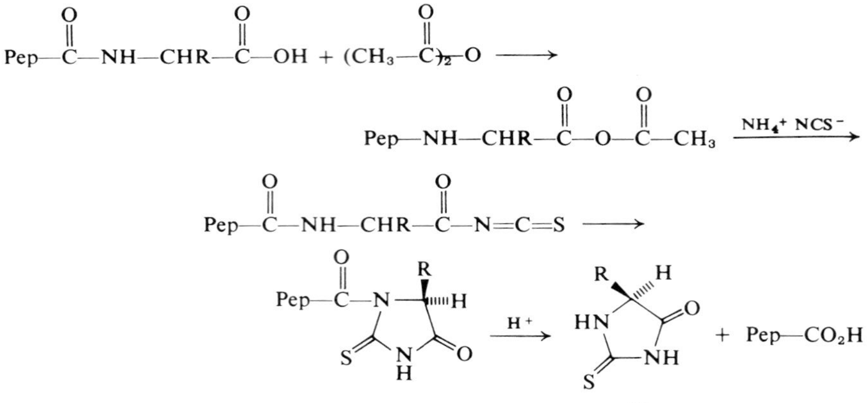

$$\text{Pep}-\overset{\overset{\displaystyle O}{\|}}{C}-NH-CHR-\overset{\overset{\displaystyle O}{\|}}{C}-OH + (CH_3-\overset{\overset{\displaystyle O}{\|}}{C})_2O \longrightarrow$$

$$\text{Pep}-NH-CHR-\overset{\overset{\displaystyle O}{\|}}{C}-O-\overset{\overset{\displaystyle O}{\|}}{C}-CH_3 \xrightarrow{NH_4^+ \ NCS^-}$$

$$\text{Pep}-\overset{\overset{\displaystyle O}{\|}}{C}-NH-CHR-\overset{\overset{\displaystyle O}{\|}}{C}-N=C=S \longrightarrow$$

$$\text{Pep}-\overset{\overset{\displaystyle O}{\|}}{C}-\underset{\underset{\displaystyle H}{N}}{\overset{R}{N}}\cdots H \ \ \underset{S}{\ } \ \ N-O \xrightarrow{H^+} \ \ \underset{S}{HN}\overset{R \quad H}{\ }\underset{NH}{\ }O \ + \ \text{Pep}-CO_2H$$

(2)

The Edman Degradation from the amino terminus:

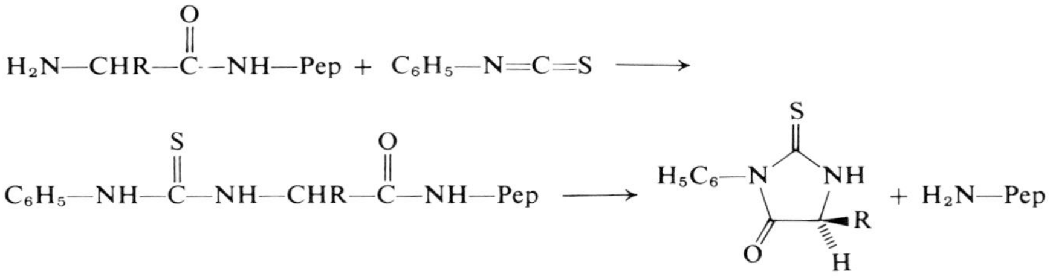

$$H_2N-CHR-\overset{\overset{\displaystyle O}{\|}}{C}-NH-\text{Pep} + C_6H_5-N=C=S \longrightarrow$$

$$C_6H_5-NH-\overset{\overset{\displaystyle S}{\|}}{C}-NH-CHR-\overset{\overset{\displaystyle O}{\|}}{C}-NH-\text{Pep} \longrightarrow \ \ H_5C_6-N\overset{S}{\ }NH \ \underset{O \quad H}{\overset{R}{\ }} \ + \ H_2N-\text{Pep}$$

(3)

Scheme 2   The Stark carboxyl-terminal peptide degradation.

$$\text{Pep}-\overset{\overset{\displaystyle O}{\|}}{C}-NH-CHR-\overset{\overset{\displaystyle O}{\|}}{C}-OH + \underset{\underset{\displaystyle SR'}{|}}{H_2\overset{+}{N}=C}-NH_2 \xrightarrow{EDC}$$

$$\text{Pep}-\overset{\overset{\displaystyle O}{\|}}{C}-NH-CHR-\overset{\overset{\displaystyle O}{\|}}{C}-\underset{\underset{\displaystyle NH}{\overset{\displaystyle \|}{C-SR'}}}{NH} \xrightarrow{base}$$

$$\text{Pep}-\overset{\overset{\displaystyle O}{\|}}{C}-NH-CHR-\overset{\overset{\displaystyle O}{\|}}{C}-NH-C\equiv N + R'SH \xrightarrow{OH^-} \text{Pep}-CO_2H + \ \ HN\overset{NH}{\ }NH \ \underset{O \quad R}{\ }$$

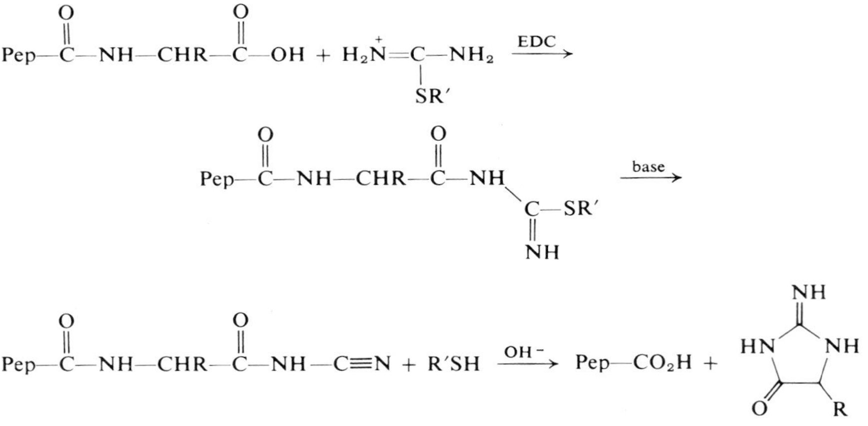

(4)

Scheme 3   The Tarr carboxyl-terminal peptide degradation (EDC, ethyldiaminopropyl-carbodiimide).

reducing agents have been used in this manner to reduce specifically the carboxyl-terminal amino acid to an amino alcohol. Chibnall and Rees [31] found that esters of peptide carboxyl groups can be reduced by LiBH$_4$ without concomitant reduction of peptide bonds. The carboxyl-terminal amino alcohols resulting from this procedure can be isolated and identified after hydrolysis of the reduced peptide. Bailey [32] extended this idea to a sequential degradation by cyclizing the carboxyl-terminal amino alcohol after LiBH$_4$ reduction to an imidate ester, which opens in acids to a new ester on which the LiBH$_4$ procedure can be repeated (Scheme 4). The requirement for acid-catalyzed ester formation and solvents (tetrahydrofuran, acetonitrile) generally incompatible with solubility of larger proteins would appear to be a drawback to the general applicability of these procedures. However, the existence of modern solid-phase sequencing methods, as well as some newer methods for making peptide methyl or ethyl esters (trimethyl- or triethyloxonium fluoroborate [29,33]), suggests that this procedure, in which the yields were really quite good, might be appropriately reexamined. Other hybride reducing agents, such as sodium borohydride [34] and sodium dihydrobis(2-methoxyethoxy) aluminate [35], have also been investigated, but the generality of these reagents is not established, and the advantages over LiBH$_4$ are not clear.

The next methods to be discussed in this section, unlike the previous method, are incapable of extension to sequential processes; on the other hand, two of these methods, the Akabori hydrazinolysis and the Matsuo tritiation

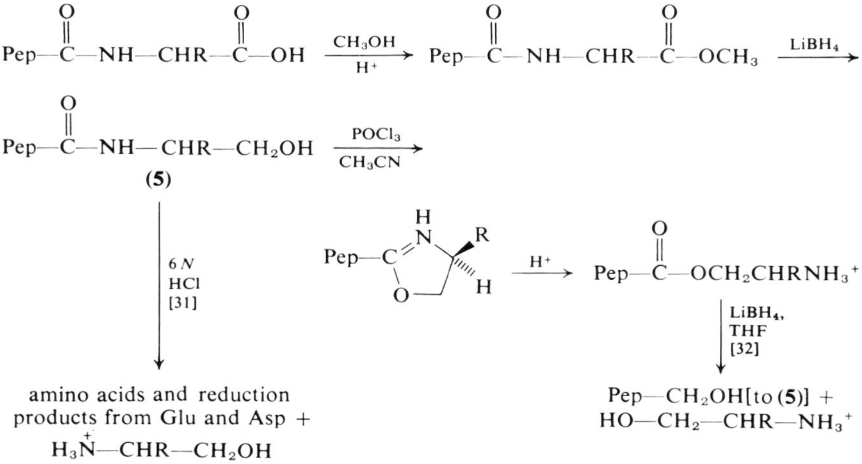

Scheme 4 The LiBH$_4$ procedure (THF, tetrahydrofuran).

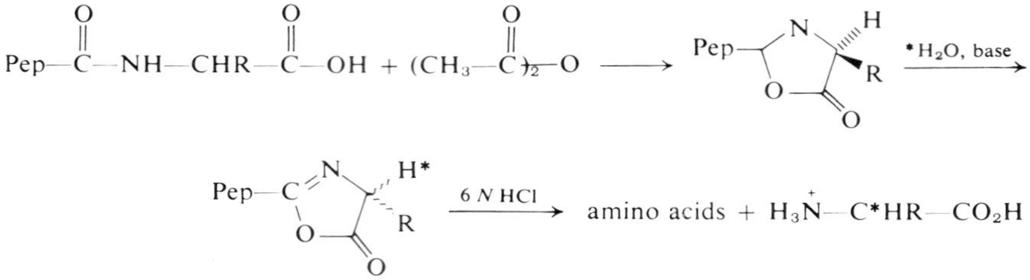

$$\text{Pep}-\overset{\overset{\displaystyle O}{\|}}{C}-NH-CHR-\overset{\overset{\displaystyle O}{\|}}{C}-OH \ + \ (CH_3-\overset{\overset{\displaystyle O}{\|}}{C})_2-O \ \longrightarrow \ \text{Pep} \begin{array}{c} N \\ \diagdown \end{array} \overset{H}{\underset{R}{\diagdown}} \ \overset{*H_2O,\ base}{\longrightarrow}$$

$$\text{Pep}-C \overset{N}{\underset{O}{\diagdown}} \overset{H^*}{\underset{R}{\diagdown}} \ \overset{6\ N\ HCl}{\longrightarrow} \ \text{amino acids} \ + \ H_3\overset{+}{N}-C^*HR-CO_2H$$

Scheme 5   The Matsuo tritium labeling procedure.

method, are currently the two most widely used chemical methods for identification of the carboxyl-terminal amino acid residue.

In the hydrazinolysis procedure [36–38], the peptide is exhaustively hydrazinolyzed with pure, freshly distilled hydrazine; only the carboxyl-terminal amino acid does not form a hydrazide and is readily separated from the other hydrazides. Among the difficulties with this method are the hydrazinolysis of glutamine and asparagine, the partial destruction of arginine, and the occasional difficulty of obtaining a reagent of satisfactory quality [38]. Although it has been noted that freshly opened bottles of hydrazine are evidently free of explosion hazard, it would be unwise to minimize the potential difficulties with the safety of this method.

The Matsuo tritium labeling method [39,40], like the Khorana procedure illustrated above, not only is useful, but has the aesthetic appeal of turning a reaction that is a nuisance in peptide synthesis into a reliable analytical tool. This method, illustrated in Scheme 5, involves the formation of a carboxyl-terminal oxazolinone (azlactone), which is then radiolabeled in tritiated water and base. Subsequent hydrolysis yields a labeled amino acid characteristic of the carboxyl terminus. Although it has been noted [41] that β-linked aspartic acid and γ-linked glutamic acid residues will also be labeled, these residues will in fact interfere in any carboxyl-terminal method, since they are in reality additional carboxyl termini. On the other hand, the extent to which the acetic anhydride base reagent will promote an $\alpha \rightarrow \omega$ shift is not known.

Recently, a new method was introduced which uses the terminal labeling concept in conjugation with a proteolytic enzyme. In this procedure [42] a carboxyl-terminal triazine is formed with the interesting reagent dimethyl-biguanide; the triazine is released along with other amino acids by a protease from *Streptomyces griseus* and can be identified (Scheme 6). The generality of this method has not yet been explored, and it appears to require rather large amounts of protein; however, it was applied to intact enzymes.

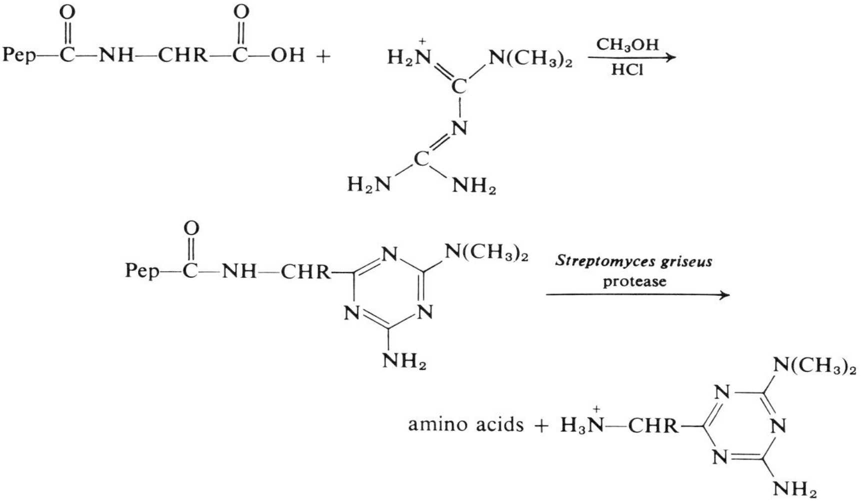

Scheme 6 The triazine procedure.

ONE-ELECTRON METHODS

In 1955, Boissonas [43] presented an electrochemical procedure in which oxidation of the carboxyl terminus leads to fragmentation with loss of CO_2 to a methylolamine (**7**) (Scheme 7). Selective hydrolysis of (**7**) to a new carboxyl-terminal amino acid was claimed, but, in view of the existing literature on the subject [44,45] as well as our own experience with similar species, a selective hydrolysis of (**7**) is unlikely, and hydrolytic conversion of this species to a C-terminal amide is in fact what generally occurs. These observations may explain the fact that Boissonas was unable to proceed further than two or three residues into a peptide chain using this intriguing approach.

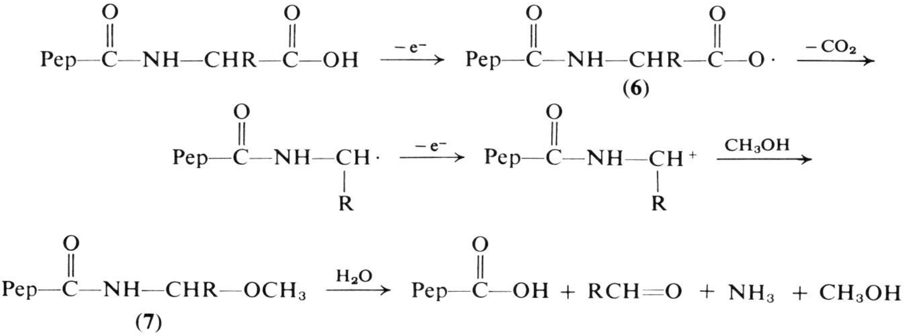

Scheme 7 The Boissonas electrochemical procedure (see text).

Related to this idea is a suggestion [46] that *tert*-butyl peresters should thermolyze via a similar process involving radical (**6**). Thus, we [47] synthesized the *tert*-butyl perester of hippuric acid and found that this compound, when heated to 80°C in water, undergoes a facile thermolysis with liberation of CO_2; compound (**8**) was isolated in high yield. The evident lack of availability of a specific hydrolysis of (**8**) (i.e., conditions that do not also cleave ordinary peptide bonds) with acyl–nitrogen cleavage rather than alkyl–nitrogen cleavage means that this result may not be useful for a sequential degradation. Furthermore, the synthesis of C-terminal *tert*-butyl peresters of peptides and proteins in solvents in which the proteins are soluble presents a real problem. Although these problems are in principle surmountable, it appears that there is other chemistry which seems to be considerably more promising for the development of a sequential degradation.

$$C_6H_5-\overset{\overset{\displaystyle O}{\|}}{C}-NH-CH_2-OH$$

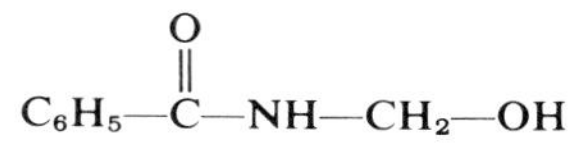

(**8**)

MIGRATION TO ELECTRON-DEFICIENT NITROGEN

The idea of using the series of molecular rearrangements in which an alkyl group migrates from a carbonyl to electron-deficient nitrogen (the Curtius, Lossen, Wawzonek, and related rearrangements) is one of the oldest that has been applied to the carbonyl-terminal degradation of peptides and, in fact, antedates the Edman procedure. In 1936, Bergmann and Zervas [48] ingeniously conceived the idea of synthesizing a carboxyl-terminal azide (9) and allowing this to rearrange with subsequent loss of the carbonyl-terminal residue as an aldehyde (Scheme 8). The hydrolysis of (**10**) to the amide of the

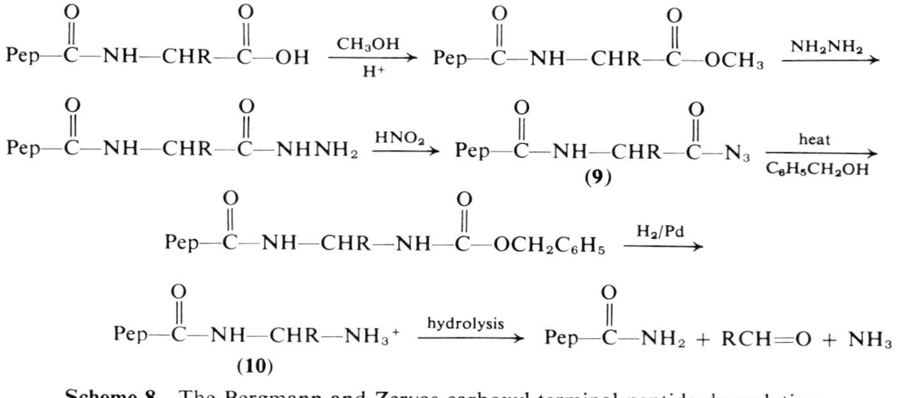

Scheme 8 The Bergmann and Zervas carboxyl-terminal peptide degradation.

new carboxyl terminus should be compared with Boissonas' claim for selective hydrolysis of (7) to an acid. This mode of cleavage would appear to bode ill for a sequential process based on this chemistry, although, as we show below, there are evidently ways of dealing with this problem. There are also several other problems in this azide procedure, not the least of which is the number of reactions required to synthesize the azide and the fact that protic solvents, especially water, readily attack acyl azides. If one could overcome these problems, the results of Bergmann and Zervas indicate that the Curtius rearrangement itself would be most satisfactory as a degradative procedure. In fact, it is interesting that one must take special precautions to avoid this rearrangement when using acyl azides as intermediates in peptide synthesis [49]. We shall show below how we are putting this reaction to use in a carboxyl-terminal analytical procedure.

Wieland and Fritz [50] used the related Lossen rearrangement in a carboxyl-terminal peptide degradation, as shown in Scheme 9. In this scheme, (12) arises presumably by attack of (10) on the isocyanate intermediate in the Lossen rearrangement. Many of the same features can be cited as problems in this method that were noted in the azide method above: the synthesis of the carboxyl-terminal O-acylhydroxamic acid and the ultimate hydrolysis of symmetrical urea (12) to a C-terminal amide. On the other hand, the rearrangement itself is smooth and quantitative and appears to be an otherwise extremely attractive process.

A NEW CARBOXYL-TERMINAL END-GROUP PROCEDURE

Against the background of the methods described above, we began development of a carboxyl-terminal degradation with the benefit of some interesting new reagents and methods. From Scheme 9, it is clear that a direct and

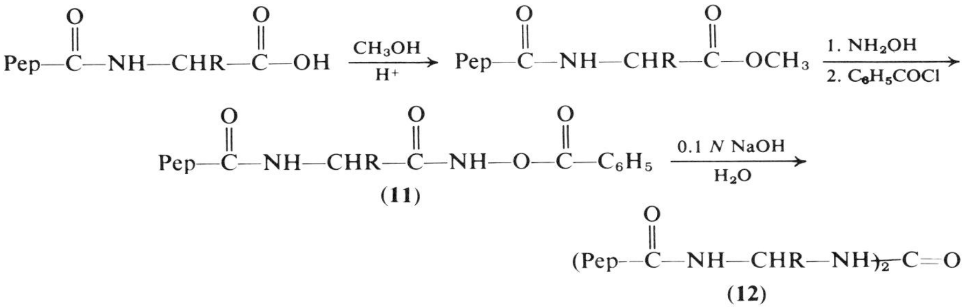

Scheme 9 The Wieland and Fritz carboxyl-terminal peptide degradation.

quantitative synthesis of O-acylhydroxamic acid (11) would be a significant advance in the implementation of the use of the Lossen rearrangement in a C-terminal peptide degradation. Moreover, such a synthesis, to be maximally useful, would have to be carried out in solvents in which most proteins are soluble, namely, water, and would have to be compatible with the presence of denaturants such as 8 M urea or guanidine hydrochloride. Such a direct synthesis would require a water-soluble species of the form NH_2X, in which X is a potential leaving group, as well as some method of forming an amide bond between NH_2X and the carboxyl groups on proteins. The water-soluble carbodiimides appeared to be ideally suited for the latter purpose. The work of Carpino [51] on the synthesis of O-substituted hydroxylamines was the initial investigation which suggested that compounds of the form NH_2X might be readily available; subsequent work [52–55] has provided related and, in some cases, more convenient routes to these compounds. The reagent NH_2X has to be nucleophilic enough to attack the carbodiimide–carboxylic acid adduct, but the X group has to be sufficiently electron withdrawing that it is a reasonable leaving group. The early work of Hauser [56] attests to the effect of leaving group on the rate of the Lossen rearrangement. We thus surveyed a wide variety of possible derivatives NH_2X, encountering several common problems. (1) These materials are often unstable. Thus, O-acetyl-hydroxylamine rather rapidly loses hydroxylamine hydrolytically or attacks another molecule of itself in water [57]; O-methanesulfonylhydroxylamine deflagrates upon drying. (2) The materials are not sufficiently nucleophilic to attack the carbodiimide–carboxylic acid O-acylisourea adduct (which under-goes competing spontaneous hydrolysis) at a reasonable rate. O-2-Nitro-phenylhydroxylamine-4-sulfonate and hydroxylamine-O-sulfonic acid are victims of this problem. (3) The materials are insufficiently water soluble to be useful. O-Mesitylenesulfonylhydroxylamine and O-2,4-dinitrophenyl-hydroxylamine typify this problem. The latter reagent, although not useful in our degradative procedure, has been shown by us [58] and others [59,60] to be a useful synthetic reagent in organic chemistry, serving as a source of "electro-philic nitrogen" (e.g., amination of enolates or amines) or as a means whereby aldehydes can be smoothly converted to nitriles in high yield [58,61].

To summarize our rather detailed survey of O-substituted hydroxylamines, we have settled provisionally on O-pivaloylhydroxylamine (OPHA) (13) as the most satisfactory reagent for use in our degradative scheme. This reagent is readily prepared [62] and used as its hydrochloride. The tert-butyl group

$$H_2N—O—\overset{\overset{\displaystyle O}{\|}}{C}—C(CH_3)_3$$

OPHA

(13)

provides sufficient steric hindrance that the hydrolytic destruction of the reagent is prevented for the most part, whereas the reagent is sufficiently water soluble to be useful. In fact, this material readily reacts with carboxylic acids in the presence of carbodiimides at pH 3.5 to form O-pivaloylhydrox-amic acids (14a), which readily ionize with a pK_a of 6.0–6.5 (depending on the

$$
\underset{\text{(14a)}}{\text{Pep}-\overset{\overset{\textstyle O}{\|}}{C}-NH-\underset{\underset{\textstyle R_C}{|}}{CH}-\overset{\overset{\textstyle O}{\|}}{C}-NHO-\overset{\overset{\textstyle O}{\|}}{C}-C(CH_3)_3} \quad\rightleftharpoons
$$

(Pep = remaining peptide)

$$
\underset{\text{(14b)}}{\text{Pep}-\overset{\overset{\textstyle O}{\|}}{C}-NH-\underset{\underset{\textstyle R_C}{|}}{CH}-\overset{\overset{\textstyle O}{\|}}{C}-\overset{-}{N}O-\overset{\overset{\textstyle O}{\|}}{C}-C(CH_3)_3}
$$

nature of R_c). At pH 8.5, (14) is thus completely in the form of its anion, which subsequently undergoes a smooth and quantitative Lossen rearrangement (Fig. 1). By the procedure shown in Scheme 10, this reaction has been used to degrade small and large peptides with a variety of carboxyl-terminal

TABLE 1

Results of the Carboxyl-Terminal Residue Analysis

Peptide[a,b]	Degradation (%)	Peptide[a,b]	Degradation (%)
Ac·Gly·Asn	71	Ac·Ala·Leu·Gly	77
Ala·Ser	72	Ac·Gly·Leu·Tyr	92
Ac·Ala·Glu	40 (60)[c]	Ac·Pro·Phe·Gly·Lys(Ac)	94
Ac·Ala·Asp	40 (50)[c]	Ac·Met·Arg·Phe·Ala	92
Pro·Gly	77	Ac·Phe·Asp·Ala·Ser·Val	93
Gly·Met	81	Ac·Leu·Trp·Met·Arg·Phe	94
Ac·Ala·Pro	82	⎰Ac·γ-Glu·Cys(SO₃H)·Gly	
		⎱ (oxidized acetylglutathione)	
Gly·Phe	87	Glu	95
Ac·Gly·Leu	90	Gly	75
Bz·Gly·Arg	93	Insulin A chain[d]	75
		(C-terminal Asn)	
Gly·Trp	86	Insulin B chain[d]	99
		(C-terminal Ala)	

[a] Abbreviations: Ac, acetyl; Bz, benzoyl; see text for definition of degradation yield.
[b] In all of these peptides, the only residues lost were the C-terminal ones, *except as noted in the text.*
[c] Yields in parentheses result from carrying out the coupling step at pH 0.75.
[d] Carried out in 8 M urea.

Step 1:

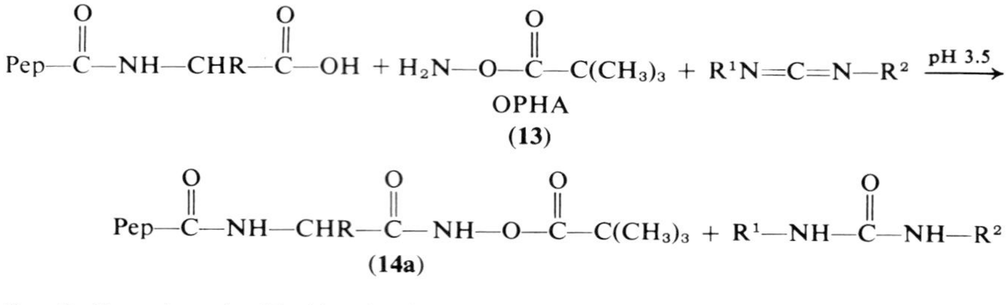

$$\text{Pep—C(=O)—NH—CHR—C(=O)—OH} + \text{H}_2\text{N—O—C(=O)—C(CH}_3)_3 + \text{R}^1\text{N}=\text{C}=\text{N—R}^2 \xrightarrow{\text{pH 3.5}}$$

OPHA

(13)

$$\text{Pep—C(=O)—NH—CHR—C(=O)—NH—O—C(=O)—C(CH}_3)_3 + \text{R}^1\text{—NH—C(=O)—NH—R}^2$$

(14a)

Step 2: Quench carbodiimide with formate buffer

Step 3: Raise pH to 8.5; **(14a)** ⟶ **(14b)**

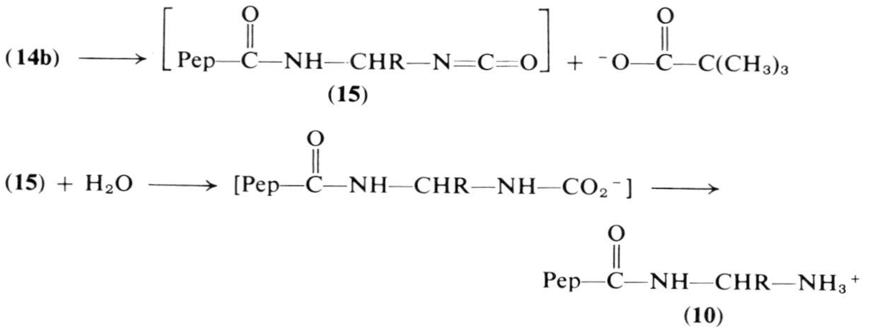

$$\textbf{(14b)} \longrightarrow \left[\text{Pep—C(=O)—NH—CHR—N}=\text{C}=\text{O} \right] + {}^{-}\text{O—C(=O)—C(CH}_3)_3$$

(15)

$$\textbf{(15)} + \text{H}_2\text{O} \longrightarrow [\text{Pep—C(=O)—NH—CHR—NH—CO}_2{}^-] \longrightarrow$$

$$\text{Pep—C(=O)—NH—CHR—NH}_3{}^+$$

(10)

Step 4: Hydrolyze **(10)** and identify aldehyde or amino acid; analyze

Scheme 10   A new carboxyl-terminal end-group procedure.

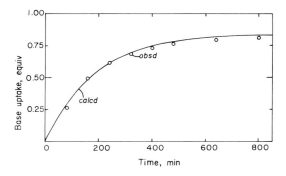

Fig. 1   The Lossen rearrangement of N_α-acetylglutamino-O-pivaloylhydroxamate as monitored by base uptake on the pH-stat at pH 8.5, 50°C. The points are observed, and the line is calculated for a first order rate constant of 0.00541 min^{-1}.

residues, as shown in Table 1 [63]. Degradation yields are assessed by differential amino acid analysis; it would clearly be possible to identify the aldehyde produced by one of several methods [64]. We have done this on isolated occasions, but we have preferred to concentrate our resources to date on the development of the degradative chemistry. This procedure has been operated at the lower limit of our analysis facilities (ca. 50–100 nmoles), but there appears to be no reason why problems should be encountered at lower peptide levels.

Several advantages are evident in this procedure. The procedure is a one-vessel, aqueous-solution method which can also be used in 8 M urea or guanidine hydrochloride. It readily detects carboxyl-terminal amino acids with or without side-chain functionality, including Arg, Lys, Pro, Ser, Asn, and Gln. Although the procedure is currently limited to peptides that can be analyzed by difference, the submicro techniques for aldehyde identification [64] might remove this restriction. Because of the large excess of (13) employed, protection of protein nucleophiles appears to be unnecessary except as indicated below.

We have encountered some problems in this procedure, which include the following. The amino-terminal amino acid is lost routinely to the extent of about 26%; this loss also occurs in the presence of OPHA only and does not occur if the peptide is acetylated prior to degradation. The interfering reaction is evidently amination of the amino terminus; OPHA is a weak animating agent [62].

Internal tyrosine is lost to the extent of about 30%, and this loss can be duplicated by the presence of OPHA only. The loss incurred during the pH 3.5 coupling process can be reversed by dithionite prior to amino acid analysis, but that which occurs at pH 8.5 during the Lossen rearrangement (about 12%) cannot be so reversed. These losses do not interfere with a successful C-terminal analysis [47]. Tryptophan evidently undergoes serious side reactions of a similar character, and internal tryptophan is lost to the extent of 50%. Since carboxyl-terminal tryptophan is lost nearly quantitatively, the procedure still provides some evidence for carboxyl-terminal tryptophan, although an investigator would probably wish to have confirming data from another method.

In unprotected peptides, internal serine is not lost, but serine is lost to the extent of about 50% when the peptide is acetylated prior to degradation. This loss can be duplicated by incubation of the acetylated peptide alone at pH 8.5 for a time equivalent to that of the Lossen rearrangement. Quantitation of the NH_3 liberated on acid hydrolysis gives the result [serine + NH_3] = 1.0 residue-equivalent, which suggests that the side reaction is the elimination of acetic acid from O-acetylserine. The anhydroserine thus produced yields, of course, pyruvic acid and ammonia on hydrolysis.

Two minor problems, the loss of glycine in only 75% and the sluggish rate of the Lossen rearrangement (Fig. 1), appear to be related. The former presents no problem for analysis of the C-terminal glycine but was puzzling, since glycine is generally one of the most well-behaved amino acids. The latter problem is an annoyance only because one has to wait overnight for the results; the Lossen rearrangement itself, however, can be monitored by a pH stat and left unattended. If one considers the factors controlling the rate of the Lossen rearrangement, not only the nature of the leaving group but also the nature of the migrating group will have an important influence. In fact, the migrating group in the degradation is an α-amidoalkyl group, the superiority of whose migratory aptitude accounts for our ability to use a leaving group as poor as pivalate. However, of all the migrating groups represented by the various amino acids, that embodied in glycine is the poorest because it is primary, whereas all of the others are secondary. This consideration suggests at first that the glycine degradation yields can be raised by merely allowing longer times for the degradation, but in fact longer times do not improve the yield. Evidently, we are facing a competition between two reactions [Eq. (1)].

$$
\begin{array}{c}
\underset{\text{Pep}}{\overset{\displaystyle O}{\overset{\|}{C}}}-NH-CH_2-\overset{\displaystyle O}{\overset{\|}{C}}-NHO-\overset{\displaystyle O}{\overset{\|}{C}}-C(CH_3)_3 \xrightarrow[\text{Lossen}]{\text{(a)}}
\end{array}
$$

$$\text{(b)} \downarrow \begin{array}{c}[OH^-]\\ \text{hydrolysis}\end{array} \qquad \underset{\text{Pep}}{\overset{\displaystyle O}{\overset{\|}{C}}}-NH-CH_2-\overset{+}{N}H_3 + CO_2 + {}^-O_2C-C(CH_3)_3$$

$$
\underset{\text{Pep}}{\overset{\displaystyle O}{\overset{\|}{C}}}-NH-CH_2-\overset{\displaystyle O}{\overset{\|}{C}}-NHOH + {}^-O_2C-C(CH_3)_3 \tag{1}
$$

Glycine is especially unfavorable on both counts because of its poor migrating group and because it has no side chain to provide steric hindrance to dead-end path (b). Clearly, more hindered leaving groups of lower pK_a should be developed to avoid this problem, and work to this end is in progress.

The problems with Asp and Glu (Table 1) have been traced to a side reaction in the coupling process rather than the Lossen rearrangement [eq. (2)]. Species (16) results from attack of a carbodiimide on a C-terminal dicarboxylic amino acid which has previously coupled with an equivalent of OPHA [68]. Imide (17) can and does reopen with OH$^-$ at pH 8.5 at any of the three possible carbonyl groups; opening at C_ω leads to the desired degradation, but this unfortunately is not the major pathway that is observed [58].

The hydrolysis of (10) has been found to occur with alkyl–nitrogen cleavage under conditions that do not affect peptide bonds; thus, an amide is produced after the degradation. This finding appears to present a serious problem for

$$Pep-\overset{O}{\overset{\|}{C}}-NH-\underset{\underset{\underset{O}{\overset{C}{/}}\diagdown O-\underset{\underset{NHR^2}{|}}{C}=NHR^1}{\overset{|}{(CH_2)_n}}}{CH}-\overset{O}{\overset{\|}{C}}-NH-O-\overset{O}{\overset{\|}{C}}-C(CH_3)_3 \longrightarrow$$

(16)

$$Pep-\overset{O}{\overset{\|}{C}}-NH-\underset{\underset{\underset{O}{\overset{\|}{C_\omega}}}{\overset{(CH_2)_n}{|}}\diagdown N-OC-C(CH_3)_3}{CH}-\overset{O}{\overset{\|}{C_\alpha}}\qquad (2)$$

(17)

the extension of the very simple chemistry of Scheme 10 to a sequential method. To overcome this problem, we must (1) convert (10) to an acid directly, perhaps by condensation with 1,2-dicarbonyl compounds and cyclization to an acylimidazole; (2) find a selective hydrolysis of primary amides to acids; or (3) develop a method for continuing the degradation on primary amides themselves. We have not yet investigated idea (1), we have had some progress with (2), but idea (3) appears now to be most promising, as we indicate below.

DETERMINATION OF β-ASPARTYL AND γ-GLUTAMYL LINKAGES IN PROTEINS

One can envision several other uses of this and related chemistry in both analysis [69] and modification [70] of proteins. One way in which we have used these ideas is the determination of ω linkages in copolymers containing glutamic acid or aspartic acid. [58]. The basis of this determination is readily seen in Scheme 11. Under conditions used for the Lossen rearrangement, the rearrangement of the side-chain O-pivaloylhydroxamates is quite slow from migratory aptitude considerations (see above) when the glutamic or aspartic acids are present in their normal α linkage. After amino acid analysis of the rearranged protein, one retains most of the Asp and Glu residues, and those that do rearrange are readily assayed as the corresponding diamino acids derived from (19a) or (19b). On the other hand, dicarboxylic amino acids in ω linkage are in reality new carboxyl termini which readily rearrange. Such a rearrangement gives net total loss of these residues. Thus, for example, the number of β-linked aspartic acid residues in a protein is the total number

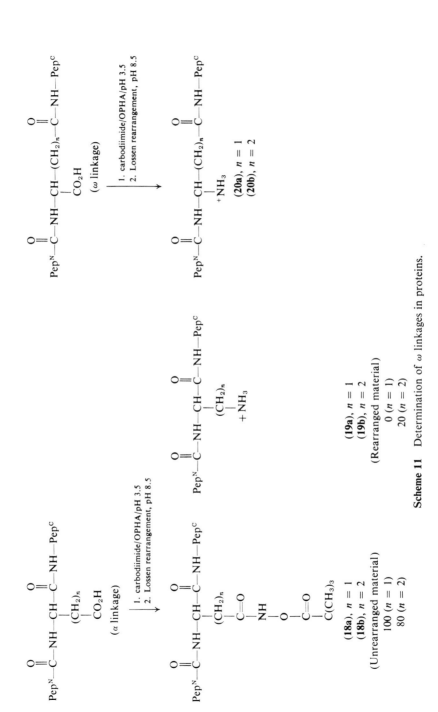

Scheme 11 Determination of ω linkages in proteins.

lost less the number of 2,3-diaminopropionic acids formed. The idea of using a Lossen rearrangement for this analysis is an old one [71], but the attractive thing about this specific chemistry is that the side-chain O-pivaloylhydroxamates required for the Lossen rearrangement are synthesized under mildly acidic conditions (pH 3.5) in which transpeptidation (a → ω shifts) is eliminated [72]. The chemistry in Scheme 11 has been used to successfully detect large amounts of β-aspartic acid linkage in copolymers of 9% Asp and 91% N_γ-4-hydroxybutylglutamine which had a history of treatment under conditions known to promote transpeptidation; the same sequence detected no β-aspartic linkages in copolymers of identical constitution in whose handling transpeptidation condition were avoided. The method thus appears to be qualitatively sound; its value as a quantitative method will hinge on the results of a study of synthetic peptides containing known linkage types. In Scheme 11 it is interesting that the degradation of ω linkages results in a specific cleavage of polypeptide chains after hydrolysis of (20).

DEGRADATION OF PEPTIDES ON SOLID SUPPORTS VIA A C-TERMINAL AZIDE

The Bergmann and Zervas degradation [48] via the azide shown in Scheme 8 would be most attractive if one could synthesize the azide in one step and prevent its hydrolytic destruction. The availability of solid supports to which one may covalently link proteins and peptides now enables one to work in reagents and solvents with which these species are normally incompatible. These supports have the additional advantage of permitting ready separations of the peptide from both reagents and fragments that are produced in the degradation. Finally, these materials tend to prevent intermolecular reactions which might otherwise occur if the peptides were not immobilized. Thus, we developed the synthesis and degradation of C-terminal azides of immobilized peptides as shown in Scheme 12. There are several keys to the success of this work. First is the availability of porous glass supports (Controlled Pore Glass, Corning); the styrene divinylbenzene resins used in solid-phase peptide synthesis are not compatible with certain solvents which we wished to use in this procedure. Second is the recent work with diarylphosphorylazides [73], which permit the quantitative conversion of carboxyl groups to acylazides. The di-p-nitrophenylphosphorylazide is shunned in peptide synthesis [74] because its use is often accompanied by extensive racemization. Since racemization is of no concern to us, the superior reactivity of this reagent makes it ideally suited for our purpose. As our first results in Table 2 indicate, the method appears to be promising indeed as a carboxyl-terminal procedure [75]. One

All peptides are bound to activated Controlled Pore Glass, 550 Å pore size, 70 m²/gm surface area, 80/120 mesh with the following type of linkage:

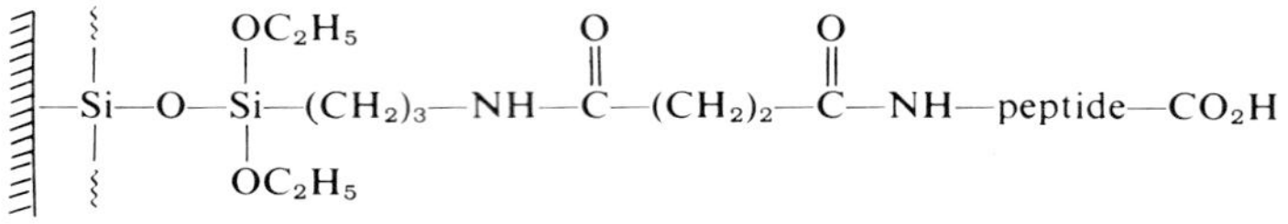

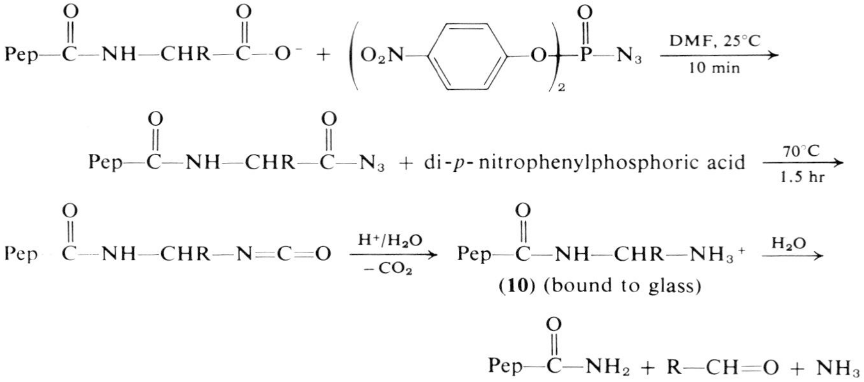

$$-\overset{\underset{|}{O}}{\underset{\underset{|}{O}}{Si}}-O-\overset{\underset{|}{OC_2H_5}}{\underset{\underset{|}{OC_2H_5}}{Si}}-(CH_2)_3-NH-\overset{\underset{\|}{O}}{C}-(CH_2)_2-\overset{\underset{\|}{O}}{C}-NH-peptide-CO_2H$$

$$Pep-\overset{\underset{\|}{O}}{C}-NH-CHR-\overset{\underset{\|}{O}}{C}-O^- + \left(O_2N-\!\!\left\langle\!\!\bigcirc\!\!\right\rangle\!\!-O\right)_2\overset{\underset{\|}{O}}{P}-N_3 \xrightarrow[10\ min]{DMF,\ 25°C}$$

$$Pep-\overset{\underset{\|}{O}}{C}-NH-CHR-\overset{\underset{\|}{O}}{C}-N_3 + di\text{-}p\text{-}nitrophenylphosphoric\ acid \xrightarrow[1.5\ hr]{70°C}$$

$$Pep\ \overset{\underset{\|}{O}}{C}-NH-CHR-N\!=\!C\!=\!O \xrightarrow[-CO_2]{H^+/H_2O} Pep-\overset{\underset{\|}{O}}{C}-NH-CHR-NH_3^+ \xrightarrow{H_2O}$$

(10) (bound to glass)

$$Pep-\overset{\underset{\|}{O}}{C}-NH_2 + R-CH\!=\!O + NH_3$$

Scheme 12   The azide carboxyl-terminal peptide degradation on a solid support (DMF, dimethylformamide).

should note the successful degradation by this method of Pro, Asp, and Trp, as well as the other amino acids.

It might be noted that another conceivable degradation pathway through the azide would be via a photochemically generated nitrene. However, the Curtius reaction appears not to occur by way of nitrene intermediates, and a variety of other products from this reaction might reasonably be expected [76]. For these reasons we have not investigated a photochemical method.

EXTENSION OF THE SOLID-PHASE PROCEDURE
TO A SEQUENTIAL METHOD

As was pointed out above, extension of our methods to a sequential degradation at first sight appears to be barred by the production of the amide of the penultimate residue after one cycle of degradation. We have spent considerable effort attempting to optimize the use of nitrosonium ion in its various guises (N_2O_4, HNO_2, $NO^+BF_4^-$, $NO^+PF_6^-$, $NO^+HSO_4^-$, etc.) to selectively, remove this primary amide without complete success. More recently, however we have achieved a successful degradation of immobilized peptide primary

TABLE 2

Degradation of Peptides by the Solid-Phase Azide Method

Peptide[a]	Degradation product[a]	Degradation (%)
Gly*(1.0)·Asp(0.95)	Gly*(1.0)·Asp(0.06)	94
Ala*(1.0)·Ser(0.93)	Ala*(1.0)·Ser(0.09)	91
Gly*(1.0)·Glu(0.93)	Gly*(1.0)·Glu(0)	100
Ala*(1.0)·Pro(1.0)	Ala*(1.0)·Pro(0.05)	95
Ala*(1.0)·Leu(0.97)·Gly(0.98)	Ala*(1.0)·Leu(0.99)·Gly(0.05)	95
Phe*(1.0)·Asp(0.97)·Ala(0.97)·	Phe*(1.0)·DAPA(0.50)·Ala(O.99)·	100
Ser(0.86)·Val(0.97)	Ser(0.80)·Val(0)[b]	100
Gly*(1.0)·Met(0.78)	Gly*(1.0)·Met(0.09)	69–91[c]
Ala*(1.0)·Ile(1.04)	Ala*(1.0)·Ile(0.01)	99
Gly*(1.0)·Leu(0.65)·Tyr(0.84)	Gly*(1.0)·Leu(1.01)·Tyr(0.08)	76–92[c]
Glu(0.97)·Gly*(1.0)·Phe(1.01)	Glu(0.15)·Gly*(1.00)·Phe(0)·	
	DABA(0.60)[b]	100
Gly*(1.0)·Trp(0.98)[d]	Gly*(1.0)·Trp(0.07)	93
Gly*(1.0)·His(0.98)	Gly*(1.0)·His(0.01)	98
Ala*(1.0)·Arg(0.95)[d]	Ala*(1.0)·Arg(0.10)	84–91[c]
Bradykinin (C-terminal Arg)[d]	Loss of 0.92 Arg	92
Insulin A chain (C-terminal Asn)	Loss of 0.97 Asn	97

[a] Relative compositions before and after degradations are given. The asterisked residue is the one used for normalization of composition; controls showed that negligible peptide is lost from the glass support during degradation. About 0.5 μmole of peptide was degraded in each case.

[b] DAPA, 2,3-diaminopropionic acid; DABA, 2,4-diaminobutyric acid. The rearrangement products are expected from the Curtius rearrangement of the side chains of the respective dicarboxylic amino acids.

[c] Uncertainties in yields reflect uncertainties in the amino acid analysis of starting peptide.

[d] Carbobenzoxylated prior to degradation.

amides using phenyliodosyl acetate [77] in essentially quantitative yield without destruction of peptide bonds or removal of the peptide from the glass supports. Although the generality of this procedure requires much more work to be demonstrated, the initial results are most encouraging.

CONCLUDING REMARKS

Despite efforts that antedate by many years the development of the Edman degradation, no fully satisfactory sequential carboxyl-terminal peptide degradation exists, although considerable progress has been made in the development of carboxyl-terminal end-group determinations. There are, however, many promising ideas under development, and there is much interesting chemistry to be explored in this fertile area. The rational design and execution of these new methods depend on a mechanistic understanding of the chemistry involved, as well as on many interesting new reagents that only recently would have been considered biochemically irrelevant. Furthermore, application of many of the ideas discussed in this Chapter extends beyond sequential degradation to protein modification, specific polypeptide cleavages, analysis of unusual peptide bonds, and no doubt other interesting and as yet unforeseen applications which we look forward to exploring in the future.

ACKNOWLEDGMENTS

We wish to acknowledge the helpful discussions and collaboration of Mr. F. E. DeBons, Mr. John Capecchi, and Dr. Francis Cardinaux of Cornell; Mr. Roger Piasio of the Corning Glass Company; and Professor Harold Scheraga. We are grateful to the National Institute of General Medical Sciences for support of our research.

REFERENCES

1. F. Sanger and E. O. P. Thompson, *Biochem. J.* **53**, 353 (1953); F. Sanger and H. Tuppy, *ibid.* **49**, 463 (1951).
2. V. M. Ingram, *Biochim. Biophys. Acta* **28**, 539 (1958); **36**, 402 (1959).
3. C. Nolan and E. Margoliash, *Annu. Rev. Biochem.* **37**, 727 (1968). M. O. Dayhoff, P. J. McLaughlin, W. C. Berker, and L. T. Hunt, *Naturwissenschaften* **62**, 154 (1975).
4. T. C. Vanaman, K. L. Brew, and R. L. Hill, *J. Biol. Chem.* **245**, 4583 (1970).
5. G. M. Edelman, *Sci. Amer.* **223**, 34 (1970).
6. P. Edman, *Acta Chem. Scand.* **4**, 277 and 283 (1950); W. Konigsberg, *in* "Methods in Enzymology" Vol. 25, Part B, p. 326. Academic Press, New York, 1972.
7. P. Edman and G. Begg, *Eur. J. Biochem.* **1**, 80 (1967).
8. S. Stein, P. Böhlen, W. Dairman, W. Leimgruber, and M. Weigele, *Science* **178**, 871 (1972).
9. W. R. Gray, *in* "Methods in Enzymology" (C. H. W. Hirs, ed.), Vol. 11, p. 139. Academic Press, New York, 1967; C. Gros and B. Labouesse, *Eur. J. Biochem.* **7**, 463 (1969).
10. F. Sanger, *Biochem. J.* **39**, 507 (1945).
11. C. W. Bevan, T. A. Emokpae, and J. Hirst, *J. Chem. Soc. C* p. 238 (1968); D. A. Sutton, S. E. Drewes, and U. Welz., *Biochem. J.* **130**, 589 (1972).
12. G. R. Jacobson, M. H. Schaffer, G. R. Stark, and T. C. Vanaman, *J. Biol. Chem.* **248**, 6583 (1973).
13. J. Lens, *Biochim. Biophys. Acta* **3**, 367 (1949).
14. H. Fraenkel-Conrat, J. I. Harris, and A. L. Levy, *Methods Biochem. Anal.* **2**, 359 (1955). R. P. Ambler, *in* "Methods in Enzymology" (C. H. W. Hirs and S. Timasheff, eds.), Vol. 25, Part B, pp. 143 and 262. Academic Press, New York, 1972,
15. R. Hayashi, S. Moore, and W. H. Stein, *J. Biol. Chem.* **248**, 2296 (1973).
16. R. Hayashi, S. Aibara, and T. Hata, *Biochim. Biophys. Acta* **212**, 359 (1970).

17. R. Hayashi, Y. Bai, and T. Hata, *J. Biochem.* (Tokyo) **77**, 1313 (1975); *J. Biol. Chem.* **250**, 5221 (1975); Y. Bai and R. Hayashi, *FEBS Lett.* **56**, 43 (1975).
18. P. A. Hargrave and F. Wold, *Int. J. Pept. Protein Res.* **5**, 85 (1973).
19. R. G. Duggleby and H. Kaplan, *Anal. Biochem.* **65**, 346 (1975).
20. H. G. Khorana, *J. Chem. Soc.* p. 2081 (1952).
21. P. Schlack and W. Kumpf, *Z. Phys. Chem.* **154**, 125 (1926).
22. S. Tibbs, *Nature (London)* **168**, 710 (1951).
23. S. G. Waley and J. Watson, *J. Chem. Soc.* p. 2394 (1951).
24. G. W. Kenner, H. G. Khorana, and R. J. Stedman, *J. Chem. Soc.* p. 673 (1953).
25. S. Yamashita, *Biochim. Biophys. Acta* **229**, 301 (1971).
26. L. D. Cromwell and G. R. Stark, *Biochemistry* **8**, 4735 (1969); G. R. Stark, *in* "Methods in Enzymology" (C. H. W. Hirs and S. Timasheff, eds.), Vol. 25 Part B, p. 326. Academic Press, New York, 1972.
27. M. J. Williams and B. Kassell, *FEBS Lett.* **54**, 353 (1975).
28. H. Darbre and M. Rangarajan, *in* "Solid Phase Methods in Protein Sequence Analysis" (R. A. Laursen, ed.), p. 131. Pierce Chemical Co., Rockford, Illinois, 1975.
29. G. R. Stark, *Adv. Protein Chem.* **24**, 302 (1970).
30. G. R. Tarr, *in* "Solid Phase Methods in Protein Sequence Analysis" (R. A. Laursen, ed.), p. 139. Pierce Chemical Co., Rockford, Illinois, 1975.
31. A. C. Chibnall and M. W. Rees, *Biochem. J.* **48**, xlvii (1951); *Chem. Struct. Proteins, Ciba Found. Symp., 1952* p. 70 (1953).
32. L. Bailey, *Biochem. J.* **60**, 173 (1955); "Techniques in Protein Chemistry." Am. Elsevier, New York, 1967.
33. S. M. Parsons and M. A. Raftery, *Biochemistry* **8**, 4199 (1969).
34. T. Hamada and O. Yonemitsu, *Biochem. Biophys. Res. Commun.* **50**, 1081 (1973).
35. A. K. Saund, B. Prashad, A. K. Koul, J. M. Bachawat, and N. K. Mathus, *Int. J. Pepto Protein Res.* **5**, 7 (1973).
36. S. Akabori, K. Ohno, and K. Narita, *Bull. Chem. Soc. J.* **25**, 214 (1952).
37. V. Braun and W. A. Schroeder, *Arch. Biochem. Biophys.* **118**, 241 (1967).
38. W. A. Schroeder, *in* "Methods in Enzymology" (C. H. W. Hirs and S. Timasheff, eds.), Vol. 25, Part B, p. 138. Academic Press, New York, 1972.
39. H. Matsuo, Y. Fujimoto, and T. Tatsuno, *Biochem. Biophys. Res. Commun.* **22**, 69 (1966).
40. D. N. Holcomb, S. A. James, and D. N. Ward, *Biochemistry* **7**, 1291 (1968).
41. G. Ramponi, G. Cappugi, and P. Nassi, *Biochem. Biophys Res. Commun.* **41**, 642 (1970).
42. K. Maekawa and E. Kuwano, *Z. Anal. Chem.* **276**, 121 (1975).
43. R. A. Boissonas, *Helv. Chim. Acta* **35**, 2226 (1952).
44. H. Hellman and G. Opitz, "α-Aminoalkylierung." Verlag-Chemie, Weinheim, 1960.
45. J. Ugelstad and J. deJonge, *Recl. Trav. Chim. Pays-Bas* **76**, 919 (1957).
46. We acknowledge and thank Professor Paul D. Bartlett for this suggestion.
47. F. E. DeBons and G. M. Loudon, unpublished work.
48. M. Bergmann and L. Zervas, *J. Biol. Chem.* **113**, 341 (1936).
49. See, for example, p. 1388 of A. M. Felix and R. B. Merrifield, *J. Am. Chem. Soc.* **92**, 1385 (1970).
50. T. Wieland and H. Fritz, *Chem. Ber.* **86**, 1186 (1953); *Chem. Struct. Proteins, Ciba Found. Symp., 1952* p. 146 (1953).
51. L. A. Carpino, C. A. Giza, and B. A. Carpino, *J. Am. Chem. Soc.* **81**, 955 (1951); L. A. Carpino, *ibid.* **82**, 3133 (1960).

52. G. Zinner, *Arch. Pharm. (Weinheim, Ger.)* **293**, 42 (1960).
53. G. Zinner, G. Nebel, and M. Hitze, *Arch. Pharm. (Weinheim Ger.)* **303**, 317 (1970).
54. von A. O. Ilvespaä and A. Marxer, *Chimia* **18**, 1 (1964).
55. F. D. King and D. R. M. Walton, *Synthesis* p. 788 (1975).
56. C. R. Hauser and W. B. Renfrow, Jr., *J. Am. Chem. Soc.* **59**, 121 (1937).
57. W. P. Jencks, *J. Am. Chem. Soc.* **80**, 4581 (1958).
58. M. J. Miller, Ph.D. Dissertation, Cornell University, Ithaca, New York (1976).
59. M. Kim and J. D. White, *J. Am. Chem. Soc.* **97**, 451 (1975).
60. Y. Tamura, J. Minamikawa, K. Sumoto, S. Fujii, and M. Ikeda, *J. Org. Chem.* **38**, 1239 (1973).
61. M. J. Miller and G. M. Loudon, *J. Org. Chem.* **40**, 126 (1975).
62. W. N. Marmer and G. Maerker, *J. Org. Chem.* **37**, 3520 (1972).
63. M. J. Miller and G. M. Loudon, *J. Am. Chem. Soc.* **97**, 5295 (1975).
64. The methods we have used include 2,4-dinitrophenylhydrazine, a resynthesis of the aldehyde to an amino acid [65], and the production of a colored adduct of the aldehyde with 4-amino-3-hydrazino-5-mercapto-1,2,4-triazole [66]. A third method possible is the use of dansyl hydrazide [67], which will yield a fluorescent dansylhydrazone.
65. B. Pereyra, O. O. Blumenfeld, M. A. Paz, E. Henson, and P. M. Gallop, *J. Biol. Chem.* **249**, 2212 (1974). We thank Professor W. R. Gray, The University of Utah, for calling this to our attention.
66. R. G. Dickinson and M. W. Jacobsen, *Chem. Commun.* p. 1719 (1970).
67. R. Chayen, R. Dvir, S. Gould, and A. Harell, *Anal. Biochem.* **42**, 283 (1971).
68. Species **16**, of course, may result also from initial coupling at the side-chain carboxyl followed by a process similar to that shown in eq. 2 taking place at the activated α-carboxyl.
69. P. M. Gallop, S. Seifter, M. Lukin, and E. Meilman, *J. Biol. Chem.* **235**, 2619 (1960).
70. P. Bodlaender, G. Feinstein, and E. Shaw, *Biochemistry* **8**, 4941 (1969); G. Feinstein, P. Bodlaender, and E. Shaw, *ibid.* p. 4949.
71. See Discussion, *Chem. Struct. Proteins, Ciba Found. Symp., 1952* p. 149 (1953).
72. S. A. Bernhard, A. Berger, J. H. Carter, E. Katchalski, M. Sela, and Y. Shalitin, *J. Am. Chem. Soc.* **84**, 2421 (1962).
73. T. Shiori and S. Yamada, *Chem. Pharm. Bull.* **22**, 855 (1974).
74. S. Yamada, N. Ikota, T. Shioiri, and S. Tachibana, *J. Am. Chem. Soc.* **97**, 7174 (1975).
75. M. E. Parham and G. M. Loudon, unpublished results.
76. W. Lwowski, "Nitrenes," Chapter 6. Wiley (Interscience), New York, 1970.
77. K. Swaminathan and N. Venkatasubramanian, *J. Chem. Soc., Perkin Trans, 2* p. 1161 (1975).

5

Interactions of Selected Antitumor Antibiotics with Nucleic Acids

J. W. Lown

INTRODUCTION

Chemotherapy continues to play a key role in the control of neoplastic diseases since more effective procedures seem unlikely to result from further improvements in early diagnosis or in surgery or radiotherapy [1]. Immuno-chemistry holds some promise for a rational approach to the combating of cancer but is still in its infancy [2]. Chemical agents are often effective in the treatment of leukemia and solid tumors (and metastases therefrom) inoperable by surgery or radiation therapy. In some forms of cancer (Burkitt's lymphoma, choriocarcinoma, and Wilm's tumor), regression and long-term tumor-free survival can be achieved only by chemotherapy [1]. Antitumor agents currently in clinical use fall into the following categories: (a) alkylating agents, (b) antibiotics, (c) antimetabolites, and (d) miscellaneous, e.g., platinum complexes [3]. Many pieces of evidence indicate significant selectivity of site within the cell, e.g., nucleic acids, nucleic acid polymerases, or other enzymes such as those involved in cell wall production [4]. Extensive information is available on the interactions between drugs and proteins, especially the detailed information on the interaction between substrates, metabolic inhibitors, and enzymes, but there is less detailed information on the interaction of drugs and nucleic acids [5]. Not until this challenge is met will we be able to account for the essential chemical features of most antineoplastic agents.

This review has the modest aim of illustrating how, by using a combination of chemical modification of recognized antitumor antibiotics and physical methods, especially the fluorescence characteristics of intercalative dyes, one can begin to understand some of the ways in which these agents interact with nucleic acids.

Antitumor Agents That Are Inhibitors of Nucleic Acid Synthesis

Present-day cancer relies very heavily on inhibition of nucleic acid synthesis or interference with mitosis as a means of discriminating selectively against fast-growing tumor cells. There is much evidence (inactivation of viruses, antimitotic, cytological, and mutagenic effects) from *in vitro* and *in vivo* reactions and physicochemical alterations indicating that DNA is the principal cell target site for many antitumor agents [1,6–8]. The complex of processes involved in DNA replication, transcription to mRNA, and translation to proteins on the ribosomal system offers various possibilities for drug action. Some antibiotics, cytostatics, carcinogens, and mutagens find their points of attachment here [9–15].

Antibiotics may interfere with the synthesis of nucleic acids in several different ways. Drugs may inhibit nucleic acid synthesis at the level of nucleotide metabolism or may interact with double-stranded DNA to block normal access of DNA polymerase and thus its replication. The different chemical modes of interaction with DNA include the following: (a) covalent cross-linking of bases of DNA, either interstrand or intrastrand, by polyfunctional agents, e.g., nitrogen mustards and mitomycin C, which may lead to unequal partition of chromosomes between daughter cells [12]; (b) intercalation between the base pairs in DNA of planar polycyclic molecules (e.g., actinomycin, acridines), leading to frame-shift mutation [16, 17]; a secondary binding known as apposition may also occur [12,13]; (c) alkylation which occurs preferentially on guanine units and can cause sublethal damage such as depurination, leading to mutations [18]; (d) strand scission of DNA by antibiotics, which results in fragmentation of chromosomes and inhibition of DNA synthesis [19].

This review concentrates on agents that impair the template function of DNA in the cell.

Hitherto, interstrand cross-linking of DNA has been demonstrated by techniques such as cesium chloride gradient density centrifugation [19, 20] or by the hyperchromicity of cross-linked denatured DNA [21]. However, such techniques are unwieldy, and there is no convenient procedure (apart from radioactive labeling) for detecting the concomitant and more common alkylation, which does not result in cross-linking. Similarly, single-strand scission of DNA may be detected by sedimentation velocity changes or by using radiolabeled DNA [22], but this method cannot be adapted to the investigation of the chemical mechanism, nor can it detect, for example, simultaneous strand scission and covalent cross-linking.

Certain fluorescence procedures employing the intercalative dye ethidium described here complement existing techniques but also have the distinct advantages of sensitivity (typically 0.5 μg of DNA can be employed), simplicity, rapidity, and adaptability and lend themselves to an investigation of various aspects of interaction of antibiotics with DNA [22a].

Ethidium Fluorescence Assays for Examination of DNA–Antibiotic Interactions

It was observed initially by Le Pecq and Paoletti [23] that ethidium bromide, which is widely used as a histological stain for nucleic acid-containing structures in cells, intercalates into duplex nucleic acids (Fig. 1). Ethidium is a phenan-

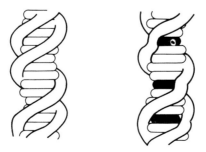

thridine trypanocide which is also active against bacteria [24]. It inhibits DNA and RNA synthesis by binding reversibly to DNA [25]. There are two modes of binding: (a) primary, by intercalation, and (b) secondary, by external attachment. The intercalation process shows no base preference, and the primary sites are saturated when one drug is bound for every four or five nucleotides. Binding is also independent of pH within the range where the double helix is stable [4]. During primary binding the fluorescence shows a 25-fold enhancement. This phenomenon has been attributed to the removal of quenching by the polar solvent as the dye is absorbed in the hydrophobic region of the duplex [4,23].

The basis of a representative assay developed by Morgan and Paetkau which permits the detection of covalent cross-linking of DNA is given in

Fig. 1   Native DNA (left) and DNA containing intercalated molecules (right).

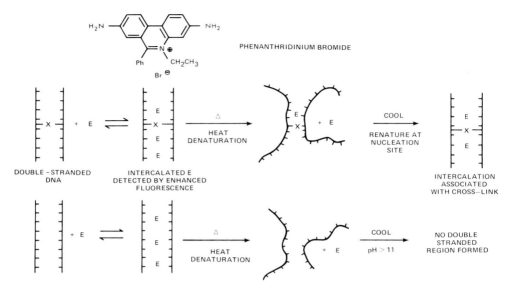

Fig. 2 Ethidium fluorescence assay for detecting covalent cross-linking of DNA (E, ethidium; X, covalent cross-link). Excitation, 525 nm; emission, 600 nm.

Fig. 2 [26,27]. Typically, the excitation wavelength is 525 nm and the emission wavelength is 600 nm.

Native DNA at neutral pH after heat denaturation to separate the strands typically shows a 50% return of fluorescence owing to the formation of short local regions of self-complementarity [26]. If the pH is adjusted to 11.8 this hydrogen bonding in the short regions is disrupted with little effect on long duplex DNA, and the fluorescence enhancement drops to zero. However, if a covalent cross-link is introduced by chemical means this may serve as a nucleation point, so that after the heating and cooling cycle renaturation occurs rapidly at 10^7–10^8 base pairs/sec [28]. Palindromic DNA also shows quantitative return of fluorescence [26]. Covalently closed circular (CCC) DNA with topologically linked strands behaves like cross-linked DNA with respect to heating in contrast to open circular (OC) DNA in which the strands can separate with total loss of fluorescence. This property is made use of in later assays [27]. In other assays duplex and denatured DNA existing within the same molecule complicate the results [29]. Thus, the fluorescence assay at pH 11.8 is a very rapid and sensitive means of detecting cross-linking and, as will be shown, of alkylation as well. Although depurinated sites give rise to strand cleavage in alkali, 7-alkylated purine (the commonest site of DNA alkylation) rings open rapidly at pH 11–12 preventing strand scission [30]. Fuller and

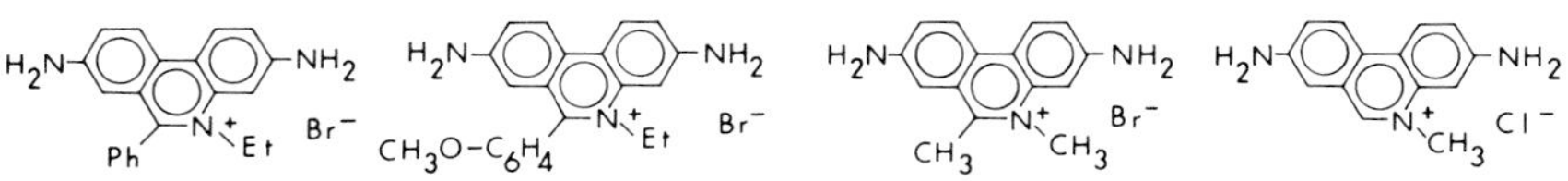

Fig. 3 Ethidium analogs and relative enhancement of fluorescence upon intercalation into λDNA (given by the number under the structures).

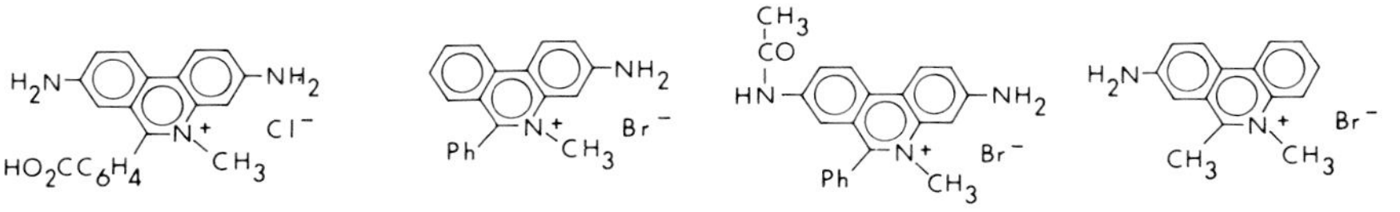

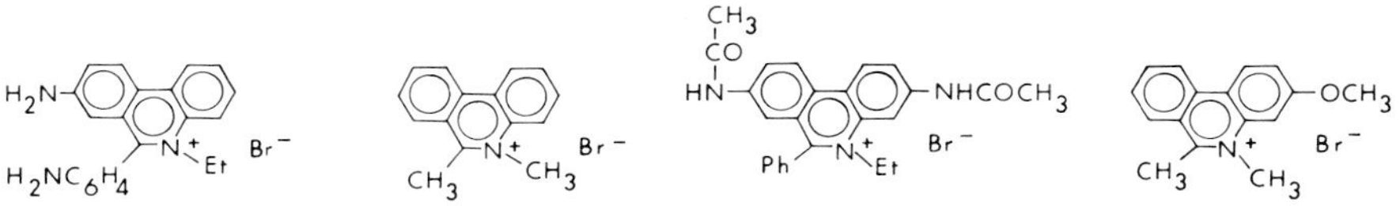

Waring carried out an X-ray diffraction and molecular model building study on the ethidium–DNA complex [31]. They concluded that the tricyclic phenanthridine ring system lies perpendicular to the helix axis through van der Waals interaction with the base pairs above and below. The phenyl and ethyl groups are perpendicular to the plane of the phenanthridine ring and lie in one of the grooves of the helix.

We prepared several analogs of ethidium to determine the substituent requirements for optimizing the fluorescence enhancement upon intercalation into DNA. The pronounced effects of changing the substituents are clearly evident (Fig. 3).

AMINOQUINONE-CONTAINING ANTITUMOR ANTIBIOTICS

A number of antitumor agents or antibiotics, including examples of clinical importance, which are otherwise very differently constituted contain the common aminoquinone moiety. These include mitomycin C (1) [32–32c], porfiromycin [32–32c], streptonigrin (2) [33], the actinomycins (4) [34] and the structurally related aurantins [35,36], and several representatives of synthetic antitumor agents of the aziridinoquinone class, e.g., Trenimon (5). Rifamycin (3), which inhibits RNA synthesis in prokaryotes, is an example of a group of macrolide antibiotics that contain a potential aminoquinone group [37]. Since

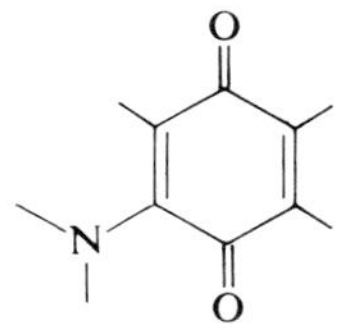

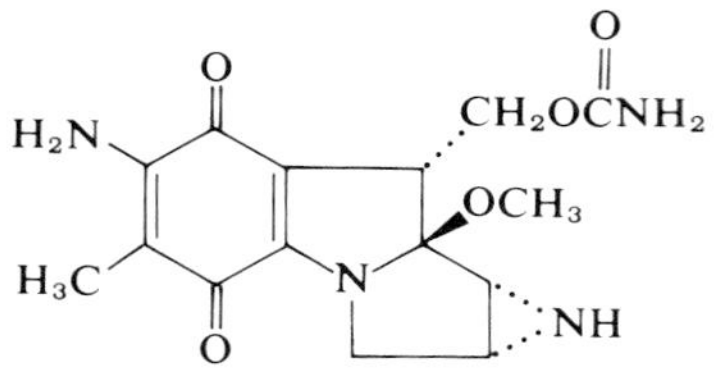

Mitomycin C
(1)

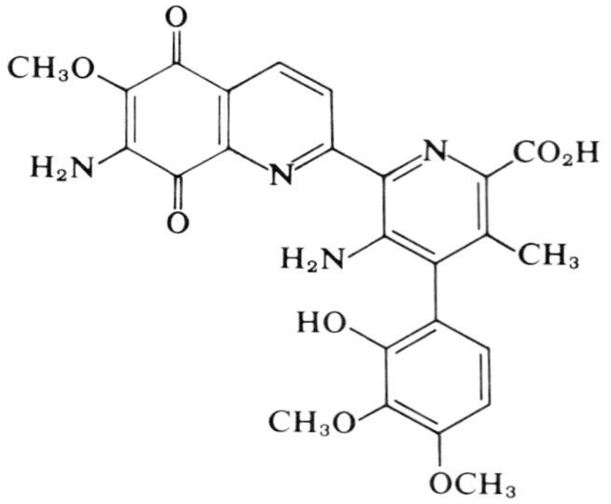

Streptonigrin
(2)

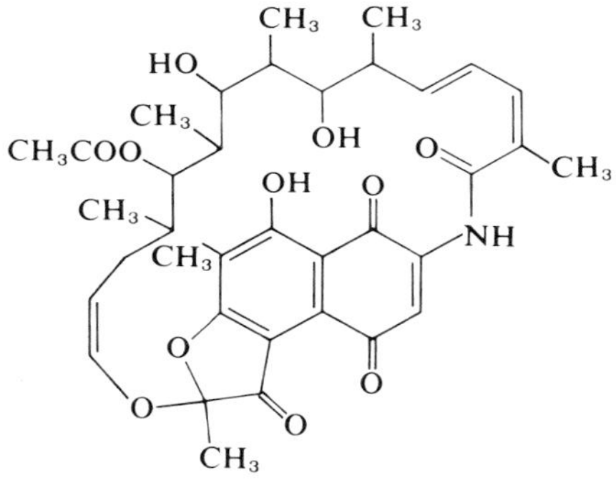

Rifamycin
(3)

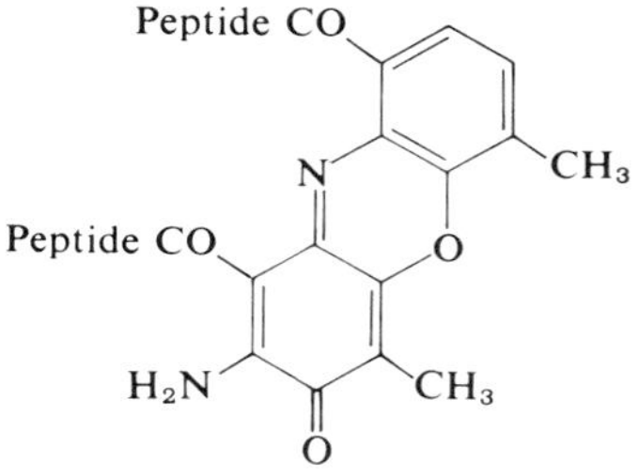

Actinomycin
(4)

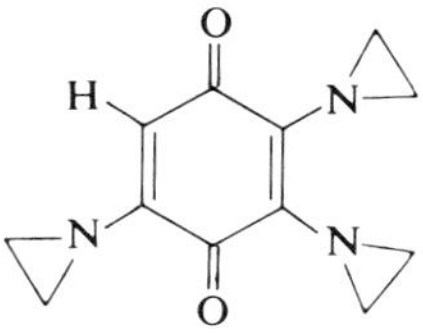

Trenimon
(5)

loss of physiological activity frequently accompanies replacement of the amino group this strongly suggests that the aminoquinone moiety is intimately associated with the expression of anticancer activity. All of these compounds act variously as inhibitors of nucleic acid synthesis.

Streptonigrin

Streptonigrin is an antitumor antibiotic which is a metabolite of *Streptomyces flocculus* [38] and inhibits DNA synthesis in tissue culture cells and bacteria. It inhibits various tumors, e.g., Carcinoma 755, Sarcoma 180, Lewis lung carcinoma, Walker 256 carcinosarcoma, and Ridgeway osteogenic sarcoma [39,40]. The human tumor, H.S. No. 1, grown in rats is particularly sensitive [41]. Also several viral tumors such as Rauscher and Friend leukemia are inhibited [22a,42]. Owing to its high toxicity (the tolerated dose is ca. 2 mg per course of treatment) it has not received widespread clinical use. Streptonigrin inhibits the growth of both gram-negative and gram-positive bacteria. The structural element necessary for antitumor activity is the aminoquinone moiety [43]. As with mitomycin, the bactericidal action is accompanied by DNA degradation [44,45]. In contrast to mitomycin, which also cross-links DNA, it seems that the lethal action of streptonigrin is more directly correlated with degradative damage to DNA; killing occurs at concentrations that do not impair DNA, RNA, or protein synthesis. White and White [46] found that both an electron source and oxygen were required for streptonigrin to exert its greatest lethal effect, suggesting that a reaction product of oxygen and intracellularly reduced streptonigrin is the lethal agent. Considerable evidence suggests that DNA is the principal target in the cell [45–48].

The evidence to date indicates that streptonigrin exerts its antitumor effects by two distinct mechanisms: (a) interference with the cell respiratory mechanism and (b) disruption of the replication mechanism.

An *in vitro* system for studying streptonigrin-induced cleavage of PM2 covalently closed circular DNA was developed [27,49].

FLUORESCENCE ASSAY FOR DETECTION OF SINGLE-STRAND SCISSION
OF PM2 DNA

Heating PM2 CCC DNA [50] with ethidium at pH 8 and cooling gives a denatured form with a large but incomplete loss of fluorescence. However, at pH 11.8 there is a quantitative return of fluorescence, presumably because of the increased mobility of the two strands relative to one another. The topological constraints permit renaturation despite the electrostatic repulsion that prevents duplex formation with homopolymers. Strand scission by a nuclease or by chemical means results in a 30% increase in fluorescence intensity before denaturation due to the removal of topological constraints on ethidium binding [51]. After heat denaturation the strands separate and, provided the pH is greater than 11.8, the fluorescence intensity falls to zero (Fig. 4). This assay has the advantages of simplicity and convenience and can be adapted to the investigation of the mechanism of DNA cleavage by the addition to the

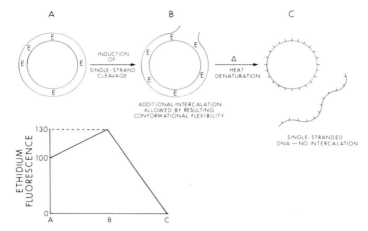

Fig. 4   Fluorescence assay for detecting single-strand scission of PM2 CCC DNA.

system of selective inhibitors. The fluorescence assay for nuclease using PM2 DNA is some 10^5 times as sensitive as the hyperchromicity assay [27].

Streptonigrin itself was found not to cleave PM2 DNA. Prior chemical reduction of the antibiotic was not possible owing to the rapid reoxidation of reduced SN in air. The reduction of streptonigrin by NADH *in situ* nicked approximately 50% of the DNA in 20 min (Fig. 5). From the known requirements of oxygen for streptonigrin-induced damage to DNA [48] the superoxide radical anion $O_2^{\cdot -}$ was a likely candidate. This was confirmed by the complete inhibition of DNA cleavage by both superoxide dismutase and catalyse. The two enzymes together exert a synergistic effect. Superoxide radical generated independently by the aerobic oxidation of xanthine in the presence of xanthine oxidase exhibits similar nicking of the DNA. Evidently, there is a slow, non-

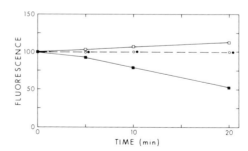

Fig. 5   Cleavage of PM2 CCC DNA by 0.4 mM streptonigrin, 0.4 mM NADH. "Before-heat" fluorescence is denoted by open symbols; closed symbols denote fluorescence after heating and cooling cycle; squares, no additional components; circles, addition of superoxide dismutase or catalase at 4 μg/ml. (**Data** from Cone *et al.* [49].)

enzymatic reduction of SN by NADH with rapid autoxidation of the hydro-quinone species as it is formed. This is also suggested by the kinetics of DNA cleavage, which continues for relatively long periods of time compared with the rapid autoxidation of hydroquinone or compared with the very short lifetime of the semiquinone. The latter has been detected by electron paramagnetic resonance (epr) methods as a short-lived species in metabolizing *Escherichia coli* and *Bacillus megaterium* treated with the drug [52]. In the cell the reduction of streptonigrin is probably enzymatically catalyzed. From studies with model 5,8-quinolinequinones related to streptonigrin, metal ions complexed to the quinolinehydroquinone moiety were required for the autoxidation step [53]. Although metal ions were not deliberately added to the antibiotic solution it is possible that traces of ions were present. Separate control experiments involving the addition of Fe^{2+}, Cu^{2+}, and Co^{2+} ions supported the model scheme for selective metal complex autoxidation of reduced streptonigrin with the production of oxidizing species. Certainly the addition of 10 mM EDTA completely inhibited the cleavage reaction and supported the idea that metal ions were required. The results suggest the mechanism shown in Eqs. (1)–(6) (see also Fig. 6). The superoxide radical undergoes rapid spontaneous dismutation so that only hydrogen peroxide is detected by the spectrophotometric assay used with the metal complexes of 5,8-dihydroxyquinoline. A team of protective enzymes is present *in vivo* to remove damaging oxidative species [54,55]. Superoxide dismutase, a metal-containing enzyme, catalyzes reaction (4), while catalase catalyzes reaction (5). The single-strand scission induced by

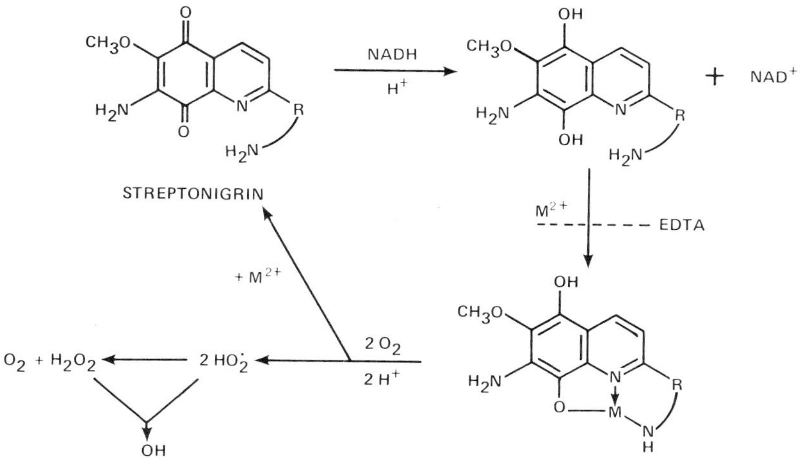

Fig. 6 Metal-ion-catalyzed autoxidation of reduced streptonigrin with production of hydroxyl radicals.

$$SN + NADH + H^+ \longrightarrow SNH_2 + NAD^+ \qquad (1)$$

$$SNH_2 + O_2 \longrightarrow SNH\cdot + HO_2\cdot \qquad (2)$$

$$HO_2\cdot \rightleftharpoons O_2^- + H^+ \qquad (3)$$

$$2O_2^- + 2H^+ \xrightarrow{\text{superoxide dismutase}} H_2O_2 + O_2 \qquad (4)$$

$$2H_2O_2 \xrightarrow{\text{catalase}} 2H_2O + O_2 \qquad (5)$$

$$O_2^- + H_2O_2 \longrightarrow OH\cdot + OH^- + O_2 \qquad (6)$$

streptonigrin is reminiscent of that produced by ionizing radiation, the chemistry of which has been examined recently [56].

It can be concluded that the superoxide radical is an obligatory intermediate in the mode of action of streptonigrin. This mechanism has also been implicated in the depolymerization of hyaluronate in synovial fluid [57]. A rational approach could be developed for modification of streptonigrin to improve its antitumor activity, for example, the addition of chemically reactive side groups to attach to the DNA.

REACTION OF SYNTHETIC STREPTONIGRIN IN ANALOGS WITH DNA

The above-mentioned studies confirm that the 5,8-quinolinequinone moiety in streptonigrin is principally responsible for the expression of antitumor and antibiotic character.

A series of substituted 5,8-quinolinequinones (6) which exhibit significant antitumor activity [58] was prepared and their interaction with PM2 CCC

(6)

DNA was investigated in the presence of a reducing agent, NADH or sodium trimethoxyborohydride. Selective enzymatic and chemical inhibitors of the induction of lesions in the DNA were used to determine the mechanism of cleavage. These studies confirmed that the mechanism of cleavage was identical to that operating with streptonigrin, i.e., via production of the superoxide ion and thence hydroxyl radicals. There was a marked effect of change of substituents on the rate of cleavage of the circular DNA as demonstrated in Figure 7, which shows a range of activity flanking that of the parent antibiotic.

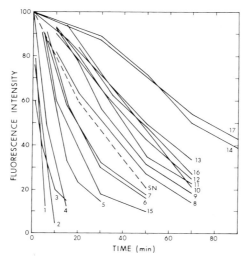

Fig. 7 Single-strand scission of PM2 CCC DNA by reduced 5,8-quinolinequinones at 37°C pH 7, with 1.0 A_{260} unit/ml of DNA, 2.0 × 10^{-4} M substrate, 6 × 10^{-4} M NADH as measured by after-heat fluorescence. Substituents: (1) 6,7-DiBr; (2) 6,7-DiCl; (3) 6,7-DiNH$_2$; (4) 6-MeO; (5) H; (6) 7-Cl-6-MeO; (7) 7-Br-6-MeO; (8) 6-NH$_2$; (9) 7-NH$_2$-6-MeO; (10) 7-N$_3$-6-MeO; (11) 6-NH$_2$-7-Br; (12) 6NH$_2$-7-Cl; (13) 6,7-diMeO; (14) 6-NH$_2$-7-MeO; (15) 6-AcNH-7-Cl; (16) 6-Me$_2$N-7-Cl; (17) 6-Me$_2$N; dashed line, streptonigrin. [Reproduced by permission of the National Research Council of Canada from J. W. Lown and S. K. Sim, *Can. J. Biochem.* **54**, 446 (1976)].

In general those agents with higher redox potential (e.g., containing two halogens) are more efficient in cleaving the DNA. However, since this process requires both reduction of the quinone and subsequent reoxidation of the hydroquinone the relationship is not a simple one.

From *in vivo* data of the tumor inhibitory activity of these substituted 5,8-quinolinequinones published by the National Cancer Institute [58], it is of interest to note that a correlation exists between $t_{1/2}$ (time to achieve 50% scission of PM2 CCC DNA) and tumor weight inhibition of Walker 256 carcinosarcoma in test animals (Fig. 8). This simple parameter may therefore be useful as a prescreening procedure for potential antitumor agents.

The Mitosanes

The mitosanes, a group of six structurally related antibiotics, were discovered by Hata *et al.* in Japan in 1956 [32] as metabolites of *Streptomyces caespitosus* and isolated by Wakaki *et al.* [32] in the form of deep violet crystals. Of these compounds, mitomycin C (R_1 = H; R_2 = CH$_3$;R_3 = NH$_2$) shows the

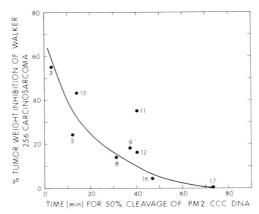

Fig. 8 Correlation of inhibition of Walker 256 carcinosarcoma (data of Driscoll *et al.* [58]) by substituted 5:8-quinolinequinones with efficiency of single-strand cleavage of PM2 CCC DNA.

broadest spectrum of antitumor activity. It is especially effective against chronic myelogenous leukemia and a variety of solid tumors, e.g., carcinomas, chorioepithelioma, reticulum cell sarcoma, and seminoma [59,60]. Mitomycin C is currently in commercial production in Japan, where it has widespread clinical application in the treatment of stomach cancer, which is prevalent in that country. The antibiotic also shows activity against certain viruses [60].

Mitomycin C is quite toxic so that the tolerated dose is approximately 40 mg per full course of treatment (i.e., it is somewhat less toxic than streptonigrin). The chemical structures of the mitosanes were elucidated by Webb and co-workers in 1962 [32]. The mitosanes constitute the first naturally occurring compounds containing an aziridine ring. They contain three known carcinostatic groups; aziridine, quinone, and urethane.

THE MODE OF ACTION OF MITOMYCIN C

Considerable evidence indicates that mitomycin C reacts primarily with DNA [60,61] in two distinct ways: (a) There is covalent cross-linking of DNA with consequent disruption of semiconservative replication; (b) DNA suffers degradation to smaller fragments.

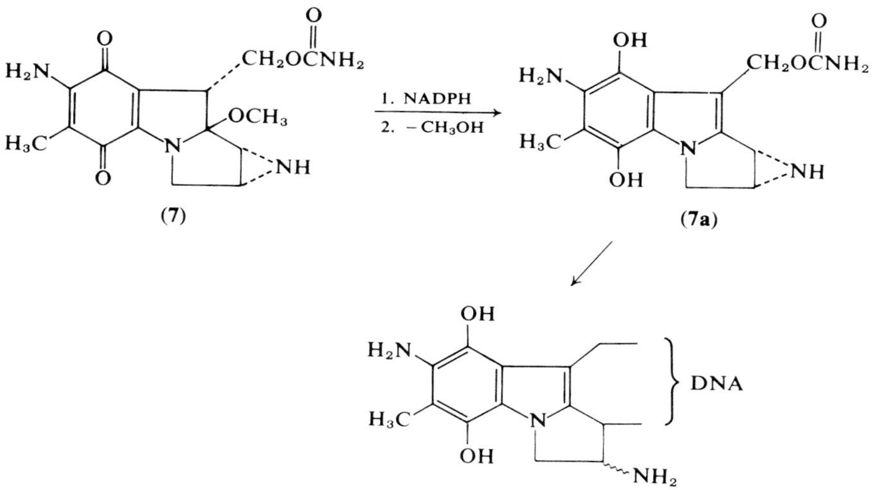

Fig. 9 Suggested reductive activation of mitomycins and subsequent covalent cross-linking of DNA.

COVALENT CROSS-LINKING OF DNA BY MITOMYCIN C

It is considered that mitomycin C is subjected *in vivo* to an initial NADPH-mediated reduction with a cellular reductase with concomitant elimination of the 9a-methoxy grouping to form the reactive and unstable (7a) (60) (Figure 9).

The covalent cross-linking of the complementary strands of DNA is envisaged as shown in Fig. 9. The existing evidence implicates the aziridine and carbamate functions, but there is some speculation that another binding site may be involved [61]. The assumption of DNA as a target site neatly explains why bifunctional agents are so much more cytotoxic for tumors than mono-functional analogs. It is proposed that cross-linking of DNA is the principal cause of mitotic inhibition [18].

Covalent cross-linking of λ phage DNA occurred rapidly on incubating the DNA with reduced mitomycin in a phosphate buffer at pH 7.2 at 22°C using sodium borohydride for reduction. Typically, with a concentration of 0.04 $\mu g/\mu l$ of mitomycin, 100% cross-linking is achieved within 4–5 min [62]. The technique employed is illustrated in Fig. 2. The results of the new technique of fluorescence enhancement are therefore in accord with the demonstration of the cross-linking of DNA by mitomycins by hyperchromicity by Iyer and Szybalski [19] and by others [63–66].

CONFIRMATION OF CROSS-LINKING BY S$_1$-ENDONUCLEASE

In order to confirm that the fluorescence enhancement assay procedure detects the formation of CLC (covalently linked complementary) DNA formed

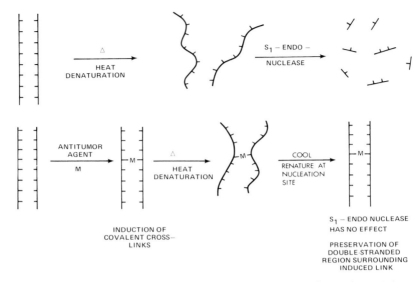

Fig. 10 Confirmation of induction of covalent cross-linking of *E. coli* DNA by anti-tumor agents employing S₁-endonuclease.

TABLE 1

Comparison of the Cross-Linking[a] of *E. coli* DNA Assayed by Ethidium Fluorescence and S₁-Endonuclease Sensitivity

	% cross-linked for run no.		
Assay	1	2	3
Ethidium fluorescence			
Before dialysis	34	48	62
After dialysis	39	51	60
S₁-Endonuclease			
After dialysis	32	51	44

[a] Runs 1, 2, and 3 contained 0.06, 0.15, and 0.2 mM mitomycin C, respectively, with a constant molar ratio of 96:1 of sodium borohydride to mitomycin.

as a result of the chemical cross-linking event, experiments were performed with the enzyme S_1-endonuclease. The latter specifically cleaves single-stranded DNA [67] and is essentially inactive to duplex DNA [68]. Since the time for the S_1-endonuclease digestion is long enough to allow renaturation of denatured λDNA, E. coli DNA was used, which has a suitable C_0t value.

The DNA treated with mitomycin C and sodium borohydride was dialyzed before treatment with the enzyme to remove excess inorganic salts and decomposed mitomycin C. The basis of the assay is outlined in Fig. 10. The results, summarized in Table 1, confirm the formation of covalent cross-links with the antitumor agent. The incomplete cross-linking with E. coli DNA as compared with T7 [26] or λDNA is probably a result of the much lower molecular weight of the E. coli DNA, which therefore requires a much larger proportion of the drug to effect complete cross-linking. There is good correlation between the CLC assay and the endonuclease assay (see Table 1). The somewhat poorer agreement with the higher-concentration run may be attributed to some slow degradation of the DNA by the enzyme at the ends and also as a result of the natural "breathing" mechanism which momentarily exposes short single-stranded regions [69].

DETECTION OF ALKYLATION WITHOUT CONCOMITANT
DEPURINATION DURING TREATMENT OF DNA WITH
ACTIVATED MITOMYCIN C

As the level of mitomycin was increased a progressive decrease in the fluorescence intensity was observed. The loss of fluorescence is linearly related to the extent of drug binding [62] (Fig. 11).

Many effective antitumor agents belong to the alkylating class, and therefore a means of quantifying this mode of action is a useful finding. Alkylating agents are used particularly in the treatment of lymphomas, especially Hodgkin's disease, and of melanoma [18]. Alkylation is a much more toxic event if cells are forced to divide a short period after administration of an alkylating agent. Thus, tissues with a high mitotic index are the most sensitive [18,70].

Poly(dG$\cdot$dC) bearing ^{14}C label in the guanine and ^{3}H in the cytosine was incubated with progressively increasing concentrations of mitomycin C in parallel with E. coli DNA. There was a progressive decrease in ethidium fluorescence. There was no loss of soluble radioactivity, and the ratio of ^{3}H/^{14}C counts was essentially constant under conditions in which increasing concentrations of mitomycin produced from 16.1 to 79.6% reduction in ethidium fluorescence (see Table 2). This confirms that there has been no detectable loss of either purine or pyrimidine bases as a result of alkylation. The fluorescence losses which accompany alkylation at N-7 of guanosine (with e.g., nitrogen mustard, dimethyl sulfate) show marked pH and time dependence.

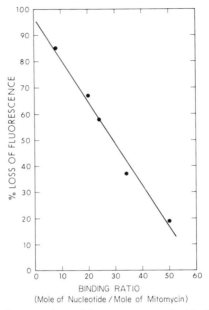

Fig. 11 Dependence of percent loss of fluorescence on the binding ratio of mitomycin to λDNA determined spectroscopically [62]. [Reproduced by permission of the National Research Council of Canada from the *Can. J. Biochem.* **54**, 110 (1976).]

This is due to base catalyzed imidazolium ring opening with concomitant change in ethidium intercalation sites [70a]. The alkylation of DNA by mitomycin C in contrast shows no such pH dependence in accord with previous findings that alkylation takes place at positions other than N-7 [65a].

TABLE 2

Radioactivity Assay for Alkylation of $[^{14}C]dG_n \cdot [^{3}H]dC_n$ by Mitomycin C [a]

Mitomycin C ($\mu g/ml$)	^{3}H(cpm)	^{14}C(cpm)	$^{3}H/^{14}C$ ratio	Decrease in fluorescence with *E. coli* DNA (%)
20	1292	1785	0.724	16.1
40	1225	1682	0.728	43.5
60	1265	1733	0.730	61.7
80	1149	1418	0.810	72.2
100	1187	1528	0.776	79.6

[a] The molar ratio of mitomycin C to sodium borohydride was 1:96 in all experiments.

DEPENDENCE OF EFFICIENCY OF COVALENT CROSS-LINKING OF
DNA BY MITOMYCIN C ON (G + C) CONTENT OF DNA

We applied the ethidium fluorescence assay to detect covalent cross-links and alkylation in three native DNA's of different (G + C) content: *Clostridium perfringens* (30%), calf thymus (40%), and *E. coli* (50%). Assuming a Poisson distribution of the links and further that one link is sufficient to permit spontaneous renaturation of the molecule, an estimate of the average number of cross-links per molecule, $m = \ln(1/Po)$ (where Po is the proportion of the molecules unlinked), was made as 0.54, 0.92, and 1.27 for the three DNA's for a mitomycin concentration of 3.0×10^{-4} M and as 0.08, 0.29, and 0.46 cross-links for a mitomycin concentration of 0.6×10^{-4} M [62]. This result, which demonstrates a clear preference for alkylation of guanine units (Fig 12), is in accord with results of previous workers in which, for example, alkylation of tRNA with porfirmoycins and subsequent hydrolysis yield demonstrable quantities of monoguanyl- and diguanylporfiromycin [60]. Examination of space-filling models of DNA and of mitomycin suggested that the best fit was obtained by postulating links between the O-6 groups of nearby guanines on opposing strands [19,60]. A new assay to measure 7-alkylation of guanine

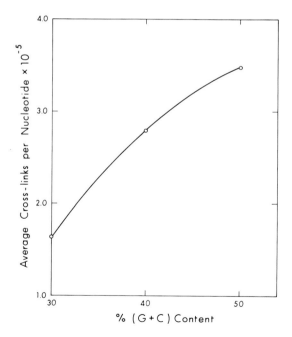

Fig. 12 Dependence of efficiency of covalent cross-linking of λDNA by mitomycin C on the (G + C) content of the DNA.

residues in DNA was described by Tomasz and applied to mitomycin C, which was shown not to attack at this site [65].

THE pH DEPENDENCE OF CROSS-LINKING AND ALKYLATION OF DNA BY MITOMYCIN C AND THE SEQUENCE OF COVALENT CROSS-LINKING

Recalling the structure of mitomycin, which contains an aziridine ring, one would expect a pH dependence of the ability to alkylate at this site. The decrease in fluorescence due to mitomycin C is strongly pH dependent with lower pH favoring more rapid alkylation (Fig. 13a). This suggests that under these conditions the aziridine function alkylates the bases of DNA first. This conclusion is in agreement with previous results in which the product, after treatment of mitomycin C with a reducing agent followed by secondary rearrangement and then reoxidation, corresponded to the structure in which the aziridine ring is opened [32c].

Similarly, the trend toward more efficient covalent cross-linking with lower pH is clear (Fig. 13b). The results, which parallel those obtained for alkylation, again suggest that the first event in cross-linking is due to attack by the acid-sensitive aziridine. Tumor cells are characterized by lower pH caused by their high anaerobic glycolytic rate and the production of lactic acid [18,70]. This fact coupled with the more reducing environment in tumor cells could contribute to the selectivity of action of mitomycin. Lowering the pH alone was found to be sufficient to promote alkylation of DNA by the aziridine ring followed by cross-linking. Freese and Cashel [71] have demonstrated that treatment of DNA under low pH conditions can induce covalent cross-links; however, careful controls showed that under the conditions of the present experiment no acid-promoted cross-linking was significant. Cross-linking is considerably slower than in the presence of the reducing agent, but given the lower pH conditions that obtain in tumor cells [70], it does raise the possibility that mitomycin may not have to be reduced even within the cell for cross-linking to occur. Mitomycin C was attached to DNA by a short pH 4.0 treatment, and the resulting DNA–antibiotic complex was dialyzed to remove unreacted and unattached mitomycin. The extent of cross-linking was determined by the CLC assay before and after dialysis. Subsequent treatment of the complex with sodium borohydride resulted in an increase in the extent of cross-linking, confirming the establishment of stepwise cross-linking (Fig. 14). This finding may be useful in chemotherapy since in many cases the DNA–drug complex has greater antitumor activity than the drug alone [70].

THE MECHANISM OF DNA DEGRADATION BY MITOMYCIN C

Mitomycin C has been observed to cause DNA degradation which accompanies covalent cross-linking, and hitherto this had been considered to be

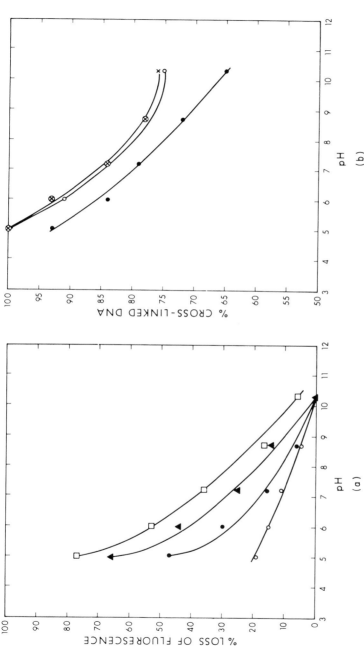

Fig. 13 (a) The pH dependence of alkylation of DNA by reduced mitomycin C. The reactions were at 22°C in 50 mM phosphate at the appropriate pH and λDNA at 1.2 A_{260}. The antibiotic concentrations were 0.6 (○), 1.2 (●), 2.0 (▲), and 4.0×10^{-4} M (□). The sodium borohydride concentrations were 1.3 (○), 2.2 (●), 3.1 (▲), and 5.3×10^{-3} M (□) [62]. (b) The pH dependence of cross-linking DNA by reduced mitomycin C. The conditions were as for (a), but the fluorescence values were obtained after heating the DNA solution. The antibiotic concentrations were 0.6 (●), 1.2 (○), 2.0×10^{-4} M (x). The sodium borohydride concentrations were 1.3 (●), 2.2 (○), and 3.1×10^{-3} M (x) [62].

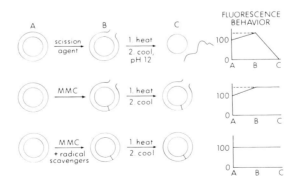

CONDITIONS	% CROSS—LINKED BY CLC ASSAY
Before dialysis	50
After dialysis, pH 7.0	50
After dialysis and NaBH₄ reduction	65

Fig. 14   Stepwise cross-linking of λDNA by mitomycin C (MMC).

due largely to the activation of intracellular deoxyribonucleases [72,73]. However, a comparison of the redox potentials of streptonigrin and mitomycin indicated that the degradation of DNA by an oxidative pathway was feasible for the latter. The assay for strand scission employed to investigate the mode of action of streptonigrin is complicated by concomitant cross-linking and also alkylation. Neither of the latter reactions can give rise to an increase in fluorescence, which can be accounted for only by nicking CCC DNA (Fig. 15). However, the loss of fluorescence normally observed on heating the nicked DNA is not observed due to cross-linking. Indeed, the OC DNA originally

Fig. 15   Ethidium fluorescence assays for detecting strand cleavage and cross-linking of PM2 CCC DNA by mitomycin C (MMC).

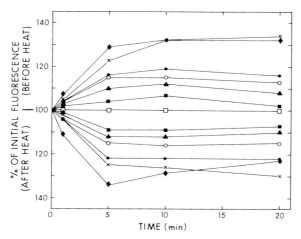

Fig. 16 Single-strand scission of PM2 CCC DNA by reduced mitomycin C, 3×10^{-4} M and $NaBH_4$, 5.3×10^{-3} M, at 22°C, pH 7. Additional components were ×, none; ○ catalase (4.1×10^{-6} M); ●, catalase (4.1×10^{-6} M), superoxide dismutase (6.1×10^{-5} M); ▲, sodium benzoate (5.0×10^{-2} M); ■, isopropyl alcohol (2.5×10^{-1} M); ◆, superoxide dismutase (3.0×10^{-5} M); □, control.

present in the CCC DNA (85% return of fluorescence after heat treatment) appears as CLC DNA. It was shown in a control experiment that sodium borohydride is without effect on PM2 CCC DNA. The results, which are summarized in Fig. 16, confirm that mitomycin C when reductively activated does induce single-strand cleavage of DNA as shown by the rise in fluorescence. This was independently confirmed by sedimentation velocity studies in which the compact CCC DNA ($s_{20,w} = 30$) was converted to material with a sedimentation rate of OC DNA ($s_{20,w} = 23$) [62]. The production of single-strand cleavage may be retarded by catalase or by a combination of catalase and superoxide dismutase or by free-radical scavengers such as isopropyl alcohol or sodium benzoate. This suggests a mechanism of cleavage similar to that operating for reduced streptonigrin and its analogs in which cleavage of PM2 CCC DNA is induced by hydroxyl radical. The pathway requires the intermediacy of mitomycin semiquinone, which has recently been detected by epr and shown to have a lifetime of several seconds [74]. In contrast to streptonigrin, mitomycin C does not inactivate the action of either of the cell protective enzymes catalase or superoxide dismutase [62]. It may be noted in connection with this proposed mechanism that the inhibitory effect of mitomycins on the synthesis of RNA is explained by Kersten and Kersten [15] as involving the quinone ring since mitomycin derivatives lacking the aziridine function and synthetic indoloquinones have been found to inhibit

TABLE 3

Correlation of Covalent Cross-Linking of λ-DNA by Bifunctional Alkylating Agents with Antitumor Activity against Various Tumors [a,b]

Structure	L1210 OD	L1210 ILS	W-256 OD	W-256 ILS	CA-755 OD	CA-755 TWI	S-180 OD	S-180 TWI	KB ED$_{50}$	λDNA cross-linking Concentration (μg/μl)	Maximum (%)	Time to reach maximum (min)
Structure 1 (H$_2$N, H$_3$C, CH$_2$OCNH$_2$, OCH$_3$, NH quinone/mitosane)	1.5	60	1.0	95	1.0	90	3.0	75	0.04	0.04 / 0.02	100 / 93	5 / 5
Structure 2 (triaziridinyl quinone)	0.1	47	0.1	96	0.04	70	0.13	49	0.01	0.4 / 0.04	100 / 100	2 / 10
Structure 3 (diaziridinyl quinone) [c]	0.5	26	—	—	0.25	60	1.25	39	< 1.0	0.01	92	10
Structure 4 (CH$_3$, CH$_3$O diaziridinyl quinone) [c]	2.0	36	2.0	95	—	—	—	—	—	0.4 / 0.04	92 / 82	60 / 120

0.01	39	—	—	0.05	80	0.13	65	4.9	0.4 0.04	90 85	25 100		
4	50	1.0	95	3.0	78	5.0	5.0	0.07	0.4 0.04	85.5 79	30 60		
2.0	7.0	0.98	84	2.0	80	3.0	63	—	0.4	55	255		
11, 22	0	—	—	11	50	25	35	—	0.4	18	120		
100	Toxic	—	—	—	—	—	—	—	0.4	0	—		

[a] Abbreviations used: OD, optimal dose; ILS, increased life span of test animals; TWI, percent tumor weight inhibition; ED$_{50}$, effective dose for 50% survival; L1210, Leukemia 1210; W-256, Walker 256 carcinoma; CA-755, carcinoma 755; S-180, Erlich ascites sarcoma 180 [78]. [b] Data taken from Akhtar et al. [78]. [c] Shows antibacterial activity against *Staphylococcus aureus*, *E. coli*, and *Streptococcus faecalis*. [d] Shows marked carcinostatic activity against Jensen sarcoma.

the synthesis of proteins and RNA in microorganisms. Similarly, Lerman and Benyumovich [76] in a study of the effects of mitomycin on protein synthesis in human neoplastic cell lines concluded that the function of the quinone ring predominates in tumor tissue.

Bifunctional Aziridinoquinones with Antineoplastic Activity

Many aziridine-containing compounds of quite different structures have useful antitumor properties.

Aziridinoquinones are of special interest because of their central nervous system antineoplastic activity [70]. The mechanism whereby these agents exert their antitumor activity is still a subject of controversy [77]. The induction of CLC sequences in λDNA by a series of ortho- and paraaziridinoquinones was examined and confirmed independently by the S_1-endonuclease assay described previously.

The aziridinoquinones display a similar strong pH dependence for mono-alkylation and for cross-linking as was found for the mitosanes. Experiments conducted with poly(^{14}C]dG-[^3H]dC) showed that alkylation is not accompanied by depurination. The aziridinoquinones also exhibit the same preference for attack on guanine residues as shown by increased cross-linking with higher (G + C) content of the DNA.

It is of interest that a fair correlation exists between the extent and rate of covalent cross-linking of DNA and antineoplastic activity, especially against Leukemia L1210 (Table 3) [78]. This parameter may prove to be useful for prescreening antitumor agents for clinical trials. It is recognized that other biological and pharmacological parameters, in addition to DNA covalent cross-linking (i.e., drug uptake, partition across membranes, metabolism, and toxicity), contribute to the ultimate effectiveness of cancer inhibitory properties.

CONCLUSIONS

What is sorely needed in cancer chemotherapy is a good guide or rationale for planning the structure of an effective cytotoxic agent. This stage will come when we have an understanding of the mechanism of action of antitumor drugs. The brief survey of our approach adopting chemical modification of recognized antitumor agents and the application of the novel techniques of fluorescence enhancement indicates that quite detailed information can be obtained on the different modes of chemical interaction of antitumor agents with macromolecular target sites. Important supportive studies are being made in an attempt to understand DNA replication, control of gene expression in animal and human tissue, and interaction of known effective anti-tumor agents with these processes [79].

Another important aspect of this problem should be considered. There is a correlation between the sensitivity of mammalian and bacterial cells to alkylating agents and their ability to repair alkylated DNA [18,80]. It is likely that malignant tissues, which are highly sensitive to alkylating agents, have lost the ability to repair DNA efficiently or the antitumor agent may interfere with the repair mechanism selectively. It should be noted that the ethidium fluorescence assays described here are readily adapted to monitor the function of the cell repair enzymes: nucleases, ligases, and polymerases [27].

REFERENCES

1. J. A. Montgomery, T. P. Johnston, and Y. F. Shealy, *in* "Medicinal Chemistry" (A. Burger, ed.), p. 680. Wiley (Interscience), New York, 1970.
2. R. T. Prehn, *Proc. Can. Cancer Res. Conf.* 10, 136 (1974).
3. "Report of the Division of Cancer Treatment," Vols. 1 and 2. Natl. Cancer Inst., Bethesda, Maryland, 1974. Quoted by L. N. Ferguson [70].
4. E. F. Gale, E. Cundliffe, D. E. Reynolds, M. H. Richmond, and M. J. Waring, "The Molecular Basis of Antibiotic Action," p. 24. Wiley (Interscience), New York, 1972.
5. T. J. Franklin and G. A. Snow "Biochemistry of Antimicrobial Action," p. 19. Chapman & Hall, London, 1975.
6. W. C. J. Ross, "Biological Alkylating Agents." Butterworths, London, 1962.
7. G. P. Wheeler, *Cancer Res.* 22, 651 (1962).
8. E. F. Gale, E. Cundliffe, D. E. Reynolds, M. H. Richmond, and M. J. Waring, "The Molecular Basis of Antibiotic Action," pp. 246–252. Wiley (Interscience), New York, 1972.
9. E. Boyland, *Biochem. Aspects Antimetab. Drug Hydroxylation, Fed. Eur. Biochem. Soc., Meet., 5th, 1968* p. 183 (1969).
10. P. Brookes and P. D. Lawley, *J. Cell. Comp. Physiol.* 64, Suppl. 1, 111 (1964).
11. H. Dellweg, *Dsch. Med. Wochenschr.* 91, 1697 (1966).
12. H. Kersten and W. Kersten, *in* "Inhibitors: Tools in Cell Research" (T. Buchner and H. Sies, eds.), p. 11. Springer-Verlag, Berlin and New York, 1969.
13. L. S. Lerman, *J. Cell. Comp. Physiol.* 64, Suppl. 1, 1 (1964).
14. D. Neubert, *Internist* 7, 435 (1966).
15. M. J. Waring, *Nature (London)* 219, 1320 (1968).
16. A. Goldstein, L. Aronow, and S. M. Kalman, "Principles of Drug Action," p. 618. Harper, New York, 1968.
17. S. Sesnowitz-Horn and E. A. Adelburg, *J. Mol. Biol.* 46, 1 (1969).
18. T. A. Connors, *Top. Curr. Chem.* 52, 148 (1974).
19. V. N. Iyer and W. Szybalski, *Proc. Natl. Acad. Sci. U.S.A.* 50, 355 (1963).
20. P. Brookes, *Cancer Res.* 26, 1994 (1966).
21. K. Hamaguchi and E. P. Geiduschek, *J. Am. Chem. Soc.* 84, 1329 (1961).
22. J. M. Saucier and J. C. Wang, *Biochemistry* 12, 2755 (1973).

22a. T. J. McBride, J. J. Oleson, and D. Woolff, *Cancer Res.* **26**, 727 (1966).

23. J. B. Le Pecq and C. Paoletti, *J. Mol. Biol.* **27**, 87 (1967).

24. F. Hawking, *Exp. Chemother.* **1**, 129 (1963).

25. B. A. Newton, *Adv. Chemother.* **1**, 35 (1964).

26. A. R. Morgan and V. Paetkau, *Can. J. Biochem.* **50**, 210 (1972).

27. A. R. Morgan and D. E. Pulleyblank, *Biochem. Biophys. Res. Commun.* **61**, 346 (1974).

28. M. Eigen and D. Porschke, *J. Mol. Biol.* **53**, 123 (1970).

29. M. J. Waring, "Biochemical Studies of Antimicrobial Drugs," p. 235. Cambridge Univ. Press, London and New York, 1966.

30. P. D. Lawley and P. Brookes, *Biochem. J.* **89**, 127 (1963).

31. W. Fuller and M. J. Waring, *Ber. Bunsenges. Phys. Chem.* **68**, 805 (1964).

32. T. Hata, Y. Sano, R. Sugawara, A. Matsumae, K. Kanamori, T. Shima, and T. Hoshi, *J. Antiobiot., Ser. A* **9**, 141 (1956).

32a. S. Wakaki, H. Marumo, T. Tomioka, G. Shimizu, E. Kato, H. Kamada, S. Hudo, and Y. Fujimoto, *Antibiot. Chemother. (Washington, D.C.)* **8**, 228 (1958).

32b. S. Wakaki, *Cancer Chemother. Rep.* **13**, 79 (1961).

32c. J. S. Webb, D. B. Cosulich, J. H. Mowat, J. B. Patrick, R. W. Broschard, W. E. Meyer, R. P. Williams, C. F. Wolf, W. Fulmor, C. Pidaks, and J. E. Lancaster, *J. Am. Chem. Soc.* **84**, 3185 (1962).

33. K. V. Rao, K. Biemann, and R. B. Woodward, *J. Am. Chem. Soc.* **85**, 2532 (1963).

34. S. A. Waksman, *Antibiot. Chemother. (Washington D.C.)* **6**, 90 (1956); H. Brockmann G. Bohnsack, B. Franck, H. Grone, H. Muxfeldt, and C. Suling, *Angew. Chem.* **68**, 70 (1956); G. E. Foley, *Antibiot. Annu.* p. 432 (1956); J. P. Cobb and D. G. Walker, *J. Natl. Cancer Inst.* **21**, 263 (1958).

35. M. B. Bitteeua, *Mikrobiologiya* **31**, 492 (1962); Y. O. Sazykin and G. N. Varisova, *Fed. Proc. Fed. Am. Soc. Exp. Biol.* **23**, 380 (1964).

36. G. P. Georgiev, O. P. Samarina, M. I. Lerman, M. N. Smirnov, and A. N. Severton, *Nature (London)* **200**, 1291 (1963).

37. P. Sensi, *Res. Prog. Org.-Biol. Med. Chem.* **1**, 337 (1964).

38. K. U. Rao and W. P. Cullen, *Antibiot. Annu.* p. 950 (1959).

39. J. J. Oleson, L. A. Calderella, K. J. Mjos, A. R. Reith, R. Thie, and I. Toplin, *Antibiot. Chemother. (Washington, D.C.)* **11**, 158 (1961).

40. H. C. Reilly and K. Sugiura, *Antibiot. Chemother. (Washington, D.C.)* **11**, 174 (1961).

41. M. N. Teller, S. F. Wagshul, and G. W. Woolley, *Antibiot. Chemother. (Washington, D.C.)* **11**, 165 (1961).

42. P. S. Ebert, M. A. Chirigos, and P. A. Ellsworth, *Cancer Res.* **28**, 363 (1968).

43. K. V. Rao, *Cancer Chemother. Rep.* **4**, 11 (1974).

44. M. M. Cohen, M. W. Shaw, and A. P. Craig, *Proc. Natl. Acad. Sci. U.S.A.* **50**, 16 (1963).

45. D. S. Miller, J. Laszlo, K. S. McCarty, W. F. Guild, and P. Hochstein *Cancer Res.* **27**, 632 (1967).

46. H. R. White and J. A. White, *Biochim. Biophys. Acta* **123**, 648 (1966).

47. N. S. Mizuno and D. P. Gilboe, *Biochim. Biophys. Acta* **224**, 319 (1970).

48. H. L. White and J. R. White, *Mol. Pharmacol.* **4**, 549 (1968).

49. R. Cone, S. K. Hasan, J. W. Lown, and A. R. Morgan, *Can. J. Biochem.* **54**, 219 (1976).

50. P. H. Pouwels, J. van Rotterdam, and J. A. Cohen, *J. Mol. Biol.* **40**, 379 (1969).

51. C. Paoletti, J. B. Le Pecq, and I. R. Lehman, *J. Mol. Biol.* **55**, 75 (1971).

52. K. Ishizu, H. H. Dearman, M. T. Huang, and J. R. White, *Biochim. Biophys. Acta* **165**, 283 (1968).

53. M. H. Akhtar, S. K. Hasan, J. W. Lown, J. A. Plambeck, and S. K. Sim, unpublished results.
54. I. Fridovich, *Acc. Chem. Res.* **5**, 321 (1972).
55. E. Friden, *Chem. & Eng. News* **52**, 42 (1974).
56. M. Dizdaroglu, C. von Sonntag, and D. Schulte-Frohlinde, *J. Am. Chem. Soc.* **97**, 2277 (1975).
57. J. M. McCord, *Science* **185**, 529 (1974).
58. J. S. Driscoll, F. G. Hazard, H. B. Wood, and A. Goldin, *Cancer Chemother. Rep.*, *Part 2* **4**, (21), 1 (1974).
59. F. R. White, *Cancer Chemotherap. Rep.* **2**, 21 (1959); K. Sugiura, *ibid.* **13**, 51 (1961).
60. W. Szybalski and V. N. Iyer, *In* "Antibiotics I" (D. Gottlieb and P. D. Shaw, eds.), pp. 211–245. Springer-Verlag, Berlin and New York, 1967.
61. H. Kersten and W. Kersten, *Mol. Biol. Biochem. Biophys.* **18**, 4 (1974).
62. J. W. Lown, A. Begleiter, D. Johnson, and A. R. Morgan, *Can. J. Biochem.* **54**, 110 (1976).
63. R. J. Cohen and D. M. Crothers, *Biochemistry* **9**, 2533 (1970).
64. A. E. Pritchard and B. E. Eichinger, *Biochemistry* **13**, 4455 (1974).
65. M. Tomasz, *Biochim. Biophys. Acta* **213**, 288 (1970).
65a. M. Tomasz, C. M. Mercado, J. Olson, and N. Chatterjie, *Biochemistry* **13**, 4878 (1974).
66. D. A. Wilson and C. A. Thomas, *J. Mol. Biol.* **84**, 115 (1974).
67. W. D. Sutton, *Biochem. Biophys. Acta* **240**, 522 (1971).
68. F. T. Morrow and P. Berg, *J. Virol.* **12**, 1631 (1973).
69. P. H. von Hippel and K. Y. Wong, *J. Mol. Biol.* **61**, 587 (1971).
70. L. N. Ferguson, *Chem. Soc. Rev.* **4**, 289 (1975).
70a. H. Hsiung, J. W. Lown, and D. Johnson, *Can. J. Biochem.* **54**, 1047 (1976).
71. E. Freese and M. Cashel, *Biochim. Biophys. Acta* **91**, 67 (1964).
72. W. Kersten, *Biochim. Biophys. Acta* **55**, 558 (1962).
73. Y. Nakata, K. Nataka, and Y. Sakamoto, *Biochem. Biophys. Res. Commun.* **6**, 339 (1961).
74. T. Nagata and K. Matsuyama, *Prog. Antimicrob. Anticancer Chemother., Proc. Int. Congr. Chemother., 6th, 1969* Vol. 2, p. 423 (1970).
75. H. Kersten and W. Kersten, *in* "Inhibitors: Tools in Cell Research" (T. Buchner and H. Sies, eds.), pp. 11–31. Springer-Verlag, Berlin and New York, 1969.
76. M. I. Lerman and M. S. Benyumovich, *Nature (London)* **206**, 1231 (1965).
77. O. C. Dermer and G. E. Ham, "Ethyleneimine and other Aziridines," p. 425. Academic Press, New York, 1969.
78. M. H. Akhtar, A. Begleiter, D. Johnson, J. W. Lown, L. McLaughlin, and S. K. Sim, *Can. J. Chem.* **53**, 2891 (1975).
79. F. M. Thompson, A. N. Tischler, J. Adams, and M. Calvin, *Proc. Natl. Acad. Sci. U.S.A.* **71**, 107 (1974).
80. J. J. Roberts, *in* "DNA Repair Mechanisms" (F. C. Schattauer, ed.), p. 41. Springer-Verlag, Berlin and New York, 1972.

CHAPTER

6

Composition and Structure of Thermal Condensation Polymers of Amino Acids

Paul Melius

INTRODUCTION

The thermal condensation of amino acid mixtures to polymeric substances was demonstrated by Fox [1] when a high proportion of aspartic acid, glutamic acid, or lysine was included in the polymerization mixture. This review covers the nature of these polymers which are referred to as "proteinoids" by Fox. The amino acid composition of the polymer in relation to the amino acid composition of the polymerization mixture, the molecular size and amino acid composition, N- and C-terminal amino acid residues, branching, and some sequence information are discussed.

The role of thermal condensation of amino acids to proteinoids in the origin of life has been extensively discussed by Fox et al. [2,3]. The origin of living organisms probably required formation of macromolecules such as proteins and nucleic acids. Miller [4] demonstrated the production of simple aliphatic amino acids and organic acids by an electrical discharge through a mixture of methane, ammonia, water, and hydrogen gases. Harada and Fox [5] found that a mixture of methane, ammonia, and water heated to 900°–1100°C in the presence of silica sand produced a complex mixture of amino acids. That the atmosphere of primitive earth was composed of methane, ammonia, water, and hydrogen gas has been suggested by Urey [6] and Oparin [7].

Proteinoids produced before the appearance of life may have had far different amino acid compositions than proteinoids produced in current laboratory experiments. For example, Miller [4] produced simple aliphatic amino acids as did Harada and Fox [5] and Oro [8] in their experiments without contact materials present. Harada and Fox [5] did produce more of the protein amino acids when they used silica sand, but they still did not produce the basic amino acids. Of course, they did not produce the sulfur amino acids, as their system did not contain any sulfur. They did find β-alanine, α-aminobutyric acid, and alloisoleucine, which are nonprotein amino acids.

Lawless et al. [9] have reported the presence of valine, alanine, glycine, proline, and aspartic acid (protein amino acids) and the nonprotein amino acids N-methylglycine, α-aminoisobutyric acid, β-aminoisobutyric acid, β-alanine, and γ-aminobutyric acid in the Orgueil meteorite. The DL racemic mixtures of the amino acids suggested a nonbiogenic origin for the amino acids. Kvenvolden et al. [10] found glycine, alanine, valine, proline, glutamic acid, and aspartic acid in the Murchison meteorite. Along with the protein amino acids they found N-methylglycine, α-aminoisobutyric acid, β-alanine, α-amino-n-butyric acid, γ-amino-n-butyric acid, β-amino-n-butyric acid, isovaline, and pipecolic acid with tentative evidence for at least β-aminoisobutyric acid, norvaline, N-methylalanine, and N-ethylglycine. Lawless [11] reported the characterization of 17 more amino acids present in the Murchison meteorite, all of probable abiotic origin. Thus, it would appear that the environment of the primitive abiotic earth could have contained a much larger number of different types of amino acids than appear in modern proteins.

It is of interest here that the six protein amino acids found in the Murchison meteorite constitute 64% of the amino acids in a ferredoxin of the anaerobic bacteria Clostridium butyricum [12]. Also, 91% of the amino acids in this organism have been synthesized in abiogenic laboratory experiments. Thus, Hall et al. [13] suggest this as one of the earliest proteins formed.

Proteinoids produced by Fox and his collaborators and also in our laboratory [14] have been found to be in the molecular weight range of 5,000–11,000 for the acidic types. The ferredoxin molecular weights are about 6,000–12,000.

From the foregoing discussion, it is apparent that there is a great deal of interest in thermal condensation of amino acids to polymers in the area of molecular evolution. The chemistry of this process is now becoming important to the nutritionist and food scientist with regard to the changes that take place in polypeptides during food preparation and food processing. For example, Sternberg et al. [15] have found in commercial foods and home-cooked foods that had not been exposed to alkali a factor, lysinoalanine, that may be toxic to rat kidney. Alkaline treatment of foods and proteins is known to produce the lysinoalanine.

Pyrolysis of individual amino acids at very high temperatures (500°C) has been carried out by Ratcliffe *et al.* [16] with analysis of products by gas chromatography and mass spectrometry. A wide variety of volatile products were identified and mechanisms suggested for the degradative reactions. Lien and Nawar [17] studied the thermal decomposition of amino acids at 180°–270°C. At 220°C no apparent change occurred (aliphatic amino acids), whereas at 250°C a light yellow color developed and at 270°C the amino acid became black and the crystalline appearance was lost. Products at the higher temperatures were NH_3, CO_2, CO, alkanes, alkenes, ketones, aldehydes, alkylamines, aromatic amines, and water.

EFFECT OF AMINO ACID COMPOSITION OF REACTANTS ON AMINO ACID CONTENT OF PROTEINOIDS

Fox and Harada [18] found that dicarboxylic amino acids exerted a protective effect against the decomposition of the neutral amino acids, particularly at temperatures from 150° to 210°C. Thus, using either or both aspartic acid and glutamic acid in the reaction mixture in a hypohydrous state results in the production of polypeptide polymers (proteinoids) rather than destruction of the amino acids. The reaction mixture was subjected to "dialytic washing(s)" or was dissolved in sodium bicarbonate and dialyzed in cellulose tubing (preparation A). Diffusible, soluble and nonsoluble, nondiffusible fractions were obtained. When a mixture of two parts of glutamic acid, two parts of aspartic acid, and one part of an equimolar mixture of the neutral and basic (BN) amino acids (2:2:1) was used for polymerization, the nondiffusible material contained 67.5% aspartic acid and 10.7% glutamic acid and all other amino acids ranged from 0.1% (threonine) to 3.0% (alanine). When 25 gm of a 2:2:1 mixture was treated at 170°C, 2.40 gm of S (insoluble proteinoid) was obtained after 2 hr and 9.41 gm was obtained after 6 hr. When a 1:5:3.5 mixture (comparable to the amino acid composition of modern proteins) was used only 0.24 gm of S was obtained, which indicates the requirement of a high proportion of acidic amino acids for high yields of proteinoid. The products were all ninhydrin-negative and biuret-positive. In the A preparation from the 2:2:1 mixture, there was a 3% decrease in glutamic acid with a 3% increase in aspartic acid; the BN remained essentially constant during 2- to 6-hour heating times. In contrast, the A fraction from the 1:1:1 mixture had a constant glutamic acid content but a 5% decrease in aspartic acid and a 5% increase in BN over the 2- to 6-hour heating periods.

When the temperature was increased from 160° to 190°C, there was an increase of BN acids but a decrease of aspartic acid. The increase in temperature also resulted in an increase in mean chain weight from about 4000 to 8000 as determined by end-group analysis.

When Fox and Harada [19] polymerized a 2:2:3 type of mixture, the percentage of alanine and lysine incorporated was twice that of alanine and lysine in the reaction mixture. The percentage of aspartic acid incorporated was almost twice the percentage of aspartic acid in the reaction mixture, whereas half as much glutamic acid was incorporated as was present in the reaction mixture. The percentage of half-cystine and glycine incorporated was also greater than the percentage of these residues in the reaction mixture, while for the remaining amino acids the percentage of residues incorporated was lower than the percentage of residues present in the reaction mixture. It would be of interest to use even greater ratios of single amino acids than were reported here.

Thus, Fox and Harada were able to incorporate all 18 amino acids common to modern proteins into their proteinoids. Tryptophan was absent from the acid hydrolyzates of the proteinoids but was found by color test and microbial assay of the alkaline hydrolyzates of the proteinoids.

As indicated before, standard protein color tests were obtained for the proteinoids. The infra red spectra indicated amide and imide bonds. The imide bond spectrum was replaced by an amide bond spectrum upon gentle alkaline treatment of the proteinoid. The proteinoids had a very low water solubility.

Phillips and Melius [20] studied the polymerization of a mixture of eight amino acids: one part aspartic acid one part glutamic acid, and one part, glycine, alanine, valine, isoleucine, and proline at 180°–190°C for 9–12 hr. These studies were initiated with the intention of obtaining information about the role and possible directing effects of the various amino acids in the polymerization process. The products under these conditions of preparation were proteinoids that were 90% diffusible. Fox defines proteinoids as macro-molecular preparations with molecular weights in the thousands containing most of the 20 amino acids found in protein hydrolyzates. We also apply the term to products obtained which do not contain most of the 20 amino acids found in proteins. Of course, there are proteins that do not contain all 20 amino acids. Unreacted amino acids were removed by preparative high-voltage electrophoresis at pH 1.9. The diffusible fraction was subjected to gel filtration on Sephadex G-25, and two major fractions were obtained.

The isolated proteinoids were acid hydrolyzed and analyzed by the gas chromatography method of Gehrke [21] for their amino acid composition. The results indicated a variation of amino acid content with polymer size. Three effects were observed. The first was a reversal of predominance of aspartic acid in going from the largest (nondiffusibles, 49 mole %) to the smallest (most retarded molecules on Sephadex column, 13.8 mole %). The second effect was an increase of glutamic acid from 9.8 mole % in the non-diffusibles to 52.0 mole % in the slow-moving proteinoids on the Sephadex column. Thus, the nondiffusible preparation of Phillips and Melius [20] is similar in aspartic acid content to polymers prepared by Fox [18,22,23]. The third effect was an almost 4-fold increase (7.3 mole %) of proline in the smallest proteinoids and the nondiffusible fraction (1.9 mole %). The small amount of glutamic acid in the nondiffusible fraction and aspartic acid in the smallest proteinoid indicates a nonrandom incorporation of amino acid into the polymers.

The fractions isolated by Sephadex G-25 chromatography in the work of Phillips and Melius [20] were subjected to high-voltage electrophoresis (HVE). The hypochlorite–starch–potassium iodide stain of Pan and Dutcher [24] was used to detect the spots, as they were ninhydrin-negative. The most electropositive (E-1) of three fractions obtained by HVE was subjected to chromatography in a butanol–acetic acid–pyridine–water system and two major fractions (EC-1 and EC-2) were obtained. The amino acid compositions E-1, EC-1, and EC-2 are presented in Table 1. It appears to be obvious from the amino acid composition of these proteinoids that nonrandom incorporation of amino acids has occurred. Fraction EC-1 is almost completely a glutamic acid polymer, and aspartic acid content is almost constant in all three fractions. A comment on homogeneity of these fractions will be made when end-group composition is discussed.

TABLE 1

Amino Acid Composition of Polymer Fractions [a]

Amino acid	Fraction (mole %)		
	E-1	EC-1	EC-2
Ala	8.3	3.5	17.1
Val	5.2	—[b]	10.6
Gly	9.6	3.7	12.2
Ile	3.8	—	8.7
Leu	4.0	—	12.2
Pro	20.9	4.0	6.8
Asp	7.7	6.2	6.1
Glu	40.6	82.3	26.5

[a] From Philips and Melius [20].
[b] The amino acids denoted by dashes are present in trace amounts.

Melius and Sheng [14] thermally polymerized a mixture of 2.5 gm of glutamic acid with 0.5 gm each of glycine, alanine, proline, phenylalanine, and leucine at 180°–190°C for 12 hr. Upon dialysis of the reaction mixture, the bulk of the polymer was found to be diffusible. When the diffusible fraction was chromatographed, all the starting amino acids were found and three ninhydrin-negative spots were located with hypochlorite–starch–iodide spray. The three fractions were labeled P₁ (smallest R_f value), P₂, and P₃ (largest R_f value). Fractions P₁, P₂, and P₃ were isolated by a preparative chromatography and analyzed for amino acids. They were homogenous by gel electrophoresis. Table 2 indicates the amino acid composition of the polymers which had molecular weights of 5000–11,500 (Table 3). Again, we can see the independence of amino acid composition of the polymer from that of the reacting mixture. Although glutamic acid and glycine did not vary much, there was a very wide variance in the other four amino acids with phenylalanine and proline having the smallest uptake into the polymer.

Harada and Fox [25] characterized some copolymers of α-amino acids with dicarboxylic amino acids or lysine. In this instance the two amino acids (copolymers) were heated with orthophosphoric acid. When aspartic acid was reacted separately with alanine, histidine, isoleucine, leucine, lysine, methionine, phenylalanine, proline, tyrosine, or valine, aspartic acid was 84–95% of the polymer formed. When glutamic acid was included to form three amino acid mixtures, the aspartic acid content decreased somewhat but still ranged from 74% to 86% of the final polymers formed. An equimolar mixture of aspartic acid, glutamic acid, and glycine resulted in an increase in glycine content with time and a concomitant decrease in aspartic acid in the polymer

TABLE 2

Amino Acid Compositions of Polymer Fractions [a]

Amino acid	Fraction (mole %) [b]		
	P₁	P₂	P₃
Ala	7.1	24.4	2.8
Gly	26.0	21.7	23.1
Glu	43.4	40.7	37.0
Leu	23.5	4.6	21.0
Phe	—	1.6	6.2
Pro	—	6.9	9.9

[a] From Melius and Sheng [14].
[b] The amino acid analyses were done by gas–liquid chromatography procedure.

TABLE 3

**Molecular Weights and Electrophoretic Mobilities of
Standard Proteins and Thermal Peptides** [a]

Protein	Molecular weight	Mobility [b]
Insulin	6,000	1.14
Cytochrome c	13,350	0.93
Lysozyme	16,800	0.89
Myoglobin	17,200	0.88
Carboxypeptidase A	34,600	0.73
Alcohol dehydrogenase	41,000	0.62
Albumin (bovine serum)	68,000	0.47
Peptide 1	5,200	1.17
Peptide 2	10,000	1.00
Peptide 3	11,500	0.96

[a] From Melius and Sheng [14].
[b] The mobilities were estimated from electrophoresis in SDS gels.

formed. A copolymer of alanine and lysine contained 50% alanine and 50% lysine.

Heinrich *et al.* [26] thermally polymerized L-lysine alone at 195°C and found that free lysine disappeared in 1 hr. Molecular size, judged by biuret color intensity and gel filtration, exceeded 100,000.

TABLE 4

**Amino Acid Compositions in Molar Percentages of Hydrolyzates
of a Systematically Altered Series of Proteinoids** [a]

Amino acid	Proteinoid number										
	1	2	3	4	5	6	7	8	9	10	11
Alanine	4.4	5.2	5.6	6.3	7.0	7.0	5.6	5.8	4.6	4.6	4.2
Arginine	3.9	4.5	4.9	5.3	4.8	5.2	3.9	4.0	4.0	3.8	3.5
Aspartic acid	40.3	30.1	20.3	13.7	7.3	6.3	5.7	4.9	4.1	3.6	3.7
Glutamic acid	13.9	13.8	12.1	10.5	8.6	7.8	8.8	8.9	8.1	7.8	7.5
Glycine	5.9	7.3	8.3	9.4	11.0	10.7	9.7	9.1	8.0	7.4	7.0
Histidine	3.9	4.3	5.0	5.3	4.8	5.2	4.3	4.8	4.6	4.4	4.0
Isoleucine	2.1	2.6	3.1	4.0	6.2	6.9	4.8	3.9	2.9	2.4	2.3
Leucine	4.8	5.3	6.7	7.9	11.0	11.2	9.5	8.1	6.5	5.8	5.3
Lysine	6.7	9.7	12.7	14.3	14.0	15.2	26.4	36.3	39.6	44.4	47.1
Proline	1.9	2.3	3.0	3.3	3.8	3.6	3.4	3.7	3.7	2.9	2.6
Valine	4.8	5.4	6.4	7.7	10.0	9.8	8.2	7.4	5.4	5.1	4.8
Alloisoleucine	2.1	2.5	3.1	3.5	5.1	5.0	3.9	3.2	2.5	2.1	2.0
Ammonia	5.8	6.9	8.6	8.3	6.2	7.1	5.2	5.8	5.4	4.6	5.5

[a] From Fox and Waehneldt [27].

Fox and Waehneldt [27] have demonstrated how the composition of the reaction mixture can be utilized to control to some extent the amount of certain amino acids in the resulting proteinoid. Tables 4 [27] and 5 [27a] indicate that aspartic acid, glutamic acid, and lysine are the most readily incorporated amino acids. These proteinoids have been used to study amino acid content in relationship to binding to polynucleotides.

MOLECULAR SIZE OF PROTEINOIDS AND RELATED AMINO ACID COMPOSITION

Molecular weights have been determined by end-group assay, sedimentation velocity, analysis in the ultracentrifuge [18], and gel electrophoresis [14,26]. Generally, the acidic types have the lowest molecular weights and the basic types have the highest molecular weights. The molecular weights of the acidic proteinoids from either the 2:2:1 or 1:1:1 amino acid mixtures of Fox and Harada [18] did not exceed 8600. The molecular weight of the polylysine

TABLE 5

Amino Acid Compositions in Molar Percentages of Two Proteinoids Compared to the Reaction Mixtures Polymerized [a]

Amino acid	2:2:1 Proteinoid		2:2:3 Proteinoid	
	Mixture (%)	Product (%)	Mixture (%)	Product (%)
Aspartic acid	42.0	66.0	30.0	51.1
Glutamic acid	38.0	15.8	27.0	12.0
Alanine	1.25	2.36	2.72	5.46
Lysine	1.25	1.64	2.72	5.38
Semi-cystine	1.25	0.94	2.72	3.37
Glycine	1.25	1.32	2.72	2.79
Arginine	1.25	1.32	2.72	2.44
Histidine	1.25	0.95	2.72	2.03
Methionine	1.25	0.94	2.72	1.73
Tyrosine	1.25	0.94	2.72	1.66
Phenylalanine	1.25	1.84	2.72	1.48
Valine	1.25	0.85	2.72	1.16
Leucine	1.25	0.88	2.72	1.06
Isoleucine	1.25	0.86	2.72	0.90
Proline	1.25	0.28	2.72	0.59
Serine	1.25	0.6	0.0	0.0
Theonine	1.25	0.1	0.0	0.0

[a] From Fox et al. [27a].

made by Heinrich *et al.* [26] exceeded 100,000. Krampitz *et al.* [28] found weights as high as 200,000 using the ultracentrifuge.

Harada and Fox [25], using end-group analysis to determine mean chain weights, found copolymers of aspartic acid with ten other amino acids, basic and neutral, in the range of 11,000–19,000. Aspartic acid and alanine formed the largest chains. When aspartic acid, glutamic acid, and the same ten amino acids were used, the weights ranged from 10,000 to 19,000, with isoleucine producing the largest polymer. In these experiments orthophosphoric acid was used in the reaction mixture.

Degree of polymerization and temperature are related; higher temperatures produced larger proteinoids for both the 2:2:1 and 1:1:1 mixture [29]. Fox suggested that the basic and neutral amino acids tend to act as chain terminators in the condensation of the dicarboxylic acids. This would explain the lower molecular weights of proteinoids compared to those of polyaspartic, polyglutamic, and copolyaspartic–glutamic acids.

N- AND C-TERMINAL GROUP COMPOSITION OF THERMAL POLYMERS

When Fox and Harada [18,19] determined the N-terminal amino acid residues of unfractionated 2:2:1 proteinoid, they found that 46% were glutamic acid, 48% BN, and 6% aspartic acid residues. Actually, aspartic acid made up 71% of the total proteinoid. The C-terminus was 91% BN acids, 8% glutamic acid and only 1% aspartic acid. The one significant difference in the 2:2:3 proteinoid was 28% aspartic acid in the C-terminus (or termini).

Phillips and Melius [20] and Phillips [30] found that their polymers were all ninhydrin-negative. Repeated efforts to prepare dinitrophenyl (DNP) derivatives of the N-terminal residues or dansyl derivatives failed. They concluded that the N-terminal residues were pyroglutamic acid. Hubbard [31] was able to identify pyroglutamyl residues at the N-terminus of Phillips' thermal peptides using rat liver pyrrolidonecarboxylyl peptidase and also dilute NaOH treatment at 60°C. The glutamate was identified as its dansyl derivative. Melius and Sheng [14], using the trifluoroacetic acid procedure of Smyth *et al.* [32], opened the N-terminal pyroglutamyl residue and identified it as the dansyl glutamate derivative in the P_1, P_2, and P_3 polymers that they had prepared.

Phillips and Melius [20], using hydrozinolysis, found only the neutral amino acids at the C-terminus of the nondiffusible, diffusible, and electrophoretically purified polymers. A trace of proline was found for the nondiffusible polymer. Melius and Sheng [14] found glycine as the C-terminus of the P_1 polymer, which contained 26 mole % glycine. The P_2 polymer, which contained a very high percentage of alanine (24.4), had an alanine C-terminal group. The P_3 polymer had the highest phenylalanine and proline content and had leucine at the C-terminus. Thus, as Fox and his co-workers have observed, the neutral amino acids appear to terminate chains.

Harada and Fox [25] found that aspartic acid was in most cases the major N-terminal amino acid in the two amino acid copolymers (MW 11,000–19,000) that they prepared. When glutamic acid was incorporated to give three component systems, aspartic acid and glutamic acid were the major N-terminal amino acids. However, in both instances very significant amounts of the BN acids were located at the N-termini contrary to the findings of Phillips, Melius, and Sheng [14,20].

In those cases where more than one amino acid was present at the C-terminus, there is the possibility of heterogeneity in the fraction or branching in the polypeptide chain. Melius and Sheng [14] and others have electrophoretic evidence of homogeneity of some polymers, and analysis of DNP-lysines from dinitrophenylated proteinoids indicates branching at some lysine residues [25,26]. Of course, Melius and Sheng [14] did not have lysine in their polymers.

The linkages of amino acids in proteinoids indicate an α link for most of the acids but a α and β bonds for aspartic acid, α and γ bonds for glutamic acid, and α and ε links for lysine. Thus, the proteinoids are much more complex in the links between individual amino acid residues than are the proteins. Fox et al. [25,33], using a procedure similar to that of Folk [34], have determined the amount of α and ε linkage in copolymers of lysine. This procedure involved treating the polymer with fluorodinitrobenzene, removing the α,ε-di-DNP-lysine, separating the α- and ε-mono-DNP-lysines, and measuring them spectrophotometrically. They found 29% α- and 71% ε links in a lysine–aspartic acid copolymer [33]. Lysine–glutamic acid copolymers gave almost exactly the same content of the two links, 55% α and 45% ε linkages. Obviously, the second amino acid can have a profound influence on the type of peptide bond that is formed. In a study of the copolymer between lysine and aspartic acid, Harada and Fox [25] found an influence of temperature on the linkages formed. At 200°C, 27% α links and 73% ε links formed, whereas at 210°C 35% α bonds and 65% ε bonds resulted.

Hennon et al. [35] prepared lysine copolymers with aspartic acid, proline, and arginine. Polymers up to 30,000 molecular weight were obtained. Lysine was incorporated more readily into copolymers with either proline or arginine to the extent that 60% or more lysine was incorporated in either case when the lysine was only 25% of the polymerization mixture. The aspartic acid was as readily incorporated as the lysine. The ratio of α-DNP-lysine to ε-DNP-lysine was 1.9 in the polylysine of Hennon et al., about 2.7 in their lysine–

proline polymers, and 0.58 in the lysine–aspartic acid polymers. It is of interest that there is a much greater α-peptide binding in the polymer of lysine containing the acidic amino acid aspartic acid, which of course correlates with the situation in proteins.

Fox and Harada [18] discovered that their proteinoids contained imide bonds from the existence of two strong infra red bands at 1720 and 1780 cm^{-1}. The imide bonds were replaced by peptide bonds by gentle alkaline treatment. Kovacs and Koenyves [36] indicated that a polyaspartic acid with a polyimide structure was converted to a polyamide by alkaline treatment. Rohlfing [37] heated aspartic acid containing proteinoids, converting polyaspartimide groupings to α- and β-carboxyl peptide bonds with a concomitant loss in catalytic activity.

According to Fox and Dose [3] the polyaspartimide link is not stable in aqueous solution and so cannot be expected to have lasted very long geologically except in a hypohydrous environment. The imide linkage would prevent branching in the polymer. No direct evidence is available as yet on the ratio of α and β linkages in aspartic or glutamic acid residues in the thermal polymers. Fox and Nakashima [38] have isolated a hexapeptide having the sequence Glu-Gly-Tyr-Glu-Tyr-Gly in which the glutamic acid residues are probably in α-peptide bonds, as they are reactive to leucine aminopeptidase. Harada and Fox [39] used a series of titration studies to characterize functional groups in acidic thermal polymers. They found the proteinoids to have an average uninterrupted chain length of 11 residues per carboxyl group. They concluded that these structures have some branching but are primarily linear. The neutral and basic amino acids tended to occupy C-terminal positions, which agrees with our investigations [20]. Phillips and Melius [20] found more than one C-terminal group in their purified preparations, which might be explained on the basis of branching at aspartic acid residues. In contrast, Melius and Sheng [14] identified only one C-terminal amino acid in their purified preparations, which contained glutamic acid but no aspartic acid.

With regard to amino acid sequence, of which very little direct evidence is available as yet, Harada and Fox [25] have degraded some alanine copolymers and have identified Ala-Ala dipeptides.

Phillips [30] has considered the unique role of glutamic acid in the thermal polymerization process. The glassy consistency of the cooled melt and its ability to be drawn into threads (when warm) indicate a possible polymeric structure. This behavior coupled with its spectral behavior indicates a greater degree of hydrogen bonding in the melt. When the melt was ground with dioxane and chloroform it was converted to a powdery solid containing glutamic acid and pyroglutamic acid. The glutamic acid must be present in the melt, as Wilson and Cannan [40] demonstrated that the lactam is heavily favored in aqueous solution at neutrality. Also, no increase in the amount of free glutamic acid in aqueous solution was observed by Phillips. The other neutral and acidic amino acids do not behave in this manner.

Recently, we utilized the glutamic reaction with soy A protein in order to couple methionine to the protein. When mixtures of the protein with 10% glutamic acid and 10% methionine were heated at 170°C for 4 hr, we apparently completely coupled the methionine to the protein. This was determined by analysis of the "dialysis wash" solutions from the reacted protein. The dialysis wash solution was free of methionine but contained some glutamic acid. These modified proteins are being nutritionally evaluated in mice. There is a concern in the production of meat analogs and other food products, where a large amount of processing occurs, about cross-linking reactions, degradation of amino acids, and their conversion to toxic products.

Another experimental finding on the prebiotic formation of amino acids in addition to Miller's [4] is the production by Fox and Windsor [41] of at least seven protein amino acids by heating formaldehyde and ammonia in an open system. Florkin [42] has indicated that the reducing primitive atmosphere and "hot dilute soup" concepts should be abandoned in favor of "a collection of open systems of the biochemical continuum of today derived from an open prebiological continuum." The hexapeptide containing Glu, Gly, and Tyr [38] was also reported by Nakashima, Fox, and Wang [43]. Professor E. Lederer at Gif-sur-Yvette, from mass spectrophotometric study of a permethylated sample from Professor Fox, has determined that the peptide is actually a stoichiometric mixture of the two tripeptides pGlu-Gly-Tyr and pGlu-Gly-Tyr (personal communication). Fox and his collaborators, from their earliest reports, have emphasized the nonrandom nature of the proteinoids. Dose and Rauchfuss [44] have also reviewed the nonrandom character of the thermal polyamino acids.

CONCLUSION

At present, we have a general idea about the nature of thermal condensation products produced from amino acid mixtures. Some information is available on amino acid composition vs. molecular weight and amino acid composition of reactant mixtures. A great deal of information is available on end groups, and the pyroglutamyl group is commonly the N-terminus of polymers containing glutamic acid. The C-terminal groups are usually the neutral amino acids, which may function as chain terminators. Aspartic acid may be present as an imide or in an α- or β-peptide bond. The only direct evidence indicates that glutamic acid forms α-peptide bonds. Lysine apparently forms α- and ε-peptide bonds. There is direct and indirect evidence for branching structures in the thermal polymers. A consistent finding of all investigators is that the polymerization of the amino acids is nonrandom. More direct sequence information for these polymeric substances is needed.

ACKNOWLEDGMENTS

I am most grateful to Dr. Sidney W. Fox for discussion of work in his laboratory, particularly the work on lysine polymers. The experiments in our laboratory were carried out by Dr. R. D. Phillips, James Yon-Ping Sheng, Grace Hubbard, and Andrew Edwards.

REFERENCES

1. S. W. Fox, *Naturwissenschaften* **56**, 1 (1969).
2. S. W. Fox, R. J. McCauley, and A. Wood, *Comp. Biochem. Physiol.* **20**, 773–778 (1967).
3. S. W. Fox and K. Dose, "Molecular Evolution and the Origin of Life," pp. 66–260. Freeman, San Francisco, California, 1972.
4. S. L. Miller, *J. Am. Chem. Soc.* **77**, 2351–2361 (1955).
5. K. Harada and S. W. Fox, *Nature (London)* **201**, 335 (1964).
6. H. C. Urey, "The Planets" Yale Univ. Press, New Haven, Connecticut, 1952.
7. A. I. Oparin, "Proiskhozhdenie Zhizny." Izd. Mosk. Rabochii, Moscow, 1924.
8. J. Oro, *in* "The Origins of Prebiological Systems" (S. W. Fox, ed.), p. 137. Academic Press, New York, 1965.
9. J. G. Lawless, K. A. Kvenvolden, E. Peterson, C. Ponnamperuma, and E. Jarosewich, *Nature (London)* **236**, 66–67 (1972).
10. K. A. Kvenvolden, J. G. Lawless, and C. Ponnamperuma, *Proc. Natl. Acad. Sci. U.S.A.* **68**, 486–490 (1971).
11. J. G. Lawless, *Geochim. Cosmochim. Acta* **37**, 2207–2212 (1973).
12. A. M. Benson, H. F. Mower, and K. T. Yasunoba, *Proc. Natl. Acad. Sci. U.S.A.* **55**, 1532 (1966).
13. D. O. Hall, R. Cammack, and K. K. Rao, *Nature (London)* **233**, 136–138 (1971).
14. P. Melius and J. Y. P. Sheng, *Bioorganic Chem.* **4**, 385–391 (1975).
15. M. Sternberg, C. Y. Kim, and F. J. Schwende, *Science* **190**, 992–994 (1975).
16. M. A. Ratcliff, E. E. Medley, and P. G. Simmonds, *J. Org. Chem.* **39**, 1481–1490 (1974).
17. Y. C. Lien and W. W. Nawar, *J. Food Sci.* **39**, 911–918 (1974).
18. S. W. Fox and K. Harada, *J. Am. Chem. Soc.* **82**, 3745–3751 (1960).
19. S. W. Fox and K. Harada, *Fed. Proc., Fed. Am. Soc. Exp. Biol.* **22**, 479 (1963).
20. R. D. Phillips and P. Melius, *Int. J. Pept. Protein Res.* **6**, 309–319 (1974).
21. C. W. Gehrke, D. Roach, R. W. Zumwalt, D. L. Stalling, and L. L. Wall, "Quantitative Gas-Liquid Chromatography of Amino Acids in Proteins and Biological Substances." Anal. Biochem. Lab., Columbia, Missouri, 1968.
22. K. Harada and S. W. Fox, *J. Am. Chem. Soc.* **80**, 2694–2697 (1958).
23. K. Karada and S. W. Fox, *Arch. Biochem. Biophys.* **86**, 274–280 (1960).
24. S. C. Pan and J. D. Dutcher, *Anal. Chem.* **28**, 836–838 (1956).
25. K. Harada and S. W. Fox, *Arch. Biochem. Biophys.* **109**, 49–56 (1965).

26. M. R. Heinrich, D. L. Rohlfing, and E. Bugna, *Arch. Biochem. Biophys.* **130**, 441–448 (1969).
27. S. W. Fox and T. V. Waehneldt, *Biochim. Biophys. Acta* **160**, 246 (1968).
27a. S. W. Fox, K. Harada, K. R. Woods and C. R. Windsor, *Arch. Biochem. Biophys.* **102**, 439 (1963).
28. G. Krampitz, S. Baars-Diehl, W. Haas, and T. Nakashima, *Experientia* **24**, 140 (1968).
29. S. W. Fox, *Encycl. Polym. Sci. Technol.* **9**, 294 (1968).
30. R. D. Phillips, Ph.D. Dissertation, Auburn University, Auburn, Alabama (1973).
31. W. L. Hubbard, M.S. Thesis, Auburn University, Auburn, Alabama (1973).
32. D. G. Smyth, W. H. Stein, and S. Moore, *J. Biol. Chem.* **237**, 1845 (1962).
33. S. W. Fox and F. Suzuki, *J. Mol. Cell. Behav. Origins Evol.* (in press).
34. J. E. Folk, *Arch. Biochem. Biophys.* **61**, 150 (1956).
35. G. Hennon, R. Plaquet, M. Dautrévaux, and G. Biserte, *Biochimie* **53**, 215 (1971).
36. J. Kovacs and I. Koenyves, *Naturwissenschaften* **14**, 333 (1954).
37. D. L. Rohlfing, and S. Fox, *Arch. Biochem. Biophys.* **118**, 127 (1967).
38. S. W. Fox and T. Nakashima, *Metab. Regul. Enzyme Action*, Fed. Eur. Biochem. Soc., *Meet. 6th 1969*, Abstract, p. 145 (1970).
39. K. Harada and S. W. Fox, *Biosystems* **7**, 213 (1975).
40. H. Wilson and R. K. Cannan, *J. Biol. Chem.* **119**, 309 (1937).
41. S. W. Fox and C. R. Windsor, *Science* **170**, 984 (1970).
42. M. Florkin, *Compr. Biochem.* **29B**, 241 (1975).
43. T. Nakashima, S. W. Fox, and C. Wang, *Arbeitstag. Extraterr. Biophys. Biol. Raumfahrtmed., Tagungsber.*, *2nd 1967* p. 233 (1968).
44. K. Dose and H. Rauchfuss, *in* " Molecular Evolution; Prebiological and Biological" (D. L. Ruhlfing and A. I. Oparin, eds.), pp. 199–217. Plenum, New York, 1972.

The Bioorganic Chemistry of Aggregated Molecules

F. M. Menger*

INTRODUCTION

Thomas Lincoln† once remarked that it is time to move when one can see the smoke from a neighbor's chimney. Chemists of similar temperament may feel that it is time to enter a new field when the reviews begin to appear. Although such feelings have merit, I hope that this review of micellar chemistry and previous reviews of the subject [1–3] will attract rather than repel. The vast amount of information published on micellar bioorganic chemistry in recent years has served more to define problems than to solve them.

Surfactants are molecules with dual character. They possess a hydrophilic group (quaternary ammonium ion, sulfate, etc.‡) and a hydrocarbon portion (usually an unbranched hydrocarbon chain of ten or more carbons). In dilute aqueous solutions surfactants are monomeric; at higher concentrations they spontaneously form stable aggregates called micelles. The transition from monomer to micelle occurs abruptly at the "critical micelle concentration" (cmc). Micelles typically contain 50–150 units which are spherically arranged so that the ionic heads are near the water and the hydrocarbon tails extend into the interior (Fig. 1). The tendency of surfactants toward self-association

---

* Recipient of a Camille and Henry Dreyfus Foundation Teacher-Scholar Grant and a National Institutes of Health Research Career Development Award.

† Abe's father.

‡ This review is restricted to ionic surfactants in aqueous solutions.

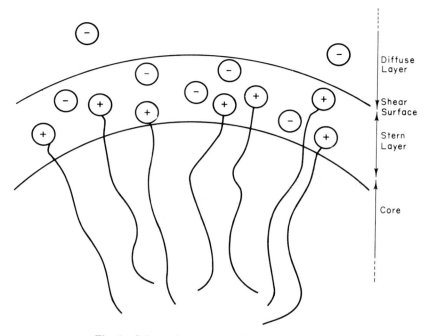

Fig. 1 Schematic representation of a micelle section.

in water stems primarily from "hydrophobic bonding" of the hydrocarbon chains [4].

Micellization of a surfactant such as dodecyltrimethylammonium bromide creates positively charged surfaces composed of the cationic "heads" (Fig. 1). Bromide "counterions," which surround the micelles, are subjected to two opposing forces: coulombic attraction and thermal agitation. Coulombic attraction gathers the bromide ions into the vicinity of the quaternary nitrogens. Thermal agitation tends to disperse the bromide ions in the water and thereby cause a more gradual bromide gradient than there would be otherwise. One portion of the ionic array, the Stern layer, deserves particular mention because some of the most interesting micellar chemistry occurs in this region. The Stern layer (Fig. 1) consists of the surfactant head groups and their strongly bound counterions and water molecules; it is probably not much thicker than the diameter of a hydrated surfactant head [5,6]. When a micelle moves through the water, the outer edge of the Stern layer constitutes the "shear surface." A useful parameter, α, has been defined as the fraction of micellar counterions that exist outside the Stern layer (i.e., in the intermicellar water). If a micelle has n ionic head groups, then the Stern layer possesses $(1 - \alpha)n$ counterions.

Organic compounds, particularly nonpolar ones, can adsorb into or onto micelles [7]. Increased solubility of a compound in an aqueous surfactant solution relative to that in pure water is a good indication of adsorption. Adsorption often alters chemical reactivity, so that rate studies of micelle-solubilized reactants provide information about micelles and micelle–substrate complexation. The bioorganic chemistry of micelles has been developed primarily by this means. Rate constants (as well as cmc, n, and α) vary with surfactant chain length, structure of the head group, counterion type, ionic strength, surfactant concentration, and the presence of an inert nonelectrolyte or a second surfactant. Since almost any aqueous reaction can be studied à la micelle using a host of surfactants and experimental conditions, it is not hard to envision the playground that micelles have provided bioorganic and physical organic chemists. In the past, most investigations have been carried out with commercially available surfactants, but this approach has fairly well burned itself out. The best future work will most likely involve surfactants that are tailor-made to test a particular point or to carry out some specific function.

In the next section I discuss important physical properties of micelles and how they are determined. This is followed by several recent examples of micellar chemistry.

PHYSICAL PROPERTIES OF MICELLES

Critical Micelle Concentration (cmc)

Formation of micelles probably occurs one monomer at a time with rate constants greater than $10^5 \, M^{-1} \, sec^{-1}$ [8]. For many purposes one may regard micellization as a fast equilibrium between monomer S and aggregate S_n [Eq. (1)]. If n is large (50–150), then the law of mass action demands that

$$nS \rightleftharpoons S_n \tag{1}$$

the micelle concentration remain small up to a certain level of surfactant and increase rapidly thereafter. The larger the value of n, the sharper the discontinuity. Equation (1) implies that only monomers and micelles are present. This is not strictly correct; premicellar association (especially dimer formation) is possible far below the cmc [3]. In any event, it is often assumed that micellar systems are monodisperse (one size) and that the monomer concentration is constant at any surfactant concentration above the cmc. Thus, adding more surfactant to a solution already above the cmc does not enlarge the micelles nor increase the monomer level [9]. Only the micelle concentration

TABLE 1

Effect of Physical and Structural Properties on the Critical Micelle
Concentration (cmc), Aggregation Number (n), and
Degree of Ionization (α)[a]

Surfactant[b]		cmc (mM)	n	α
Chain length[c]				
$C_{10}N(CH_3)_3Br$		63	52	0.18
$C_{12}N(CH_3)_3Br$		15	73	0.16
$C_{14}N(CH_3)_3Br$		3.6	107	0.15
Surfactant branching[d]				
$C_{14}OSO_3Na$		1.7	—	—
$C_{12}CH(CH_3)OSO_3Na$		3.3	—	—
$C_{11}CH(CH_2CH_3)OSO_3Na$		5.8	—	—
Surfactant head size[e]				
$C_{10}NH_3Br$		16	1100	0.07
$C_{10}NH_2CH_3Br$		20	670	0.07
$C_{10}NH(CH_3)_2Br$		12	69	0.08
$C_{10}(CH_3)_3Br$		18	48	0.13
$C_{14}N(CH_3)_3Br$		3.5	71	0.14
$C_{14}N(n\text{-}Pr)_3Br$		2.1	48	0.11
Counterion type[f]				
$C_{12}N(CH_3)_3Br$	(0.5 M NaF)	8.4	59	0.21
$C_{12}N(CH_3)_3Br$	(0.5 M NaCl)	3.8	62	0.20
$C_{12}N(CH_3)_3Br$	(0.5 M NaBr)	1.9	84	0.17
$C_{12}N(CH_3)_3Br$	(0.5 M NaIO_3)	5.1	38	0.24
Temperature[g]				
$C_{12}N(CH_3)_3Br$	(25°)C	15	43	0.17
$C_{12}N(CH_3)_3Br$	(35°C)	15	34	0.18
$C_{12}N(CH_3)_3Br$	(45°C)	17	28	0.19
$C_{12}N(CH_3)_3Br$	(55°C)	17	24	0.20
Salt concentration[c]				
$C_{14}OSO_3Na$	(0.0125 M NaCl)	1.7	120	0.12
$C_{14}OSO_3Na$	(0.050 M NaCl)	1.1	144	0.11
$C_{14}OSO_3Na$	(0.10 M NaCl)	0.32	153	0.11
Surfactant concentration[h]				
$C_{12}OSO_3Na$	(0.012 M)	—	83	0.26
$C_{12}OSO_3Na$	(0.020 M)	—	87	0.26
$C_{12}OSO_3Na$	(0.129 M)	—	110	0.26
Nonelectrolyte additive[i]				
$C_{12}OSO_3Na$	(pure H_2O)	5.7	—	—
$C_{12}OSO_3Na$	(6 M urea)	9.5	—	—
$C_{12}OSO_3Na$	(0.06 mole fraction MeOH)	7.9	64	0.19
$C_{12}OSO_3Na$	(0.12 mole fraction MeOH)	9.0	23	0.22

[a] A particularly thorough and incisive discussion of the variation
of micellar parameters with physical and structural properties appears
in Romsted [20].

increases. These assumptions permit the calculation of the micelle concentration by Eq. (2).

$$[S_n] = ([S]_{total} - cmc)/n \qquad (2)$$

Although no one has yet been able to predict cmc values by considering the various noncovalent interactions involved in micellization [10], the experimental determination of cmc values is a simple matter [11]. Conductivity [12], spectrophotometry [13], osmometry [14], refractometry [15], counterion magnetic resonance [16], and optical rotatory dispersion [17] have all been exploited for the purpose. A plot of a suitable physical property vs. surfactant concentration "breaks" at the cmc, where the solution becomes nonideal. For example, the absorbance at 610 nm of 1.0×10^{-5} M pinacyanol chloride in laurate solutions rises dramatically at the cmc of laurate (0.01 M at pH 9.59 and 50°C) [18]. A spectrophotometer is in fact unnecessary for the experiment; above the cmc the solutions are bright blue, whereas below the cmc they are a light shade of pink.

A large number of physical and structural properties affect the magnitude of the cmc (Table 1) [19,20]. Chain length is perhaps the most important factor; the cmc is roughly halved for each one-carbon increase [21]. Since micellization releases structural water surrounding the surfactant chains, the longer the chain the greater (more positive) the entropy of micellization. A double bond stiffens the chains and raises the cmc 2- to 3-fold [22]. Polar substituents on the chains and branching of any sort destabilize the micelles (elevate the cmc). Temperature effects on the cmc are rather small [23]. The cmc usually increases somewhat with temperature because (among other reasons) the entropy of micellization is less favorable when thermal agitation disorders "icelike" water adjacent to the hydrophobic portion of the monomers. Although the nature of the head certainly plays a role in micelle

---

[b] $C_{10}N(CH_3)_3Br$ refers to a surfactant bearing an unbranched chain ten carbons long.

[c] W. Prins and J. J. Hermans, *Proc. K. Ned. Akad. Wet., Ser. B* **59**, 162 (1956).

[d] L. I. Osipow, "Surface Chemistry," p. 169 Van Nostrand-Reinhold, Princeton, New Jersey, 1962.

[e] R. D. Geer, E. H. Eylar, and E. W. Anacker, *J. Phys. Chem.* **75**, 369 (1971); R. L. Venable and R. V. Nauman, *ibid.* **68**, 3498 (1964).

[f] E. W. Anacker and H. M. Ghose, *J. Phys. Chem.* **67**, 1713 (1963).

[g] L. I. Osipow, "Surface Chemistry," p. 173. Van Nostrand-Reinhold, Princeton, New Jersey, 1962; M. N. Jones and J. Piercy, *J. Chem. Soc., Faraday Trans.* **1** p. 1839 (1972).

[h] W. Hoyer, *J. Phys. Chem.* **61**, 1283 (1957).

[i] G. D. Parfitt and J. A. Wood, *Kolloid Z. Z. & Polym.* **229**, 55 (1969).

formation, no simple correlation between head group and cmc is evident. Alkyl sulfates of constant chain length give a constant cmc for a variety of divalent counterions (Mg^{+2}, Cu^{+2}, Zn^{+2}, etc.) [24]. This cmc is less than that found with univalent ions. Added salts lower the cmc as described in Eq. (3), one of the oldest empirical equations in the micelle field (C_i is the counterion concentration derived from the surfactant and added salt; m and b are positive numbers) [25]. To a first approximation, Eq. (3) is independent of

$$\log(\text{cmc}) = -m \log(C_i) - b \qquad (3)$$

the nature of the counterion, provided that the counterions bear the same charge. At high salt concentrations ($> 0.1\ M$) the cmc does indeed vary with the type of counterion; this second-order effect follows the Hofmeister "salting out" series [26] (e.g., the cmc for $C_{12}N(CH_3)_3^+$ in $0.5\ M$ NaX decreases as X changes from F^- to Cl^- to Br^-) [27]. The theoretical basis for salt-dependent cmc values is complicated. No doubt the lowering of the cmc with added salt stems in large measure from the fact that counterions screen electrostatic repulsive forces between the surfactant heads in the Stern layer. Other factors, such as alteration of water structure and monomer stability, enter the picture at high salt concentrations.

Urea and methanol both increase the cmc (Table I) despite their opposite effect on the dielectric constant of the medium. These "nonpenetrating" additives improve the water solubility of the monomer and lessen the tendency to aggregate. On the other hand, "penetrating" nonelectrolytes such as n-hexanol lower cmc values by stabilizing the micelles (perhaps by allowing the surfactant heads to separate from one another).

Aggregation Number (n)

Aggregation numbers of spherical micelles commonly fall within the range of 50–150 (Table 1). Actually, a surfactant forms a distribution of micelle sizes of which the aggregation number is the most abundant. High surfactant or salt concentrations lead to rod-shaped aggregates containing anywhere from 150 to several thousand monomers [28]. Above 10% surfactant the molecules assemble into large sheets two molecules thick ("bilayers"). To ensure a relatively constant spherical micellar size during a series of experiments, it is best not to exceed a surfactant concentration greater than ten times the cmc or a salt concentration greater than 0.1 M. Since rate constants of adsorbed additives are generally insensitive to growth and elongation of micelles, violating these limits may not be too serious for the kinetically oriented organic chemist.

Micellar size has been determined by light scattering among other methods [29]. If a particle of colloidal dimensions is much smaller than the wavelength of light, then the total intensity of light scattered by the particle is proportional to its mass times its weight concentration. In practice, the micellar mass M is calculated from Eq. (4) by extrapolating the function $Hc/R_{90°}$ to zero c (c is the surfactant concentration minus the cmc in grams per milliliter, H is a constant, and $R_{90°}$ is the intensity of light scattered at 90° in excess of that scattered below the cmc). Mass M divided by the molecular weight of the surfactant yields n.

$$\lim_{c \to 0}(Hc/R_{90°}) = 1/M \tag{4}$$

Degree of Ionization (α)

The parameter α defines the degree of ionization of the Stern layer, which, it is estimated, contains 3 M of counterions [30]. An α of unity indicates complete counterion dissociation (and a micellar charge of n), whereas an α of zero indicates one counterion per surfactant head (and a micellar charge of 0). As the α values near 0.2 show (Table 1), the majority of counterions remain bound to the Stern layer. Elevating the temperature dissociates counterions from the micelle surface and increases α. The longer the chain, the smaller the α. This is reasonable because lengthening the chain expands the micelle volume faster than the surface area. When the chain length and aggregation number increase, the surfactant head density and counterion association do likewise. Values of α also depend on the sizes of the hydrated counterion and the surfactant head. But the most striking property of α is its insensitivity to the concentration of added counterion. One might have expected, for example, that addition of sodium chloride to an aqueous solution of sodium dodecyl sulfate (SDS) would decrease α via a "common ion effect." The fact that this does not happen (Table 1) shows that the Stern layer behaves as if it were nearly saturated with counterion. Romsted [20] makes this point clearly.

Work from our laboratory illustrates the imperturbability of the Stern layer with respect to added salt [31]. We measured the rate of NH-proton exchange of dodecyldimethylamine hydrochloride in water using dynamic nuclear magnetic resonance methods. Since the proton exchange involves the surfactant head, there is no doubt that we were probing the Stern layer itself. Micellization was found to greatly enhance the exchange rate k_{ex} as well as to reduce the apparent pK_a of the protonated amine by 1.4 units. Both effects can be explained by electrostatic repulsion between the compacted ammonium ions within the Stern layer. Interestingly, addition of sodium chloride had little effect on the fast rate of exchange. If addition of salt had increased chloride binding and thereby neutralized a portion of the micellar charge, then the exchange rate would have declined.

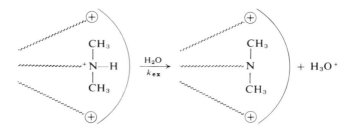

Specific ion electrodes now permit a simple and direct evaluation of α [32]. For example, to determine α for a long-chain quaternary bromide, one first prepares a standard plot of E vs. log a_{Br-} with the aid of a bromide ion electrode and a series of NaBr solutions of various concentrations [Eq. (5)]. The standard plot and the potential of a micellar solution then provide the activity of unbound Br^- (Br^- in the Stern layer is assumed not to contribute to the readings).

$$E = E_a + 2.3RT/nf(\log a_{Br}^-) \qquad (5)$$

Water Penetration, Chain Mobility, and Hydrocarbon Fluidity

Although the activity of water in the Stern layer is unknown, there is no reason to suspect that either the surfactant heads or the counterions suffer impaired hydration. The charge-transfer absorbance of a long-chain pyridinium iodide (shown below) indicates a surface polarity equivalent to that of methanol [33]. The extent of water penetration into the hydrocarbon core is still not established, although results from a variety of studies suggest that water reaches no more than the first four to six carbons [3,34]. Spin-lattice relaxation times of micellar octyltrimethylammonium bromide reveal modestly restricted motional freedom along the entire chain with maximal effects near the polar end [35]. Micelle interiors are definitely fluidlike but appear to be somewhat less fluid than hydrocarbon solvents of similar chain length [3,36].

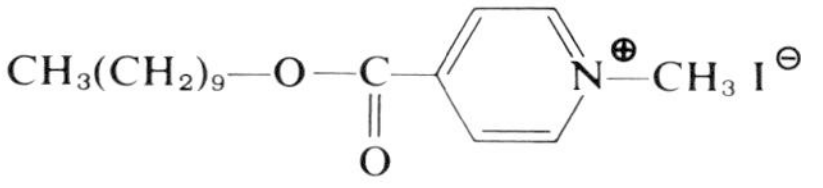

$$CH_3(CH_2)_9-O-\overset{\displaystyle O}{\underset{\displaystyle \|}{C}}-\langle\ \rangle-\overset{\oplus}{N}-CH_3\ I^{\ominus}$$

Most investigations of micelle interiors make use of a probe molecule whose ultraviolet spectrum, nmr spectrum, fluorescence, or electron spin resonance signal varies with the polarity or viscosity of the medium. Probes can, of course, disturb the micellar environment in which they are embedded, and this problem may account in part for diverse conclusions frequently found in the literature.

EXAMPLES OF MICELLAR CHEMISTRY

Duynstee and Grunwald ushered in the era of micellar chemistry with their papers on the surfactant-dependent fading of dyes [37]. Some time later we delineated the equations and assumptions involved in the "enzyme model" for micellar reactions [18]. Despite a shortcoming of the model (discussed below), the equations have been successfully applied to a variety of systems [38]. The original problem was to analyze quantitatively the hydrolysis rates of p-nitrophenyl esters at pH 9.59 as inhibited by micellar sodium laurate

$$HO^- + R-\overset{\overset{\displaystyle O}{\|}}{C}-O-\langle\!\!\!\bigcirc\!\!\!\rangle-NO_2 \xrightarrow{\text{laurate}} R-\overset{\overset{\displaystyle O}{\|}}{C}-O^- + {}^-O-\langle\!\!\!\bigcirc\!\!\!\rangle-NO_2 \tag{6}$$

[Eq. (6)]. We proposed a scheme [Eq. (7)] in which ester (E) and micelle (S_n) are in equilibrium with an ester–micelle complex (S_nE). Both the hydrolysis rate of adsorbed ester (k_2) and the binding constant (K) can be determined from Eq. (8) by plotting $1/(k_1 - k_{obs})$ vs $1/(S_n)$. [The S_n concentration,

$$S_n + E \underset{}{\overset{K}{\rightleftharpoons}} S_nE \tag{7}$$

$$\downarrow k_1 \qquad\qquad \downarrow k_2$$

$$P \qquad\qquad P$$

derived from Eq. (2), is in great excess of the ester concentration; k_1 is simply the rate constant in the absence of surfactant.] Plots based on Eq. (8) are linear and have remarkably little scatter in view of our manifold assumptions: (a) Ester does not complex with monomer; (b) ester does not perturb micellization; (c) only one ester molecule is incorporated into each micelle; (d) micellization occurs exactly at the cmc rather than over a concentration range; (e) Eq. (2) is valid. Large micelle–ester association constants were found

$$\frac{1}{k_1 - k_{obs}} = \frac{1}{k_1 - k_2} + \frac{1}{(k_1 - k_2)K(S_n)} \tag{8}$$

for long-chain esters ($K > 10^4$ M^{-1} for p-nitrophenyl octanoate). The k_2 values for the esters lie within experimental error of zero. In other words, adsorbed ester effectively resists hydrolysis. Perhaps the ester binds within the micellar core, which is void of hydroxide ion. Alternatively, ester may be adsorbed onto the Stern layer. Owing to the negative charge on the laurate micelles ($\alpha > 0$), the hydroxide ion concentration in the Stern layer is less than that in the bulk solution. Thus, ester hydrolysis within the Stern layer should be impaired. Similar reasoning leads to the expectation that quaternary ammonium surfactants might accelerate hydroxide-catalyzed reactions, and this is frequently found to be true [18,39].

The relative magnitude of k_1 and k_2 in Eq. (8) determines whether micelles catalyze or inhibit. Most micellar enhancements are less than 10^3 and many are 10^2 or less (significant but not extraordinary). Typical enhancements include cetyltrimethylammonium bromide catalysis of the hydrolysis of 2,4-dinitrophenyl phosphate dianion ($25 \times$); cetyltrimethylammonium bromide catalysis of the decarboxylation of 6-nitrobenzisoxazole 3-carboxylate ($90 \times$); and sodium dodecyl sulfate catalysis of the acidic hydrolysis of methyl orthobenzoate ($79 \times$) [40–42]. Two key questions always loom forth when one attempts to interpret micellar kinetics. What are the adsorption sites and orientations of a substrate within a micelle? Do micellar perturbations of a bimolecular reaction rate originate from a "concentration effect" or a "medium effect"? These questions are often left unanswered, but future work should rectify the situation.

The "enzyme model" for micellar reactions treats the binding of only one of the reactants in a bimolecular process [e.g., the ester in Eq. (6)]. This drawback is corrected in the "pseudophase model" developed by a group of Russian workers [43]. They assumed that a micellar solution is comprised of two phases: an aqueous phase and a micellar phase. Both reactants in a bimolecular reaction distribute themselves between the two phases much as they would between water and hexane. Partition coefficients define the micelle/water concentration ratios for the reactants. Observed rate constants are then expressed in terms of the aqueous and micellar rate constants and the two partition coefficients. A significant portion of micelle-catalysis stems from an elevated concentration of the two reactants within the micellar volume.

The ^1H nmr spectrum of 0.1 M cetyltrimethylammonium ion shows only a single large peak for the interior CH_2 groups. When 5×10^{-3} M pyrene is added to the surfactant solution, the CH_2 signal splits into two well-defined peaks of comparable size [44]. As the pyrene concentration is increased, the downfield peak remains stationary, whereas the upfield peak moves further upfield. The best explanation for these observations is that pyrene is adsorbed into the hydrocarbon core. Those methylenes that are in close proximity to the shielding region of the aromatic rings give rise to the upfield component of the CH_2 signal. 1-Pyrenesulfonic acid (PSA) broadens—but does not split— the CH_2 signal. On the other hand, PSA shifts the $N—CH_3$ peak of the surfactant much more than does pyrene. Apparently, the negatively charged

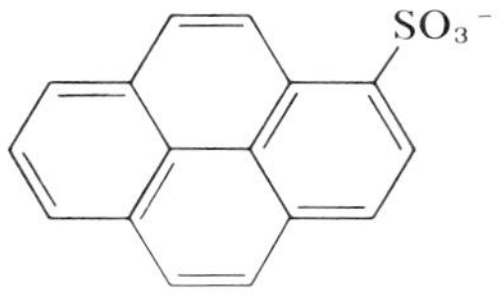

PSA

sulfonate group of PSA prefers to reside near the positive surfactant heads, thereby forcing the aromatic rings closer to the Stern layer. It was also found [44] that the fluorescence lifetime of PSA is 62 nsec in water and 140 nsec in aqueous cetyltrimethylammonium chloride. In contrast, the fluorescence lifetime of PSA in cetyltrimethylammonium bromide is 30 nsec. Clearly, the micellar environment stabilizes the excited state unless the Stern layer contains bromide ion or some other counterion which is able to quench the fluorescence. As I mentioned previously, a certain amount of caution is recommended in interpreting data like these which are based on huge probe molecules.

Hexadecyltrimethylammonium chloride (0.01 M) increases 6800-fold the quantum yield for the photochemical reaction of 4-methoxy-1-nitronaphthalene with CN^- in water [Eq. (9)] [45]. (An anionic surfactant, sodium dodecyl sulfate, lacks any special effect.) Part of the 6800-fold enhancement no doubt stems from localization of the organic substrate and CN^- in the

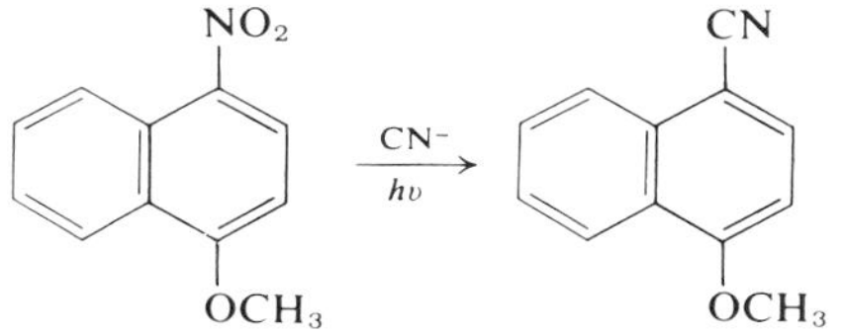

(9)

positively charged Stern layer. But the catalysis is too great for this to be the entire explanation. Very likely a "medium effect" on the photoexcited 4-methoxy-1-nitronaphthalene augments the benefits of the high CN^- concentration at the micelle surface. In support of this idea, it is found that the reaction without surfactant is favored by lowering the solvent polarity.

1,4-Dimethoxybenzene is known to inhibit photoreactions similar to Eq. (9). This fact motivated the synthesis of a series of surfactants [compounds (1)–(4)] containing 1,4-dioxybenzene units at various points along the chains [45]. The photosubstitution reaction was then carried out in "mixed micelles" comprised of hexadecyltrimethylammonium chloride and one of the four synthetic surfactants. The experiments showed that all four surfactants inhibit the reaction and that the extent of inhibition depends on the location of the 1,4-dioxybenzene group. Surfactant (4) is by far the best quencher, and (1) is the least active. The above-mentioned work brings to mind the as yet unexploited potential of regioselective micellar reactions in organic synthesis.

$$CH_3O-C_6H_4-O(CH_2)_{10}N(CH_3)_3{}^+$$
$$(1)$$

$$CH_3(CH_2)_3O-C_6H_4-O(CH_2)_{10}N(CH_3)_3{}^+$$
$$(2)$$

$$CH_3(CH_2)_7O-C_6H_4-O(CH_2)_{10}N(CH_3)_3{}^+$$
$$(3)$$

$$CH_3(CH_2)_9O-C_6H_4-O(CH_2)_4N(CH_3)_3{}^+$$
$$(4)$$

Treatment of 1-bromo-2-phenylpropane with aqueous base (pH 13, 50°C) gives three products corresponding to an E2 elimination (55%), S_N2 substitution (10%), and S_N1/rearrangement (35%) [Eq. (10)]. When the reaction

$$\begin{array}{c} \text{E2} \longrightarrow C_6H_5-\underset{\underset{CH_3}{|}}{C}=CH_2 \\[1em] C_6H_5-\underset{\underset{CH_3}{|}}{CH}-CH_2Br \xrightarrow{\quad S_N2 \quad} C_6H_5-\underset{\underset{CH_3}{|}}{CH}-CH_2OH \qquad (10) \\[1em] \text{S}_N1 \longrightarrow C_6H_5-CH_2-\underset{\underset{OH}{|}}{CH}-CH_3 \end{array}$$

is run under the same conditions except for the presence of 0.01 M cetyltrimethylammonium bromide, the product ratio changes to 95% E2, 2.5% S_N2, and 2.5% S_N1/rearrangement [46]. Preference for the micellar E2 reaction over the micellar S_N1/rearrangement may originate from an elevated hydroxide concentration in the Stern layer (only the E2 depends on hydroxide) However, this rationale cannot explain the increase in the ratio of E2 to S_N2 (both of which depend on hydroxide ion). Possibly a lower reaction-site polarity or a greater hydroxide ion basicity within the micelles favors E2 elimination over S_N2 substitution.

Nitrous acid-induced deamination of chiral secondary amines generates an intermediate [Eq. (11)] which usually partitions favorably toward alcohol of *inverted* configuration [16,47]. If the deamination is carried out on micellar 2-aminooctane (0.76 M in perchloric acid, pH 4), then the alcohol predominantly *retains* the configuration of the amine (Table 2). There is no doubt that this modified stereochemistry is micellar in origin because a "normal" inverted stereochemistry is achieved when the 2-aminooctane conjugate acid is deaminated below its cmc of 0.58 M. Remarkably, the stereochemistry reverts to inversion even above the cmc when the counterion is bromide or

TABLE 2

Stereochemistry of 2-Alkanol Formation in the Deamination of Amines

Amine	Amine concentration $(M)^a$	Counterion	Stereochemistry $(\% \text{ net})^b$
$CH_3CH_2CHCH_3$ $\mid$ $NH_3{}^+$	0.76	$ClO_4{}^-$	22.9 inv.
$CH_3(CH_2)_2CHCH_3$ $\mid$ $NH_3{}^+$	0.76	$ClO_4{}^-$	22.9 inv.
$CH_3(CH_2)_5CHCH_3$ $\mid$ $NH_3{}^+$	0.76	$ClO_4{}^-$	6.0 ret.
$CH_3(CH_2)_5CHCH_3$ $\mid$ $NH_3{}^+$	0.015	$ClO_4{}^-$	23.0 inv.
$CH_3(CH_2)_5CHCH_3$ $\mid$ $NH_3{}^+$	0.34	Br^-	22.2 inv.

^{a} The cmc of 2-aminooctane conjugate acid is 0.058 M.
^{b} Inverted, inv.; retained, ret.

acetate instead of perchlorate (Table 2). Perhaps bromide and acetate, which are tightly hydrated, associate weakly with the positive micelle. Consequently, considerable water may be forced into the Stern layer to solvate the ammonium ions, and a stereochemistry identical to that in bulk water is observed.

$$[R^*\!-\!N\!=\!N\!-\!\overset{+}{O}H_2] \underset{}{\overset{H_2O}{\longrightarrow}} \begin{cases} HO\!-\!R & \text{(inverted)} \\ R\!-\!OH & \text{(retained)} \\ [R^+] \longrightarrow R\!-\!OH + HO\!-\!R & \text{(racemic)} \end{cases} \tag{11}$$

If a poorly hydrated counterion such as perchlorate penetrates the Stern layer effectively, then the micelle will have a diminished requirement for water. In this event the intermediate in Eq. (11) may find it easiest to accept the return of its own water molecule after ejection of nitrogen. Hence, the configuration at the chiral center is retained. Alternatively, retention may be favored by perchlorate within the Stern layer, shielding the back face of the chiral carbon as the nitrogen departs.

Carbon-13 nmr spin-lattice relaxation times (T_1) are useful in assessing the degree of internal motion of alkyl chains; the more restricted the motion of a carbon, the smaller its T_1. When the T_1 method was applied to micellar 12-hydroxystearyltrimethylammonium chloride [48], it was found that the three carbons indicated below experience lower than usual relaxation times

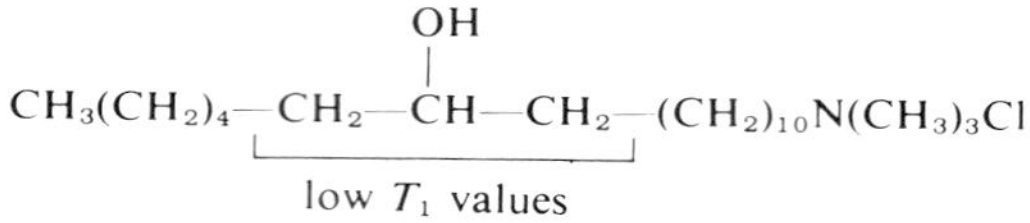

$$CH_3(CH_2)_4-CH_2-\overset{\overset{\displaystyle OH}{|}}{CH}-CH_2-\underbrace{(CH_2)_{10}N(CH_3)_3Cl}_{\text{low } T_1 \text{ values}}$$

relative to the nonhydroxylated analog. Intermolecular hydrogen bonding between hydroxy groups reduces the "segmental motion" of only a small section of the chain. Addition of a heavy-metal relaxation agent (a gadolinium complex) decreases the T_1 values of the carbons, but those near the head group are perturbed the most. This result demonstrates that the hydrocarbon chains within the micelle do not fold so as to expose the hydroxy group and its carbon neighbors to the external milieu.

The optically active sulfonate ester shown below solvolyzes in water to produce 2-octanol with 100% inverted configuration [49]. Hexadecyltrimethylammonium bromide changes neither the rate nor the stereochemistry of this reaction. However, micellar concentrations of SDS retard the solvolysis by two orders of magnitude and cause the appearance of considerable amounts of 2-octanol with retained configuration. The degree of retention depends on the relative concentration of SDS and substrate within the "mixed" micelles. If the SDS is in great excess over the substrate, one observes only 56% inversion. If the concentrations of SDS and substrate are nearly the same, the totally inverted stereochemistry returns. These data can be rationalized in terms of the water content and compactness of the Stern layer. Micelles composed mainly of SDS form a tight ionic array with the positively charged substrate, so that the need for water in the Stern layer is minimal. Consequently, the reaction rate is slowed and the stereochemistry

$$C_6H_{13}-\overset{\overset{\displaystyle CH_3}{|}}{CH}-OSO_2-\!\!\!\left\langle\!\!\!\bigcirc\!\!\!\right\rangle\!\!\!-N(CH_3)_3{}^+$$

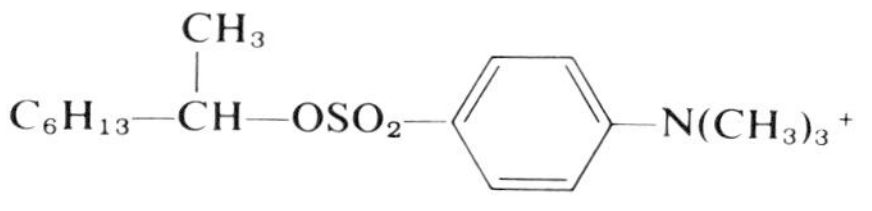

randomized. When there are large numbers of bulky quaternary nitrogen ions in the micelles, then additional water must enter the Stern layer for solvation purposes, and the stereochemistry is that found in bulk water.

Synthetic organic chemists are frequently faced with the problem of reacting a water-insoluble organic compound with a water-soluble reagent (hydroxide, permanganate, periodate, hypohalite, etc.). The use of surfactants in alleviating this problem has recently been assessed with several two-phase reactions [50]. Surfactants disperse organic liquids in water; this could conceivably generate higher yields and shorter reaction times. Micellar catalysis could further facilitate reactions involving two immiscible liquids. It was found that basic hydrolysis of α,α,α-trichlorotoluene to benzoic acid [Eq. (12)] in the presence of 0.01 M cetyltrimethylammonium bromide requires 1.5 hr at 80°C for a 98% yield. By contrast, the reaction without surfactant takes 60 hr. No product can be isolated if the reaction without surfactant is allowed to proceed for only 1.5 hr. Although general application of the method is probably limited, emulsion-micellar catalysis may find utility in large-scale reactions or in reactions of water-insoluble liquids that require a minimum exposure to an aqueous solution.

$$
\begin{array}{ccc}
\text{CCl}_3 & & \text{COOH} \\
\bigcirc & \xrightarrow{\text{OH}^-} & \bigcirc
\end{array}
\tag{12}
$$

In conclusion, I wish to cite the Andreski equation [Eq. (13)], a revealing relationship derived from a study of the social sciences [51]. Here, A is the author's ambition, K is his knowledge, and V is his verbosity. This equation would seem to have implications in the physical sciences as well, and I should probably end here before too many readers feel that the magnitude of V is excessive.

$$A/K - 1 = V \tag{13}$$

REFERENCES

1. E. H. Cordes, ed., "Reaction Kinetics in Micelles." Plenum, New York, 1973.
2. J. H. Fendler and E. J. Fendler, "Catalysis in Micellar and Macromolecular Systems." Academic Press, New York, 1975.
3. G. C. Kresheck, in "Water, A Comprehensive Treatise" (F. Franks, ed.), Vol. 4, Chapter 2. Plenum, New York, 1974.
4. C. Tanford, "The Hydrophobic Effect." Wiley (Interscience), New York, 1973.
5. D. Stigter, J. Phys. Chem. 68, 3603 (1964).
6. D. Stigter, J. Phys. Chem. 79, 1008 and 1015 (1975).
7. M. E. L. McBain and E. Hutchinson, "Solubilization and Related Phenomena." Academic Press, New York, 1955.
8. N. Muller, J. Phys. Chem. 76, 3017 (1972). P. J. Sams, E. Wyn-Jones, and J. Rassing, Chem. Phys. Letters, 13, 233 (1972).
9. T. Sasaki, M. Hattori, J. Sasaki, and K. Nukina, Bull. Chem. Soc. Jp. 48, 1397 (1975). See Fig. 4.
10. E. W. Anacker, in "Cationic Surfactants" (E. Jungermann, ed.), Chapter 7. Dekker, New York, 1970.
11. P. Mukerjee and K. J. Mysels, "Critical Micelle Concentrations of Aqueous Surfactant Systems," NSRDS-NBS 36. Superintendent of Documents, Washington, D.C., 1971.
12. R. J. Williams, J. N. Phillips, and K. J. Mysels, Trans. Faraday Soc. 51, 728 (1955); E. D. Goddard and G. C. Benson, Can. J. Chem. 35, 986 (1957).
13. M. L. Corrin, H. B. Klevens, and W. D. Harkins, J. Chem. Phys. 14, 480 (1946); K. Shinoda, J. Phys. Chem. 58, 1136 (1954).
14. W. Philippoff, Discuss. Faraday Soc. 11, 96 (1951).
15. H. B. Klevens, J. Phys. Chem. 52, 130 (1948). C. H. Arrington, Jr. and G. D. Patterson, ibid. 57, 247 (1953).

16. B. Lindman and I. Danielsson, *J. Colloid Interface Sci.* **39**, 349 (1972). H. Gustavson and B. Lindman, *J. Chem. Soc., Chem. Commun.* p. 93 (1973).
17. P. Mukerjee, J. Perrin, and E. Witzke, *J. Pharm. Sci.* **59**, 1513 (1970).
18. F. M. Menger and C. E. Portnoy, *J. Am. Chem. Soc.* **89**, 4698 (1967).
19. K. Shinoda, T. Nakagawa, B. Tamamushi, and T. Isemura, "Some Physico-Chemical Properties of Colloidal Surfactants." Academic Press, New York, 1963.
20. L. Romsted, Ph.D. Thesis, Indiana University, Bloomington (1975).
21. H. B. Klevens, *J. Am. Oil Chem. Soc.* **30**, 76 (1953).
22. L. I. Osipow, "Surface Chemistry," p. 169. Van Nostrand-Reinhold, Princeton, New Jersey, 1962.
23. J. E. Adderson and H. Taylor, *J. Pharm. Pharmacol.* **23**, 312 (1971).
24. A. Lottermoser and F. Puschel, *Kolloid-Z.* **63**, 175 (1933).
25. M. L. Corrin and W. D. Harkins, *J. Am. Chem. Soc.* **69**, 683 (1947).
26. W. P. Jencks, "Catalysis in Chemistry and Enzymology," p. 358. McGraw-Hill, New York, 1969.
27. E. W. Anacker and H. M. Ghose, *J. Phys. Chem.* **67**, 1713 (1963).
28. K. Kalyanasundaram, M. Gratzel, and J. K. Thomas, *J. Am. Chem. Soc.* **97**, 3915 (1975).
29. J. N. Phillips and K. J. Mysels, *J. Phys. Chem.* **59**, 325 (1955).
30. P. Mukerjee, *J. Phys. Chem.* **66**, 943 (1962).
31. F. M. Menger and J. L. Lynn, *J. Am. Chem. Soc.* **97**, 948 (1975).
32. J. W. Larsen and L. J. Magid, *J. Am. Chem. Soc.* **96**, 5774 (1974).
33. F. M. Menger, T. E. Thanos, and R. Bradshaw, unpublished results. see also P. Mukerjee and A. Ray, *J. Phys. Chem.* **70**, 2144 (1966).
34. F. Podo, A. Ray, and G. Neméthy, *J. Am. Chem. Soc.* **95**, 6164 (1973), and references cited therein; N. Muller *in* "Reaction Kinetics in Micelles" (E. H. Cordes, ed.), p. 8. Plenum, New York, 1973.
35. E. Williams, B. Sears, A. Allerhand, and E. H. Cordes, *J. Am. Chem. Soc.* **95**, 4871 (1973).
36. O. H. Griffiths and A. S. Waggoner, *Acc. Chem. Res.* **2**, 17 (1969).
37. E. F. J. Duynstee and E. Grunwald, *J. Am. Chem. Soc.* **81**, 4540, 4542 (1959).
38. C. A. Bunton and L. Robinson, *J. Org. Chem.* **34**, 780 (1969); G. J. Buist, C. A. Bunton, L. Robinson, L. Sepulveda, and M. Stam, *J. Am. Chem. Soc.* **92**, 4072 (1970).
39. M. T. A. Behme and E. H. Cordes, *J. Am. Chem. Soc.* **87**, 260 (1965), M. T. A. Behme, J. G. Fullington, R. Noel, and E. H. Cordes, *ibid.* p. 266.
40. C. A. Bunton and L. Robinson, *J. Org. Chem.* **34**, 773 (1969); C. A. Bunton, E. J. Fendler, L. Sepulveda, and K.-U. Yang, *J. Am. Chem. Soc.* **90**, 5512 (1968).
41. C. A. Bunton and M. J. Minch, *Tetrahedron Lett.* p. 3881 (1970).
42. R. B. Dunlap and E. H. Cordes, *J. Am. Chem. Soc.* **90**, 4395 (1968)
43. A. K. Yatsimirski, K. Martinek, and I. V. Berezin, *Tetrahedron* **27**, 2855 (1971). K. Martinek, A. V. Levashov, and I. V. Berezin, *Tetrahedron Lett.* p. 1275 (1975).
44. M. Gratzel, K. Kalyanasundaram, and J. K. Thomas, *J. Am. Chem. Soc.* **96**, 7869 (1974).
45. R. R. Hautala and R. L. Letsinger, *J. Org. Chem.* **36**, 3762 (1971).
46. C. Lapinte and P. Viout, *Tetrahedron Lett.* p. 1113 (1973).
47. R. A. Moss, C. J. Talkowski, D. W. Reger, and W. L. Sunshine, *in* "Reaction Kinetics in Micelles" (E. H. Cordes, ed.), p. 99. Plenum, New York, 1973.
48. J. M. Brown and J. D. Schofield, *J. Chem. Soc., Chem. Commun.* p. 434 (1975).
49. C. N. Sukenik, B. Weissman, and R. G. Bergman, *J. Am. Chem. Soc.* **97**, 445 (1975).
50. F. M. Menger, J. U. Rhee, and H. K. Rhee, *J. Org. Chem.* **40**, 3803 (1975).
51. S. Andreski, "Social Sciences as Sorcery." St. Martin's Press, New York, 1973.

CHAPTER

8

Recent Studies on Bioactive Compounds

Koji Nakanishi

Q*-NUCLEOSIDES [1]

The first position of the anticodons of *Escherichia coli* tRNA^Tyr, tRNA^His, tRNA^Asn, and tRNA^Asp contains the unusual Q-nucleoside, which is a 7-deazaguanosine having an unusual side chain (Fig. 1) [2]. The *trans–cis* stereochemistry of the side chain could be determined only after the synthesis of two models: (a) 3,4-*trans* and 4,5-*cis* and (b) 3,4-*cis*, the 4,5-*cis* having a phenyl group instead of the 7-deazaguanosine group. Not unexpectedly, all vicinal J values $J_{1,2}$, $J_{2,3}$, $J_{3,4}$, and $J_{4,5}$ were identical (only $J_{1,5}$ differed) due to conformational flexibility of the cyclopentene ring, and hence the stereochemical conclusion was based on chemical shift comparisons [3]. The tRNA's of various animals, in particular those of hepatoma cells, contain another modified base, Q*, generally in larger quantities than the Q-nucleoside. The Q*-nucleoside was isolated from rabbit liver tRNA [4] and was separated from Q-nucleoside by paper chromatography and finally purified by high-pressure liquid chromatography (hplc) [Poragel-PN and MeOH–H_2O (9:1)] to give a total of ~ 600 μg of pure Q*-nucleoside.

Macfarlane and co-workers have recently developed a powerful plasma desorption mass spectroctrograph (PDMS) which utilizes the fission energy of ^{252}Cf to ionize underivatized samples [5]. The ^{252}Cf fission causes a localized hot spot of temperatures up to 10,000 K on the nickel foil containing the sample. Because of the ultrarapid heating, the energy is not deposited into

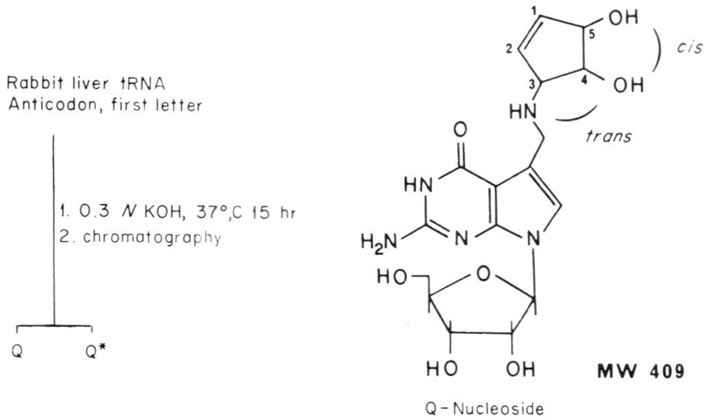

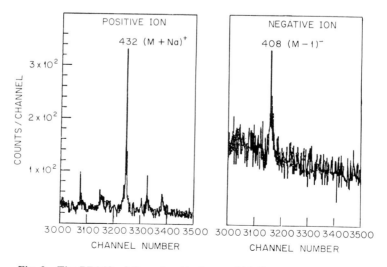

Rabbit liver tRNA
Anticodon, first letter

1. 0.3 *N* KOH, 37°,C 15 hr
2. chromatography

Q   Q*

cis

trans

HN

O

HN

H_2N N   N

HO

HO   OH   **MW 409**

Q-Nucleoside

Fig. 1 Extraction of Q- and Q*-nucleosides (left) and structure of Q nucleoside (right).

the vibrational mode leading to decomposition of sample but is consumed to desorb the molecule, thereby creating quasi-molecular cations by picking up of labile protons or Na^+ and quasi-molecular anions by loss of protons, etc. A time-of-flight mass spectrometer employing a 1.5 m tube for high sensitivity or an 8 m tube for high resolution measures the flight time, and hence the mass of both positive and negative particles by reversing the potentials of the accelerating electrostatic grid assembly. The general fragmentation pattern is similar to that obtained by *tert*-Bu CI/MS (chemical ionization mass

POSITIVE ION

$432\ (M+Na)^+$

NEGATIVE ION

$408\ (M-1)^-$

3×10^2

COUNTS / CHANNEL

2×10^2

1×10^2

3000 3100 3200 3300 3400 3500   3000 3100 3200 3300 3400 3500

CHANNEL NUMBER   CHANNEL NUMBER

Fig. 2 The PDMS of Q-nucleoside from rabbit liver tRNA, MW 409.

spectrometry) [5]. Under favorable conditions, PDMS can handle underivatized samples, e.g., pentadecapeptides, up to MW 2000–3000, and can yield high-resolution MS data at masses less than 500. The PDMS of the zwitterionic tetrodotoxin, a guanidinium-containing marine neurotoxin, gives an M + 1 peak of 320.10940 and the ^{13}C satellite peak of 321.1275 [6]; in the case of tetrodotoxin the field-desorption MS, which is so powerful in other cases, is unable to give a MW peak.

The positive and negative PDMS spectra of Q- and Q*-nucleosides are shown in Figs. 2 and 3. The difference in molecular weight of 162 between the two samples suggested that an additional hexose may be linked to Q*. The molecular weight was corroborated by an M$^+$ peak of m/e 767 for the side-chain N-acetylated permethyl (11-methyl) derivative and M$^+$ m/e 1291 for the decatrimethylsilylate. The 360 MHz paramagnetic resonance (pmr) signal (kindly measured by Dr. D. Patel, Bell Laboratories) of Q*-nucleoside in D$_2$O unexpectedly showed that three anomeric protons were present at 5.89, 4.91, and 4.43 ppm (Fig. 4) in a ratio of about 1:0.75:0.25. Since the δ and J values of the anomeric protons indicated that they were due to ribosyl, β-mannosyl, and β-galactosyl residues, respectively, the sample was hydrolyzed with acid upon which mannose and galactose were detected by paper chromatography and by gas–liquid chromatography of TMS derivatives.

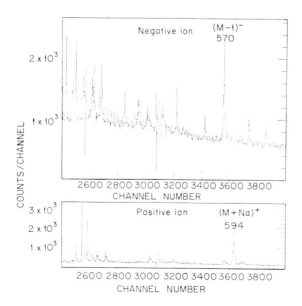

Fig. 3 The PDMS of Q*-nucleoside, MW 571. (Reprinted with permission from Kasai *et al., J. Am. Chem. Soc.* **98,** 5044 (1976). Copyright by the American Chemical Society).

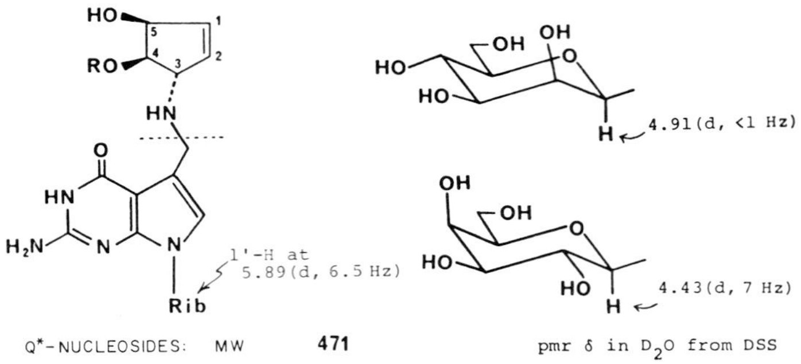

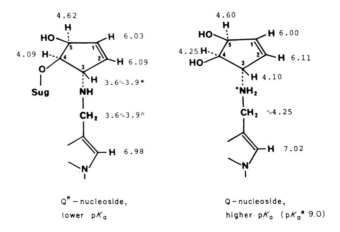

Q*-NUCLEOSIDES: MW **471**   pmr δ in D₂O from DSS

Fig. 4 Structure of Q*-nucleoside. **1**, R = H; **2**, R = β-D-mannosyl (~75%); **3**, R = β-D-galactosyl (~25%); DSS = sodium-dimethyl-2-silapentane-5-sulfonate. (Reprinted with permission from Kasai *et al.*, *J. Am. Chem. Soc.* **98**, 5044 (1976). Copyright by the American Chemical Society.)

Thus, the Q*-nucleoside sample is a mixture, which may be due to the fact that it was isolated from a mixture of tRNA's.

A comparison of side-chain pmr shifts measured in D_2O (pD 7.5) (Fig. 5) allows one to attach the sugars to C-4. Q-Nucleoside has three pK_a values, 1.1, 7.7, and 10.4, arising from the deazaguanine group and another at about 9 due to the side-chain nitrogen, the latter being measured by pmr [2]. It is noticed that the protons near the N resonate at higher fields in the Q*-nucleoside. At pD 7.5, the N in Q-nucleoside exists largely in the ammonium form, whereas the higher δ values of Q* suggest that its N is largely unproto-

Fig. 5 The 360 MHz pmr of Q- and Q*-nucleosides in D_2O, pD 7.5 (asterisk indicates same region as hexose protons).

nated. This leads to a structure having the sugars at C-4 rather than C-5, because a C-4 sugar would sterically not favor the adoption by N of an ammonium form, i.e., lower its pK_a. Assignment of the 4.62 ppm signal to 5-H is correct because in acid medium the signal underwent a downfield shift of only 0.06 ppm; if it were the 4-H, it should be shifted downfield by 0.4 ppm [2].

The Q*-nucleoside structure is unique in that it is the only tRNA base found so far to contain sugar residues in the side chain. The significance of such hydrophilic groups in anticodon position 1, in particular in hepatoma tRNA, may be related to membrane permeability. The side-chain configurations shown are based on the difference CD of Q-nucleoside hexabenzoate and guanosine tribenzoate which gives a positive exciton split CD [13] due to interaction between the 3-N- and 4-O-benzoates (unpublished).

AN *IN VIVO* REACTION PRODUCT OF BENZ[*a*]PYRENE WITH BOVINE BRONCHIAL MUCOSA [7–11]

Benz[*a*]pyrene (BP), one of the most widespread environmental carcinogens, is estimated to be emitted into the air in the United States at the rate of about 1,300 tons/year [8]. In recent years it has been shown that the microsomal hydroxylase system converts the aryl hydrocarbons into numerous oxidation products, and arene oxides have been suggested as being responsible for the carcinogenic and mutagenic activity [9]. In the case of BP, the reactive intermediate is considered to be the 7,8-diol-9,10-epoxide, two forms (see Fig. 9, I and II) of which have been synthesized [10]. Although the covalent binding of this intermediate to cellular macromolecules is a critical step in carcinogenesis, the mode of its binding was unknown. This was elucidated for the first time as follows.

Prior to studies of the *in vivo* products, the structures of adducts resulting from incubation of 7,12-dimethylbenz[*a*]anthracene-5,6-epoxide (DMBA-5,6-epoxide) with polyguanylic acid [poly(G)] in 50% acetone were determined (Fig. 6) [11]. Previous studies had shown that poly(G) was the most reactive of the homopolymers when reacted with DMBA-5,6-epoxide [12]. Careful separation of the product hydrolyzate by hplc (reverse phase, Zorbax ODS, methanol–water) gave four adducts, I–IV (Fig. 6), 0.5–1 mg each, which were subjected to circular dichroism (CD) measurements. This was fortunate since they showed that the CD spectra of I and IV had a mirror image relation, and similarly the CD spectra of II and III had a mirror image relation. This is not strictly true because the guanine–ribose links in the four adducts necessarily are all β-D, but, for practical reasons, this finding of diastereo-

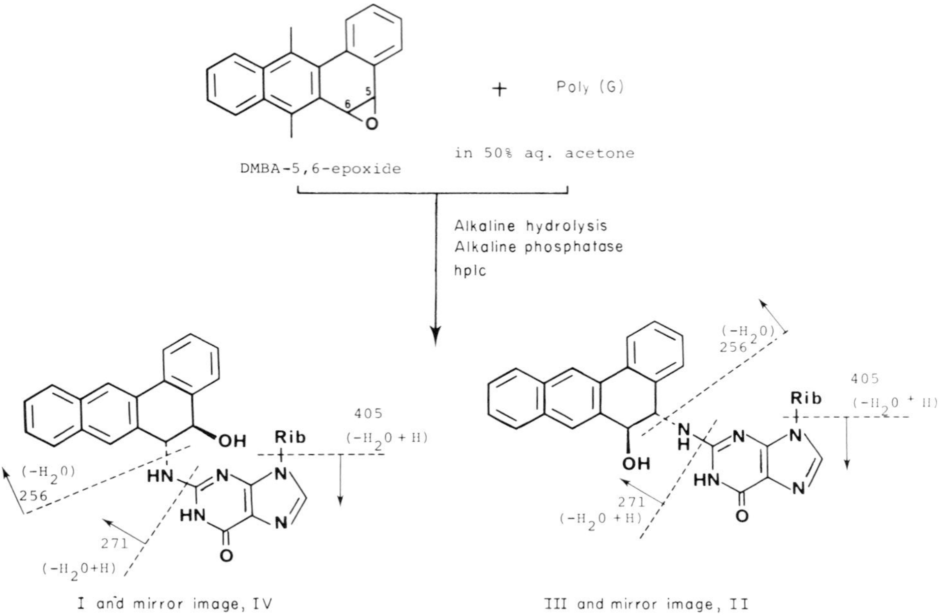

DMBA-5,6-epoxide

+   Poly (G)

in 50% aq. acetone

Alkaline hydrolysis
Alkaline phosphatase
hplc

I and mirror image, IV

III and mirror image, II

Fig. 6   Formation of DMBA-epoxide and guanosine adduct.

meric relations was very important because it allowed us to combine the
DMBA–guanine adducts obtained upon hydrolytic removal of ribose from
products I and IV and from II and III. The increased quantity resulting from
combination of bases greatly facilitated handling, especially pmr measurements.

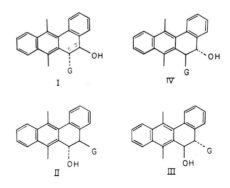

I

IV

II

III

Fig. 7   The four guanosine adducts.

Neither EI (electron impact) nor CI/MS gave the M$^+$ peak. Nevertheless, the high-resolution MS of m/e 271 peaks (Fig. 6) was very diagnostic, since the presence of the peak with an N on the DMBA moiety showed that binding should be via N-2. The pmr of adducts I–IV, the bases from I and IV and from II and III, and acetylated bases established the structures shown in Fig. 7. The absolute configurations are provisonally based on the following deductions. As shown in Fig. 8, the bases from DMBA–poly(G) adducts I and III both show split-type Cotton effects around 270 nm (CD curves of adduct bases I and IV are antipodal, and those of II and III are antipodal). Since guanine and dihydro-DMBA-5,6-diol both have ultraviolet (uv) maxima in the region 260–270 nm, it is reasonable to assume that the split CD curve originates from coupled oscillators [13]. If it is assumed that the directions of the two interacting electric transition moments at ~275 nm are as depicted in Fig. 8, the positively split CD curves lead to the absolute configurations shown for I and III; however, this is provisional until completion of more accurate linear dichroic studies by Professor J. Michl, University of Utah.

Bovine main-stem bronchial explants (16 specimens, each 1 × 1 cm, from two adult steers) were incubated with 200 μCi/ml [³H]BP (Fig. 9) for 24 hr.

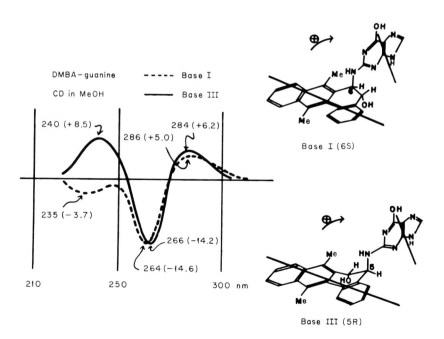

Fig. 8 The CD of bases I and III obtained from DMBA–nucleoside adducts I and III, respectively [11].

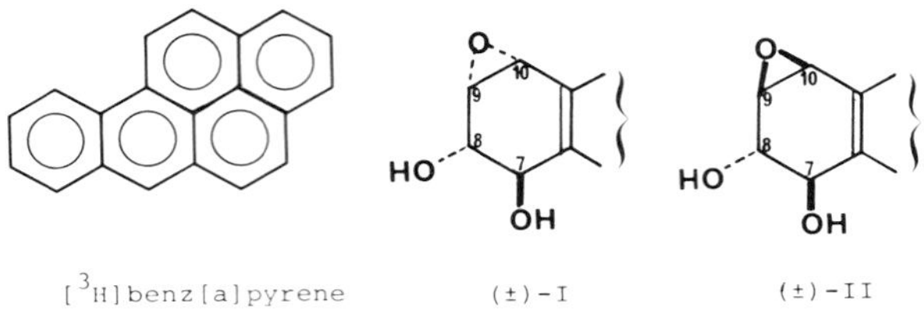

[³H]benz[a]pyrene   (±)-I   (±)-II

Fig. 9 Incubation and workup of BP with bovine glands. The procedure was as follows: (a) [³H]BP incubated with bovine bronchial explants; (b) total cellular RNA extracted and hydrolyzed; (c) hydrophobic nucleosides hydrolyzed with alkaline phosphatase; (d) modified nucleosides isolated on Sephadex LH-20; (e) adducts of poly G/racemic I and poly G/racemic II cochromatographed, Zorbax ODS, reverse phase.

The total cellular RNA was extracted, and the hydrophobic nucleosides containing the labeled BP were isolated on Sephadex LH-20 and cochromatographed with guanosine adducts obtained from *in vitro* reaction of poly(G) with racemic isomer I (Figs. 9 and 10). The radioactive peak corresponded to one of the adducts (**1a**, Fig. 10) from (±)-I but not to the adducts derived from (±)-II. The CD curves of **1a** and the other adduct **1b** were again "antipodal" (not strictly because β-D-ribose is present in both); not knowing the absolute configurations of **1a** and **1b** at this stage, we tentatively represent them as depicted in Fig. 10 [14b]. The identity of *in vivo* product with **1a** was further corroborated by hplc of the 8,9-/2′,3′-diacetonide (**2a**, Fig. 10) and

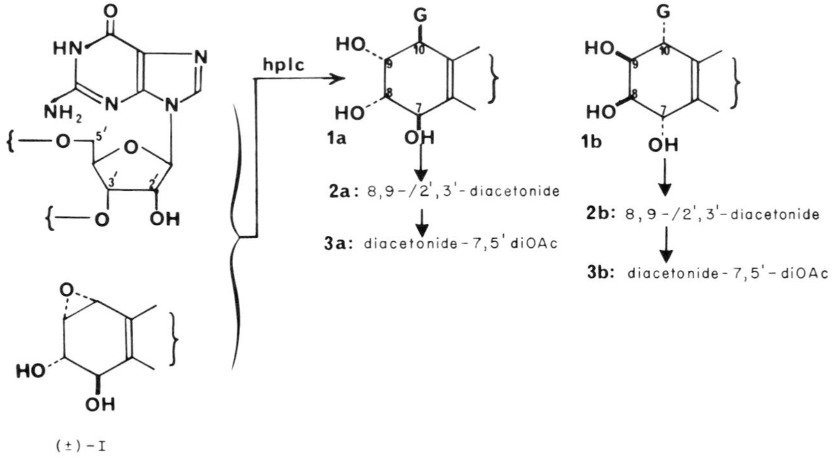

2a: 8,9-/2′,3′-diacetonide   **2b:** 8,9-/2′,3′-diacetonide

3a: diacetonide-7,5′ diOAc   **3b:** diacetonide-7,5′-diOAc

(±)-I

Fig. 10 Formation of BP–guanosine adduct *in vivo*. **1a/1b** and **3a/3b**, mirror image CD. Retention–times (hplc) of **3a**: 46.0 sec (hot); 46.8 sec (cold).

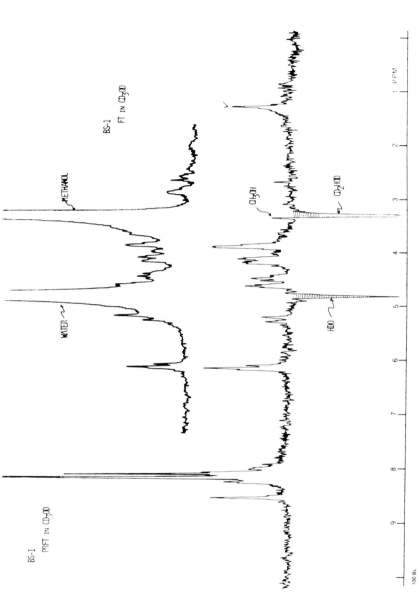

Fig. 11 The pmr of **BP**–guanosine adduct; normal (top spectrum) and **PRFT** (lower spectrum) runs. (Reprinted with permission from Jeffrey et al., J. Am. Chem. Soc. **98**, 5174 (1976). Copyright by the American Chemical Society).

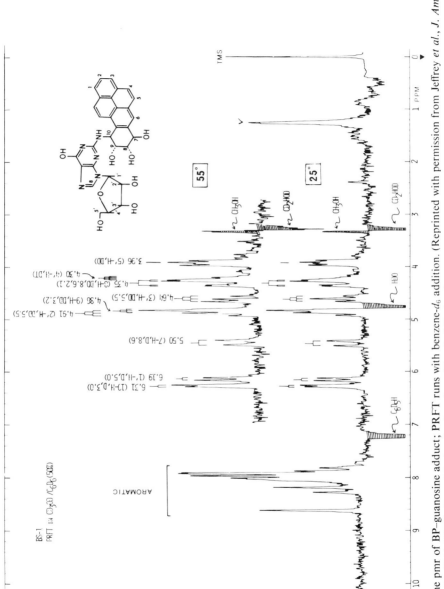

Fig. 12 The pmr of **BP**–guanosine adduct; **PRFT** runs with benzene-d_6 addition. (Reprinted with permission from Jeffrey *et al.*, *J. Am. Chem. Soc.* **98**, 5714 (1976). Copyright by the American Chemical Society.)

diacetonide-7,5'-diacetate (**3a**, Fig. 10). The CD spectra of **3a** and **3b** were antipodal as expected. It should be noted that the hplc retention time of tritiated and cold **3a** differed somewhat due to an isotope effect [14].

A detailed pmr study of the ~1 mg of adduct **1a** (Fig. 10) by partial relaxed Fourier transform $(180°-\tau-90°-T)_n$ sequence in CD_3OD/C_6D_6 dramatically clarified the spectrum. Namely, as shown in the top run of Fig. 11, the pmr in DC_3OD (containing D_2O) is hardly usable because of overlap of intense solvent peaks. When measured under PRFT (partially relaxed Fourier transform) conditions, $\tau = 2.2$ sec, $T = 6.1$ sec, 4096 scans, JEOL PS-100, the solvent peaks appear as inverted peaks because of their longer relaxation times due to the vicinal D atoms (Fig. 11, bottom spectrum). The small positive peak at ~3.5 ppm is due to the protiomethanol present in deuteriomethanol. Although more detail is seen, it is not sufficient for full analysis. Addition of 50% C_6D_6 significantly spreads out the spectrum (Fig. 12, bottom run), but a final clarification is achieved by warming the sample from ambient 25° to 55°C (Fig. 12, top run), which removes the 4.8 ppm HDO peak upfield. All coupling constants pertaining to ring A are thus measurable (Fig. 13). A comparison of J values with those of the *cis* and *trans* hydration products from I (Fig. 9) show that they are consistent with a *trans* opening of the epoxide by guanine.

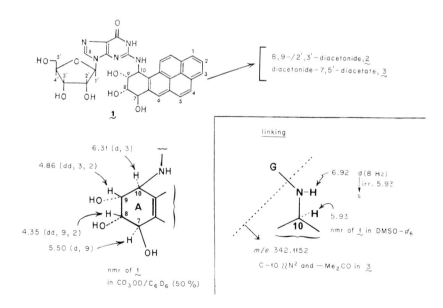

Fig. 13 Structure of BP–guanosine adduct. This also represents correct absolute configuration [14b].

The guanine must have attacked C-10 of the BP skeleton as evidenced by pmr chemical shifts of 10-H, and also by formation of the 8,9-acetonide. The point of attachment to the guanine group was determined as follows. Although weak, the high-resolution MS of 8,9-/2′,3′-diacetonide–7,5′-diacetate **3** showed a peak corresponding to the fission between the purine skeleton and an N—hence N^2 (see Fig. 13). When measured in DMSO-d_6, the pmr of adduct **1** (Fig. 13) had a doublet at 6.92 ppm (8 Hz), which collapsed to a singlet upon irradiation of the 10-H signal at 5.93 ppm and which disappeared upon addition of D_2O. The 6.92 ppm is thus clearly assignable to NH and hence the guanine is attached via N^2.

Recent studies show that isomer I (Fig. 9) is more mutagenic than isomer II (Fig. 9) and that isomer I is formed when rat liver microsomes are incubated with BP-7,8-*trans*-diol [14a]. The results outlined above are consistent with these observations and lead to the first structure determination of the *in vivo* adduct of BP and nucleic acids [14b]. However, additional reactive intermediates of BP not derived from diol epoxides apparently exist *in vivo* and remain to be elucidated.

THE β-CARBOLINES ISOLATED FROM HYDROLYSIS OF HUMAN CATARACTOUS LENS PROTEINS [15]

With aging, proteins in the central region of the human lens slowly aggregate to give species [16] of molecular weight greater than 50×10^6. Concomitantly, a yellowing of lens is observed, and its generation has been correlated with an increase in a yellow fluorescence [17] which is found primarily in the insoluble protein fraction [17,18]. It is generally believed that such aggregation and accumulation of covalently bound fluorescent material cause significant light scattering and may be a contributing factor in the development of senile cataract.

Approximately 750 mg of insoluble protein were isolated from 50–100 cataractous lenses obtained from the eye bank or morgue. This was hydrolyzed with 4 N barium hydroxide for 24 hr at 110°C and the fluorescent products were partially purified by preparative paper electrophoresis at pH 1.9 (Fig. 14). The principal fluorescent product(s), designated "C_1" (or IIA,B) in Fig. 14, was also obtained by proteolytic digestion of succinylated, reduced, and carboxymethylated protein from old human cataractous lenses [17]. Substance "C_1" is very unstable and decomposes to two fluorescent products, IA and IB by further base treatment, oxidation, or even dry storage at −80°C. As compared to IIA,B, the stable IA and IB products have a much higher mobility toward the negative pole at pH 1.9. From the ~750 mg of

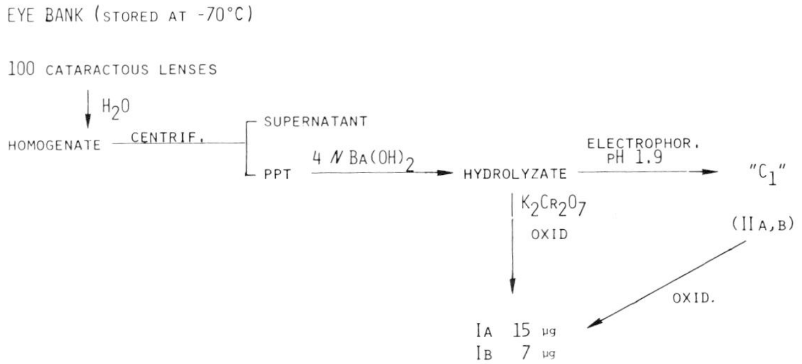

EYE BANK (STORED AT -70°C)

100 CATARACTOUS LENSES

Fig. 14 Isolation of I and II from cataractous lenses. The " C_1 " is obtained from OH^- hydrolysis and oxidation of cataractous lenses but not from young lenses, and from acid hydrolysis of tryptophan-containing proteins but not from OH^- hydrolysis of tryptophan-containing proteins.

insoluble protein, there was obtained ∼15 μg of IA and 7 μg of IB (yield based on uv absorption after characterization of structure). Substance " C_1 " is indeed derived from cataractous lenses since products IA and IB were not formed from young human and calf lenses or bovine serum albumin upon *base* treatment and subsequent oxidation. However, they are formed if tryptophan-containing proteins are subjected to *acid* hydrolysis.

The uv and CI/MS spectra suggested IA and IB to be β-carbolines, the full identity of which was established by synthesis (Fig. 15). The amount of IIA,B (or " C_1 ") was so minute that for IIA only the fluorescence data and behavior on thin layer chromatography (tlc) and electrophoresis of free IIA and its methyl ester were available, whereas for IIB only tlc and electrophoretic data were known. However, knowing the products IA and IB, the logical precursors IIA,B were synthesized, and all properties noted were in agreement.

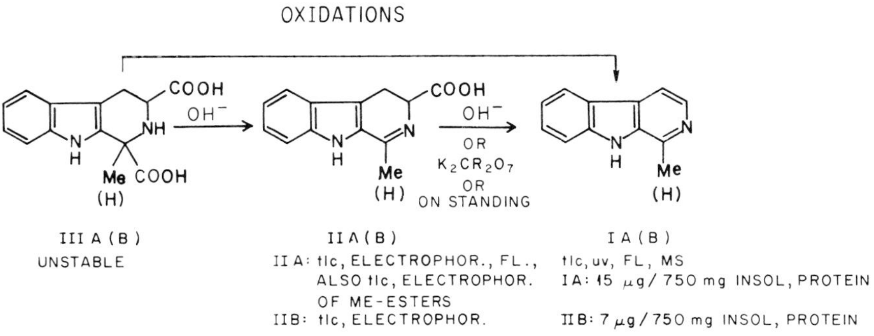

OXIDATIONS

Fig. 15 Structures of compounds I, II, and III. The (B)'s correspond to structures with (H)'s.

Synthetic II was also readily converted to I. Authentic IIB (with 1-H) was much less stable than IIA (with 1-Me), a fact that is probably related to the smaller amounts of IB (7 μg) relative to IA (15 μg). Compounds IIIA and IIIB were detected only by their fluorescence on electrophoretic plate (pH 1.9; less mobility than IIA,B toward negative pole). However, synthetic IIIA had the same electrophoretic mobility as the natural products and also gave IIA upon base hydrolysis; hence, it is the likely precursor of IIA.

The mechanism by which the hydro-β-carboline structures are formed in the lens is not understood, but the dicarboxylic structure III is particularly attractive since it can act as a cross-linking unit derived from tryptophan (and α-keto acid, which in turn is derivable from oxidative deamination of a terminal alanine and glycine residue). The β-carboline formation could be induced by photochemistry, free radicals, or enzyme deficiency, a difficult problem which remains to be solved.

AN ANTISICKLING AND DESICKLING
AGENT: DBA [19]

The DBA molecule (3,4-dihydro-2,2-dimethyl-2*H*-1-benzopyran-6-butyric acid) shown in Fig. 16 is a potent agent against sickle-cell disease and has no acute toxicity against male mice; the intraperitoneal and oral LD_{50} are judged to be 710 and > 3000 mg/kg, respectively. However, these and the following data are only *in vitro* results. Comprehensive hematological studies are in progress at several institutes, whereas clinical and pathological investigations have yet to be carried out.

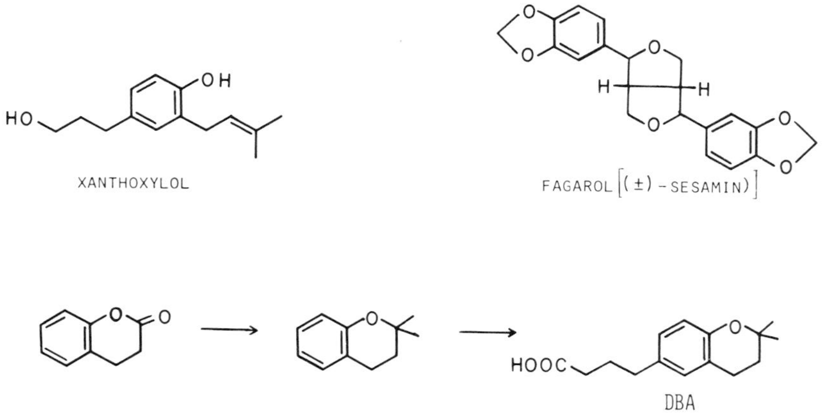

Fig. 16 *Fagara xanthoxyloides:* Xanthoxylol and fagarol; synthesis of DBA.

Sickle-cell anemia [20] is due to a genetic molecular defect in which just 2 of the 574 amino acid residues in normal hemoglobin (Hb) are different; namely, the sixth amino acid from the amino terminal in the two β chains is glutamic acid in Hb and valine in HbS (sickle-cell hemoglobin) [21]. This disease is restricted to the black population; a higher estimate of the homozygotes in the United States is 54,000 [22]. At present there is no satisfactory method for medicinal prevention or treatment.

Deoxygenation of sickle-cell blood by physical (exercise, bubbling nitrogen, etc.) or chemical (reduction) means transforms the normal round erythrocytes into sickled shapes, clots the vessels, and impairs oxygen transport. The roots of the common West African tree *Fagara xanthoxyloides* are used in West Africa as chewing sticks to clean the teeth. It has reputed antimicrobial and also antisickling activity [23], but the latter claim has been disputed [24]. In preliminary bioassays carried out on several compounds isolated from the roots, it appeared that the known xanthoxylol exhibited antisickling activity. It was therefore planned to convert the primary hydroxyl to a carboxyl by reactions with HI, then KCN, and hydrolysis. However, the HI treatment resulted in cyclization to the chroman so that the final product was DBA, the agent that turned out to possess potent *in vitro* activity against HbS. Once this was found, DBA could be made efficiently in 90% overall yield by Friedel–Crafts acylation of 2,2-dimethylchroman (obtained from dihydrocoumarin) with succinic anhydride followed by Clemmensen reduction. ^{14}C-Labeled DBA was prepared by using [1,14-^{14}C]-succinic anhydride. Interestingly, subsequent bioassays showed that xanthoxylol itself was hardly active! Fagarol or ($\pm$)-sesamin (Fig. 16) which is present in *Fagara* roots as well as in a variety of plants and which is used as a synergist (mixed-function oxidase inhibitor) in insecticidal applications of pyrethroids, also possesses antisickling activity although it is weaker than that of DBA (unpublished results). Fagarol should be at least partly responsible for the reputed [23] antisickling effect of *Fagara* roots.

The *in vitro* activities of DBA are shown in Tables 1 and 2. In the antisickling experiments (Table 1), the percent sickling in the absence of DBA was 90%. Hence, the antisickling effect was already observed with 1 mM DBA, and inhibition of sickling was almost complete at 13 mM. In the desickling studies (Table 2), the patient had only 2% normally shaped cells after deoxygenation (the rest being 69% sickle cells and 29% cells of varied shape); the effect of DBA was already apparent after 30 min, the effective concentration being 2–3 mM. Unlike other antisickling agents, e.g., cyanate [25], DBA does not shift the oxygen half-affinity curve, a result which suggests that DBA may not interfere with the delivery of O_2 to tissues. In addition, DBA inhibits the gelling of HbS at concentrations of 2 mM.

Studies of [^{14}C]DBA incorporation showed that DBA does not covalently

TABLE 1
Antisickling Effect of DBA[a]

Patient	DBA (mM)	% normal cells after deoxygenation
A	3.8	87
	8.0	93
	13.0	95
B	1.0	48
	6.0	64

[a] The sickle-cell suspensions were incubated for 1 hr with DBA and deoxygenated by reduction, evacuation, or bubbling N_2 and CO_2. The control blood was left oxygenated and contained 98% (patient A) and 78% (patient B) normally shaped blood cells, respectively.

bind to the hemoglobin. Carboxymethyl cellulose chromatography (Fig. 17) of globin from the lysate of cells incubated with [^{14}C]DBA and [^{3}H]leucine, the latter being used as internal control to measure incorporation into globin, revealed that the ^{14}C counts remained just above background and nonspecifically distributed throughout the column. Thus, unlike the internal control of [^{3}H]leucine, the DBA activity seems to be due to noncovalent binding to hemoglobin or interaction with the red cell membrane.

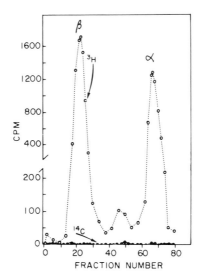

Fig. 17   Carboxymethyl cellulose chromatograph of hemoglobin.

TABLE 2

Desickling Effect of DBA[a]

Concentration (mM)	% normally shaped cells after 30 min
Control	2
10	92
6	80
3	61
1.8	61
0.9	22

[a] The sickle-cell suspensions were first deoxygenated and then treated with various amounts of DBA.

GONYAUTOXINS-II AND -III FROM THE EAST COAST TOXIC DINOFLAGELLATE [26]

Blooms caused by the toxic dinoflagellate *Gonyaulax tamarensis* give rise to the outbreak of toxic shellfish on the North Atlantic coasts of America and the United Kingdom, a hazard that forces both fish markets to be closed. The nature of the east coast poisoning is similar to that of the west coast paralytic shellfish poisoning caused by *Gonyaulax catenella*, the major toxin of which is saxitoxin (STX). Recently, the structure of STX has been established as **1** (Fig. 18) by X-ray crystallography [27] and subsequently confirmed [28]. In

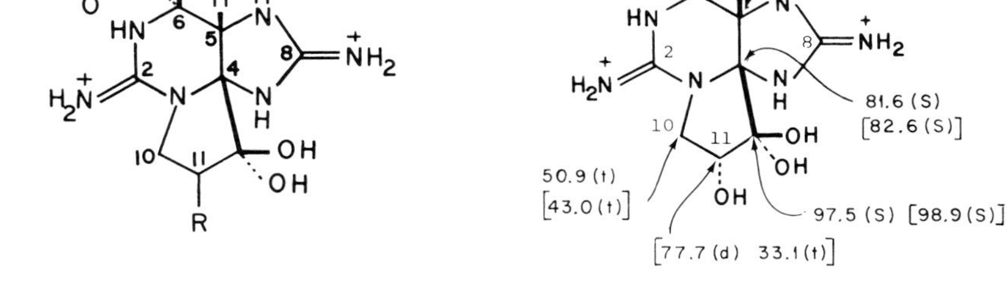

Fig. 18 Structures of STX, GTX-II, and GTX-III. **1**, R = H, STX; **2**, R = α-OH, GTX-II; **3**, R = OH, GTX-III; **4**, GTX-II. Data for STX (11-H) shown in brackets. (Reprinted with permission from Shimizu *et al.*, *J. Am. Chem. Soc.* **98**, 5414 (1976). Copyright by the American Chemical Society.)

1975, Shimizu *et al.* [29] succeeded in isolating STX and six new toxins from both the infested soft-shell clams *Mya arenaria* and cultured *G. tamerensis*. All of them are neurotoxins which block the Na^+ channel, the toxicity level being ~ 5000 mouse units per milligram, or similar to that of tetrodotoxin.

The two major new toxins, which we name gonyautoxin-II and -III (GTX-II and GTX-III) can be represented by structures **2** and **3** (Fig. 18), respectively. Gonyautoxin-II and -III were obtained as an extremely hygroscopic powder. In neutral water they slowly undergo equilibration to form a 3:1 mixture, which is separable into the two components by tlc; the equilibrium is remarkably enhanced by a trace of base such as sodium acetate. They are less basic than STX (**1**, Fig. 18), the relative mobility to the negative pole at pH 8.7 being STX, 1.00; GTX-II, 0.56; and GTX-III, 0.28.

The combined amount of the GTX-II and GTX-III mixture was ~ 2 mg. Therefore, after tlc separation of isomers, the entire supply of GTX-II was used for carbon magnetic resonance (cmr) measurements in a microprobe (JEOL). The data were most revealing since they showed that all carbons were very similar to those of STX [28] except two (Fig. 18), i.e., C-11, which is lower by 44.1 ppm, and C-10, which is lower by 7.9 ppm. The proton-noise-decoupled and off-resonance-decoupled spectra are shown in Fig. 19. A comparison of the two spectra shows that C-11 is a doublet (in STX it is a triplet) and C-12 is a triplet (the same in STX). The cmr data suggested that most probably GTX-11 has an additional hydroxyl group at C-11. The molecular weight could not be measured in spite of several attempts by various methods including PDMS [5].

In order to confirm the structure, GTX-II was gently oxidized with 0.1% H_2O_2 and the products were separated by passage through BioGel P-2 (Fig. 20). A similar oxidation with 0.8% H_2O_2 had been carried out by Rapoport and co-workers on STX [30], which yielded the lactam corresponding to **6** (Fig. 20) (no 11-OH), the key degradation product used in their structural studies. The two oxidation products from GTX-II were assigned structures **5** and **7** on the basis of their physical constants and cyclization to lactams **6** and **8**, respectively, by leaving in 1 N HCl for 18 hr (see Fig. 20). The pmr could be measured only for **5**-(Fig. 20) because of insufficient quantity of other products. Since the pmr of acid **5** in D_2O was partly overlaid by the HDO peak, the partial relaxed Fourier transform method was used again (Fig. 21; see also Fig. 12) at 55°C, which revealed a multiplet at 4.70 ppm having a 3:2 relative intensity compared to the 5.04 ppm —CH_2OH peak. As expected, **5** was optically active and showed CD Cotton effects (Fig. 20) at wavelengths corresponding to the aromatic nucleus; the relatively small CD intensities suggested that the chiral center is not directly attached to the nucleus. Similar oxidation and workup on STX gave the four products corresponding to **5–8** (Fig. 20) (no 11-OH). The structure of **7** (Fig. 20)

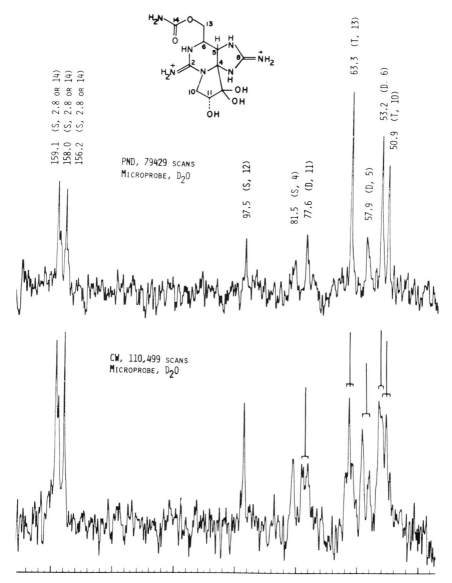

Fig. 19 Micro-cmr of GTX-II, proton decoupled and undecoupled. (Reprinted with permission from Shimizu *et al.*, *J. Am. Chem. Soc.* **98**, 5414 (1976). Copyright by the American Chemical Society.)

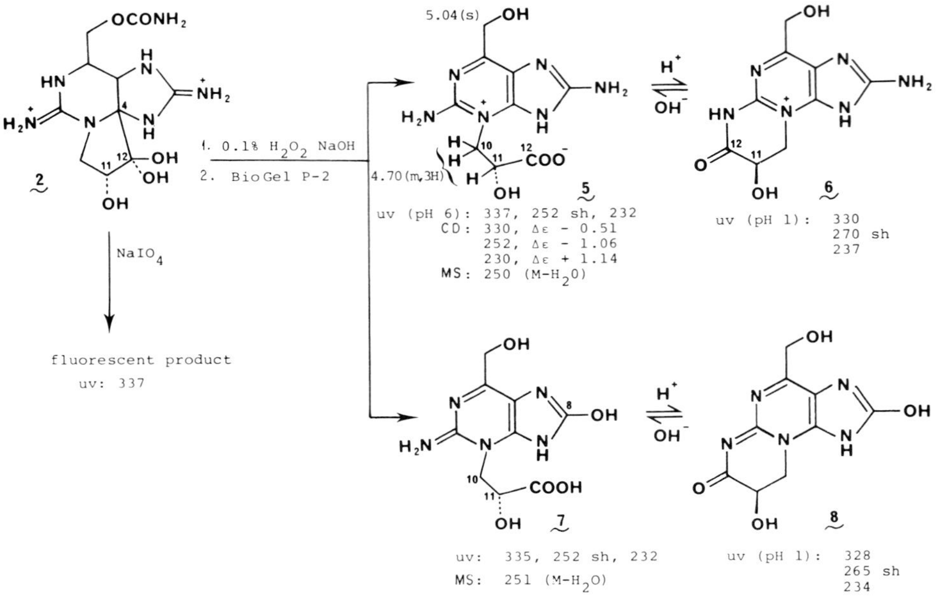

Fig. 20 Degradation of GTX-II.

having an 8-OH instead of 8-NH$_2$ is based on the uv of all eight products obtained from GTX (**5–8**, Fig. 20) and STX ($\sim$2 nm blue shift of all peaks in OH series) and some MS data. The difference in tautomeric structures between the NH$_2$ and OH series is to explain the fact that product **5** (Fig. 20) from STX (no 11-OH) had its 10-H and 11-H at 4.40 and 2.70 ppm, whereas in product **7** (Fig. 20) (no 11-OH) these signals were at the higher field of 3.39 and 2.39 ppm, respectively. Finally, GTX-II could be oxidized slowly with NaIO$_4$ to afford a fluorescent product having a uv identical to that of product **5** (Fig. 20); this is presumably a product resulting from cleavage between 11/12 followed by aromatization. The STX was resistant to NaIO$_4$.

Although the pmr signal of GTX-II was too complex to be of positive use in structural studies, once structure **2** (Fig. 22) was derived then all peaks could be correlated. The minor component is represented by the 11-epimer **3** (Fig. 22). It is known that in STX the C-12 hydrate partly exists in the keto form and that C-11 becomes deuterated upon exposure to D$_2$O [31]. Hence the 11-OH in GTX-II could readily epimerize along the sequence shown in Fig. 22. Gonyautoxin-III does not adopt the 11-keto form shown, because its pmr signal showed neither the clear quartet due to a 10-methylene nor the clear singlet due to a 12-H. Molecular models suggest the α side to be less

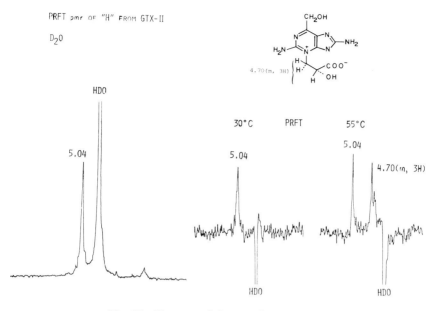

Fig. 21   The pmr of GTX-II degradation product.

Fig. 22   Equilibria of GTX-II and -III.

hindered, and this is the reason for assigning the α-OH to the major isomer **2**. Similar to STX, the GTX keto group exists in the hydrated form because it is flanked by powerful electron-attracting guanidinium groups.

ACKNOWLEDGMENTS

All the work described here was carried out in collaboration with various other laboratories, and I am most grateful for this opportunity to work in interdisciplinary areas bridging natural-products chemistry with other areas. The work in the Chemistry Department was done by Drs. V. Balogh-Nair, J. Dillon, V. P. Gullo, and H. Kasai, and nmr spectra were measured by Mr. I. Miura. The studies were supported by Grants NIH CA 11572 and AI 10187.

REFERENCES

1. H. Kasai, K. Nakanishi, R. D. Macfarlane, D. F. Torgerson, Z. Ohashi, J. A. McCloskey, H. J. Gross, and S. Nishimura, *J. Am. Chem. Soc.* **98**, 5044 (1976).
2. H. Kasai, Z. Ohashi, F. Harada, S. Nishimura, N. J. Oppenheimer, P. F. Crain, J. G. Liehr, D. L. von Minden, and J. A. McCloskey, *Biochemistry* **14**, 4198 (1975).
3. T. Oghi, T. Goto, H. Kasai, and S. Nishimura, *Tetrahedron Lett.* p. 367 (1976).
4. H. Kasai, Y. Kuchino, K. Nihei, and S. Nishimura, *Nucleic Acid Res.* **2**, 1931 (1975).
5. D. F. Torgerson, R. P. Skowronski, and R. D. Macfarlane, *Biochem. Biophys. Res. Commun.* **60**, 616 (1974); R. D. Macfarlane and D. F. Torgerson, *Science* **191**, 920 (1976).
6. Private communication from Prof. R. D. Macfarlane, Department of Chemistry, Texas A & M University.
7. A. M. Jeffrey, K. W. Jennettee, S. H. Blobstein, I. B. Weinstein, F. A. Beland, R. G. Harvey, H. Kasai, I. Miura, and K. Nakanishi, *J. Am. Chem. Soc.* **98**, 5714 1976; B. Weinstein, A. M. Jeffrey K. W. Jennette, S. H. Blobstein, R. G. Harvey, H. Kasai, and K. Nakanishi, *Science* **193**, 592 (1976).
8. Committee on the Biologic Effects of Atmospheric Pollutants, "Particulate Polycyclic Organic Matter," p. 30. *Nat. Acad. Sci.*, Washington, D.C., 1972.
9. D. M. Jerina and J. W. Daly, *Science* **185**, 573 (1974); P. Sims and P. L. Grover, *Adv. Cancer Res.* **20**, 165 (1974); P. Sims, P. L. Glover, A. Swaisland, K. Pal, and A. Hewer, *Nature (London)* **252**, 326 (1974).
10. D. G. MaCaustland and J. F. Engel, *Tetrahedron Lett.* p. 2549 (1975); H. Yagi, O. Hernandez, and D. M. Jerina, *J. Am. Chem. Soc.* **97**, 6881 (1975); F. A. Beland and R. G. Harvey, *J. Chem. Soc., Chem. Commun.* p. 84 (1976).
11. A. M. Jeffrey, S. H. Blobstein, I. B. Weinstein, F. A. Beland, R. G. Harvey, H. Kasai, and K. Nakanishi, *Proc. Natl. Acad. Sci. U.S.A.* **73**, 2311 (1976).
12. S. H. Blobstein, I. B. Weinstein, D. Grunberger, J. Weisgras, and R. G. Harvey *Biochemistry* **14**, 3451 (1975).
13. N. Harada and K. Nakanishi, *Acc. Chem. Res.* **5**, 257 (1972); D. W. Urry, *in* "Spectroscopic Approaches to Biomolecular Conformation" Chapter III., *Am. Med. Assoc.*, Chicago, Illinois, 1970.
14. Dr. A. M. Jeffrey, private communication.

14a. E. Huberman, L. Sachs, S. K. Yang, and H. V. Gelboin, *Proc. Natl. Acad. Sci. U.S.A.* **73**, 607 (1976).

14b. Recently structure **1** in Fig. 13 has been shown to represent the correct absolute configuration: K. Nakanishi, H. Kasai, H. Cho, R. G. Harvey, A. M. Jeffrey, K. W. Jennette, and I. B. Weinstein, *J. Am. Chem. Soc.* **99**, 258 (1977).

15. J. Dillon, A. Spector, and K. Nakanishi, *Nature (London)* **259**, 422 (1976).

16. A. Spector, S. Li, and J. Sigelman, *Invest. Ophthalmol.* **13**, 795 (1974); J. A. Jedziniak, J. H. Kinoshita, E. M. Yates, and G. D. Benedek, *Exp. Eye Res.* **20**, 367 (1975).

17. A. Spector, D. Roy, and J. Stauffer, *Exp. Eye Res.* **21**, 9 (1975).

18. K. Satoh, M. Bando, and A. Nakajima, *Exp. Eye Res.* **16**, 167 (1973).

19. D. E. U. Ekong, J. I. Okogun, V. U. Enyenhi, V. Balogh-Nair, K. Nakanishi, and C. Natta, *Nature (London)*, **258**, 743 (1975).

20. H. Abramson, J. F. Bertles, and D. L. Wethers, eds., "Sickle Cell Disease." Mosby, St. Louis, Missouri, 1973.

21. L. Pauling, H. A. Itano, S. J. Singer, and I. C. Wells, *Science* **110**, 543 (1949); J. A. Hunt and V. M. Ingram, *Nature (London)* **184**, 640 (1959).

22. J. V. Neel, *in* "Sickle Cell Disease" (H. Abramson, J. F. Bertles, and D. L. Wethers, eds.), p. 4. Mosby, St. Louis, Missouri, 1973.

23. E. A. Sofowora, W. A. Isaac-Sodeye, and L. O. Ogunkoya, *Lloydia* **38**, 169 (1975).

24. G. K. Honig, N. R. Farnsworth, C. Ferenc, and L. N. Vida, *Lloydia* **38**, 387 (1975).

25. A. Cerami and J. M. Manning, *Proc. Natl. Acad. Sci. U.S.A.* **68**, 1180 (1971).

26. Y. Shimizu, L. J. Buckley, M. Alam, Y. Oshima, W. E. Fallon, H. Kasai, I. Mirua, V. P. Gullo, and K. Nakanishi, *J. Am. Chem. Soc.* **98**, 5414 (1976).

27. E. J. Schantz, V. E. Ghazarossian, H. K. Schnoes, F. M. Strong, J. P. Springer, J. O. Pezzanite, and J. Clardy, *J. Am. Chem. Soc.* **97**, 1238 (1975).

28. J. Bordner, W. E. Thiessen, H. A. Bates, and H. Rapoport, *J. Am. Chem. Soc.* **97**, 6008 (1975).

29. Y. Shimizu, M. Alam, Y. Oshima, and W. E. Fallon, *Biochem. Biophys. Res. Commun.* **66**, 731 (1975). Since publication of this paper which deals with the isolation of STX and three toxins, three additional toxins have been isolated.

30. J. L. Wong, M. S. Brown, K. Matsumoto, R. Oesterlin, and H. Rapoport, *J. Am. Chem. Soc.* **93**, 4633 (1971).

31. J. L. Wong, R. Oesterlin, and H. Rapoport, *J. Am. Chem. Soc.* **93**, 7344 (1971).

Conformational Analysis of Oligopeptides by Spectral Techniques

Fred Naider and Murray Goodman

ABBREVIATIONS

The following abbreviations are used in this review.

Amine Protecting Groups

Boc	*tert*-Butoxycarbonyl
MEEA	2-Methoxy[2-ethoxy(2-ethoxy)]acetyl
Nps	*o*-Nitrophenylsulfenyl
Z	Benzyloxycarbonyl

Carboxyl Protecting Groups

OBzl	Benzyl ester
OBu^t	*tert*-Butyl ester
OEt	Ethyl ester
OMe	Methyl ester
Mo	Morpholinamide
(OPEG)$_{6000}$	A polyethylene glycol C-terminal protecting group ($M_v \sim 6000$)

Solvents

EtOH	Ethanol
HCONMe$_2$	Dimethyl formamide
F$_3$AcOH	Trifluoroacetic acid

F$_3$EtOH	Trifluoroethanol
Cl$_2$AcOH	Dichloroacetic acid
EG	Ethylene glycol
TMP	Trimethyl phosphate
HFIP	Hexafluoroisopropanol
HFA	Hexafluoroacetone sesquihydrate
DMSO	Dimethyl sulfoxide
CHCl$_3$	Chloroform
CDCl$_3$	Deuterochloroform

Spectroscopic Techniques

uv	Ultraviolet
CD	Circular dichroism
nmr	Nuclear magnetic resonance
ir	Infrared
ORD	Optical rotatory dispersion

INTRODUCTION

It is well known that both chemical and physical factors affect the activity of biologically important molecules. The conformation of a biomolecule is often a most important factor in determining activity and specificity. Investigations of model polypeptides have been extremely useful for the interpretation of conformational studies of proteins and for the elucidation of the interaction between proteins and other macromolecules. Using model polypeptides, important information has been obtained about the effects of composition, sequence, chain length, and solvation on secondary structure.

Additional information can be derived from the study of well-characterized oligopeptides. Such investigations eliminate the effects of polydispersity on the conformations of synthetic polypeptides. From the conformational analysis of a series of oligopeptides, the minimum number of residues necessary for the formation of stable secondary structures can be determined. The extensive current interest in the structure–activity relationships of short peptide hormones makes conformational investigations particularly pertinent. Since cyclic and linear peptide hormones are composed of several amino acid residues, interpretation of the complex spectra requires an increasing number of appropriate model compounds.

Over the past two decades, there has been a major effort to understand the structure of linear oligopeptides in solution. Conformational studies have shown that the solvent, temperature, and peptide concentration are critical parameters. A recent review has integrated research from several laboratories [1], and theoretical and experimental data for a number of amino acids, diamides, and oligopeptides are discussed. In this chapter, we focus attention on important developments in this field, while emphasizing work from our own laboratories. We trace the application of the various spectroscopic techniques to deduce conformations of oligopeptides and examine critically their strengths and weaknesses. Finally, we speculate on the direction of future studies.

SYNTHESIS OF OLIGOPEPTIDES

The conformational analysis of linear oligopeptides requires compounds that are both chemically and optically pure. Much of our earlier work involved the application of classical peptide coupling procedures to the preparation of glutamate-, aspartate-, and alanine-containing peptides and copeptides [2–7]. The synthetic procedure most likely to result in optically pure peptides couples a blocked peptide fragment to a urethane-blocked amino acid [Eq. (1)]. Unfortunately large peptide fragments ($n > 5, 6$) are

$$
R'-O-\overset{O}{\overset{\|}{C}}-\overset{H}{\overset{|}{N}}-\overset{H}{\overset{|}{\underset{R}{C}}}-\overset{O}{\overset{\|}{C}}-OH + H-\left[\overset{H}{\overset{|}{N}}-\overset{H}{\overset{|}{\underset{R}{C}}}-\overset{O}{\overset{\|}{C}}-OR''\right]_n \xrightarrow[\text{agent}]{\text{coupling}}
$$

$$
R'-O-\overset{O}{\overset{\|}{C}}-\left[\overset{H}{\overset{|}{N}}-\overset{H}{\overset{|}{\underset{R}{C}}}-\overset{O}{\overset{\|}{C}}\right]_{n+1}-O-R'' \quad (1)
$$

often insoluble in solvents normally employed in peptide synthesis. They also exhibit decreased nucleophilicity as compared to shorter analogs, and coupling is therefore difficult. Thus, fragment coupling (with the fragments as short as possible) has been employed for the synthesis of oligopeptides. However, the possibility of racemization (see Goodman and Glaser [8] for review) remains a major concern. It is generally accepted that oxazolinone formation during carboxyl activation or coupling results in the loss of optical integrity at the α-carbon. Scheme 1 depicts racemization via this route.

In our earlier synthesis of the glutamate oligomers, we employed *p*-nitrophenyl active ester [9] and modified azide procedures [10,11]. More recently, we have used the mixed-anhydride procedure developed by Anderson *et al.* [12]. Toniolo and his associates [13–16] have applied the Rudinger–Honzel modification [10] of the acylazide [17] to their synthesis of a variety of aliphatic homo-oligopeptides. Our original attempts using the azide procedure often resulted in low yields. Both the mixed-anhydride and the azide procedures as employed by Toniolo, however, resulted in chemically and

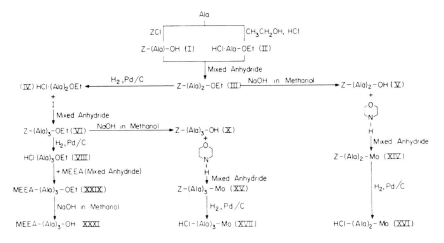

Scheme 1

chromatographically homogeneous compounds in yields of 50–95%. Minimum purification was required to obtain the monodisperse oligomers. A typical synthetic scheme is shown in Fig. 1.

Using molar rotation measurements at the sodium D line (see later sections for detailed discussion), we ascertained that little if any racemization occurs when the above coupling methods are utilized. Finally, the oligomers that we prepare are blocked on both their amine and carboxyl termini. Thus, we can explore the conformational preferences of the fully protected peptides in a wide variety of organic solvents. Changes in the blocking group can be utilized to increase solubility in helix-supporting solvents and may, in certain cases, even confer limited water solubility on the oligomers [7]. Since the

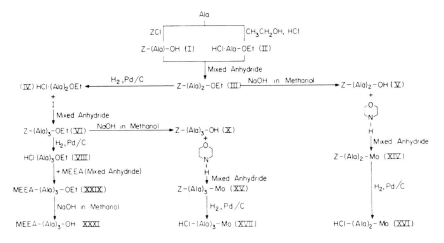

Fig. 1 Synthetic scheme for the preparation of alanine oligopeptides [7].

interior of globular proteins is relatively nonpolar, it is important to measure oligopeptide conformations in organic and organic–aqueous solvent mixtures as well as in simple aqueous solutions.

OPTICAL ACTIVITY MEASUREMENTS

Many workers have noted that the helix-to-coil transition in polypeptides is accompanied by marked changes in the optical rotation at the sodium D line [18–20]. Brand *et al.* [21] and Doty and Geiduschek [22] recognized that the optical rotation of a polypeptide is composed of a number of separable contributions. Among these contributions are the configurational optical activity of individual chiral centers [18] and the contribution arising from the presence of secondary and tertiary structure [18–20,23–26]. Using these concepts, we began a series of studies by which we attempted to predict the molar rotation of an oligopeptide from values determined using various model compounds [27–29]. Our rationale in these investigations was that the total molar rotation of a peptide could be described by Eq. (2), where $[\phi]_M$ is

$$[\phi]_M = [\phi]_{\text{configurational}} + [\phi]_{\text{conformational}} \qquad (2)$$

the molar rotation of the peptide, composed of both configurational and conformational contributions. The end residues of a peptide are chemically distinct from internal residues and solvate differently. It is therefore possible to expand Eq. (2) to Eq. (3), where $[\phi]_{\text{end}}$ is the contribution from the N- and

$$[\phi]_M = [\phi]_{\text{end}} + n'[\phi]_{\text{int}} + [\phi]_{\text{conformational}} \qquad (3)$$

C-terminal residues, $[\phi]_{\text{int}}$ represents optical activity from internal residues, and n' is the number of internal residues. Equations similar to Eq. (3), which include possible interactions between adjacent residues, have also been developed. The form of all of these equations should be linear for plots of $[\phi]_M$ vs. n' in the absence of secondary structure. The model compounds (**1**)–(**3**) were synthesized. Compounds (**1**) and (**2**) represent terminal and

OMe		OEt		OMe

Z—Glu—Gly—Gly—OMe   Z—Gly—Gly—Glu—Oet   Z—Gly—Glu—Gly—OMe

(**1**)   (**2**)   (**3**)

compound (**3**) internal residues in a homooligopeptide series. From optical rotations of these model compounds we are able to predict the molar rotation for the following series in solvents such as Cl_2AcOH, F_3AcOH, dioxane, and $CHCl_3$:

OMe ⌈ OMe ⌉ OEt
    | | | |
Z—Glu—⟨Glu⟩— Glu—OEt
          ⌊   ⌋n'

In the acidic solvents, experimental data give straight lines for $[\phi]_M$ vs. n' since no secondary structures exist in these solvents. Agreement with predicted values is close, despite the fact that our models are simplified and do not completely account for interresidue interactions. In contrast, marked deviations from linearity were observed when similar studies were conducted in dioxane, $CHCl_3$, or m-cresol. These deviations usually begin at a chain length of five or higher and can be attributed to the onset of some secondary structure.

Since the early study of the glutamates, we have applied this approach to oligopeptides containing aspartic acid [28], alanine [29,30], and methionine residues [31]. Cooligomers of alanine with γ-methyl-L-glutamic acid and of aspartic acid with glutamic acid have also been studied [29,32]. Toniolo and Bonora [13–15] have extended this approach to peptides containing isoleucine, valine, phenylalanine, and leucine residues. We believe that optical rotation at the sodium D line remains a good initial measurement for a conformational analysis of a new oligopeptide series, although these measurements must be substantiated by other, more sensitive methods (see later sections).

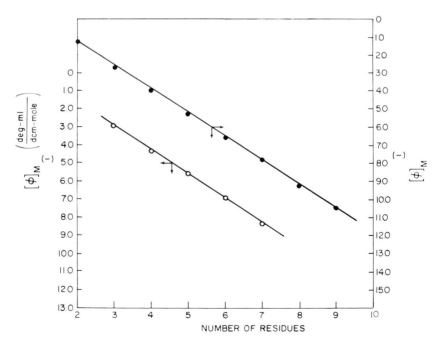

Fig. 2   Molar rotations in F_3AcOH of the MEEA(Ala)$_n$-Mo(○) and the Z-Ala$_n$-Mo(●) oligopeptide series [7].

Optical purity in short peptides is a prerequisite for the formation of stable secondary structures. In a solvent that does not support secondary structure, chromatographically and chemically pure peptides exhibit linear plots of $[\phi]_M$ vs. n' (Fig. 2) (as noted above), affirming the absence of racemization during peptide synthesis. In addition, deviations from linearity in solvents that permit stable secondary conformations give preliminary evidence for the critical number of residues necessary for helical or associated forms. In this manner we observed that γ-methyl glutamate oligopeptides begin to form secondary structures at the nonamer in F_3EtOH and $HCONMe_2$. These observations have since been confirmed and refined using CD and nmr investigations. It is not possible to distinguish between helical and associated structures by molar rotation studies; nevertheless, this method provides an important measurement for the investigation of the conformation of peptides in solution.

ULTRAVIOLET SPECTROSCOPY

The far-ultraviolet absorption spectrum of polypeptides was shown to depend on secondary structure [33,34]. Hypochromism of the $\pi \rightarrow \pi^*$ transition of the amide chromophore as compared to the spectra of simple amides and the appearance of a shoulder at 205 nm provide evidence for the existence of polypeptide helices [35]. We found that the onset of helical structure in a homologous series of oligopeptides can be studied by measuring spectral changes in the 185–230 nm range. For spectra of the γ-methyl L-glutamate oligopeptides obtained in F_3EtOH, the molar extinction coefficient at 190 nm begins to decrease at the nonapeptide [36]. By the tridecapeptide, the value of the molar extinction coefficient is quite similar to that measured for the corresponding high polymer. In contrast, analogous measurements on β-methyl aspartate oligopeptides show no evidence of hypochromicity until the undecamer, and a large difference remains between the molar extinction coefficient of the tetradecamer and that of the high polymer (Fig. 3). These conclusions are consistent with information obtained from optical rotation studies.

Although the hypochromic effect can be useful in assigning the nature of the conformations of these peptides in solution, we believe that assignments made using CD are much more definitive. The disadvantage of the uv technique is that assignments must often be based on measurements at 190 nm. This limits the number of solvents that can be utilized and also restricts the nature of the protecting groups that can be employed. We have spent considerable time attempting to correct for contributions to the molar extinction coefficient from the benzyloxycarbonyl protecting group [36,37]. These corrections

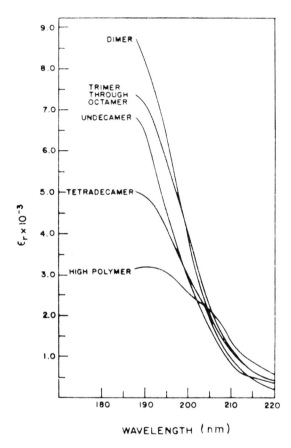

Fig. 3   Molar extinction coefficients per residue as a function of wavelength for peptides derived from β-methyl L-aspartate in F_3EtOH [36].

are not completely satisfactory and are not necessary for the CD investigations.

Toniolo has employed uv studies extensively in the study of oligopeptides which contain aliphatic residues and which tend to assume β structures in solution. On the basis of hyperchromism of the $\pi \rightarrow \pi^*$ transition of the peptide chromophore and a 4–5 nm red shift of the λ_{max}, he concluded that Boc-Ile$_n$-OMe oligomers begin to be associated at the heptamer [13]. As discussed in the next section, similar assignments can be made using CD. We feel that the major benefit of uv investigation is that they can be utilized to investigate the secondary structure of racemic peptides when CD and ORD cannot be used.

CIRCULAR DICHROISM INVESTIGATIONS

Circular dichroism is a powerful technique for examining the secondary structure of oligopeptides. There are marked differences in the shape, intensity of the extrema, and λ_{max} for Cotton effects associated with disordered, helical, and β conformations. Although care must often be exercised in interpreting an individual CD spectrum, analysis of a series of oligopeptides is simplified by the fact that lower oligopeptides serve as ideal model compounds to compare with higher members of a series. We initially applied CD to investigate the conformations of γ-ethyl glutamate oligopeptides [38,39].

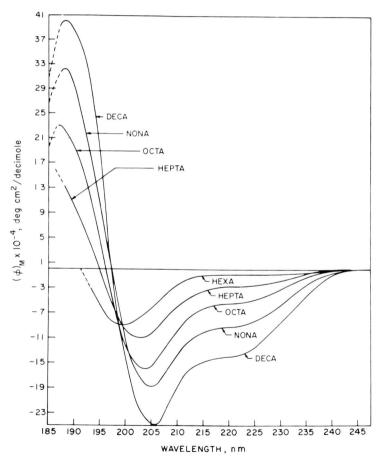

Fig. 4 Circular dichroism spectra of benzyloxycarbonyl γ-ethyl L-glutamate oligomers in 2,2,2-trifluoroethanol [39].

As seen in Fig. 4, major changes occur in the spectral patterns as the oligo-peptide chain length is increased. The spectra of the lower peptides are composed of a broad weak band at about 235 nm, a weak positive band centered at 218 nm, and a relatively strong negative band near 190 nm. The higher peptides exhibit negative intensity between 220 and 225 nm, a trough near 205 nm, and a positive band near 190 nm. The latter features are quite similar to the predicted and observed CD patterns for polypeptides in α-helical conformations [40–44]. Careful analysis of changes in the λ_{max} of the peak centered near 190 nm in the lower peptides (which splits into positive and negative contributions in the higher peptides) and of the intensity of the 222 nm shoulder allows us to conclude that helical structures begin forming at the heptamer in F_3EtOH and TMP.

Recently, Caspers *et al.* [45] reported similar findings for a series of γ-methyl L-glutamate oligopeptides. In an interesting extension of the studies in organic solution they also present evidence that the glutamate oligopeptides undergo a conformational change between the hexamer and octamer at an air–water interface. Relying primarily on surface pressure measurements, they suggest that individual molecules are helical on the surface.

The unequivocal assignment of the structure of the glutamate oligomers by CD indicates that it is a superior tool as compared to molar rotation and ultraviolet measurements. The higher sensitivity of CD is confirmed by the detection of helical species at the heptamer, whereas the formation of secondary structures was judged to occur at the nonapeptide using optical rotation or hypochromicity measurements.

In addition to the glutamate oligomers, we have also carried out an extensive CD investigation of the conformations of four different series of alanine oligopeptides [30]. We found that the major tendency of alanine oligopeptides in F_3EtOH is to change from random to associated forms (β) as the chain length increases (Fig. 5). Thus, as the chain length of the Z-Ala$_n$-Mo oligopeptides increases, an abrupt change in spectral pattern occurs at the octamer and nonamer. This change, which results in a broad trough centered near 215 nm and a strong positive band at 190 nm, is typical of CD spectra reported for β conformations in solution [46]. The tendency to aggregate is largely a function of the low solubility of the alanine peptides. Even at a concentration of 1 mg/ml, a solution of Z-Ala$_n$-Mo is nearly saturated, a situation that may lead to secondary structures typical of those assumed in the solid state. Since most oligopeptides exist in the β conforma-tion in the solid state, it is not surprising that oligomers nearly at their limiting solubility tend to form β structures. We found that dilution of certain alanine oligopeptide solutions disrupts the β structures and results in spectral patterns that resemble those of the α-helical glutamates (Fig. 6), although both the $n \rightarrow \pi^*$ and $\pi \rightarrow \pi^*$ peaks are significantly shifted toward the blue.

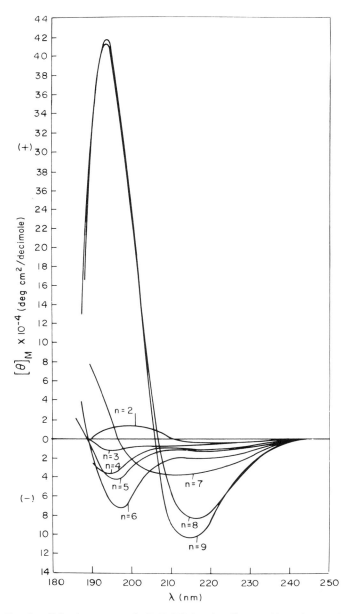

Fig. 5 Circular dichroism spectra in F_3EtOH for the oligopeptide series Z-Ala$_n$-Mo [7].

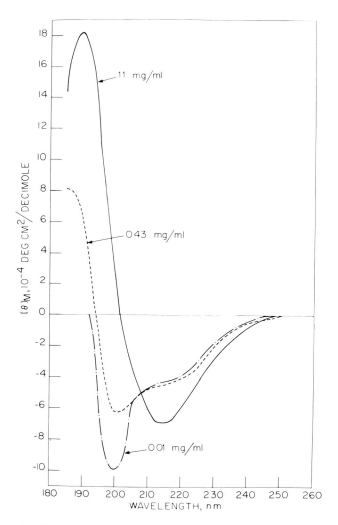

Fig. 6 Circular dichroism spectra in F_3EtOH for MEEA-$(Ala)_9$-Mo as a function of concentration [7].

Most importantly we observed that in solvent mixtures which disrupt intermolecular but not intramolecular hydrogen bonds (99% F_3EtOH/1% H_2SO_4) the alanine oligopeptides begin to form helices at the hexamer or heptamer stage.

Our CD studies show that both the solvent employed and the oligopeptide concentration strongly influence conformational preferences. In examining any new series of oligopeptides (especially those with limited solubility in helix-supporting solvents) it is premature to assign a definite preference for β structure before detailed concentration studies are carried out. Furthermore, since helices may begin to form at the hexamer or heptamer and since the spectral patterns for these compounds may be intrinscially different from previously studied oligopeptides, it is essential to have higher members of an oligomer series. For example, in an analysis of the Boc-Met$_n$-OMe series, it would have been impossible to make definitive conformational assignments without reference to the corresponding nonapeptide [47].

Careful comparison of the alanine oligopeptides in the four different series also gave indications that the protecting groups have an important influence on the overall conformation assumed by the oligomers. Furthermore, using oligomers from the MEEA-Ala$_n$-Mo series, we gained insights into the structure of alanine oligopeptides in aqueous solution. In particular, we were able to infer from the nature of the equilibrium CD patterns of MEEA-Ala$_8$-Mo in F$_3$EtOH/H$_2$O (40:60%) that protected decaalanine would assume a β conformation in water. Both the apparent influence of the protecting group on the conformation of an oligopeptide and the existence of decaalanine in a β conformation in water were recently confirmed by Bayer, Mutter, and their co-workers [48]. In their study they employed a polyethylene glycol protecting group ($M_v \sim 6000$) on the carboxyl terminus and demonstrated clearly that the removal of the Boc group from Boc-Ala$_{10}$-(OPEG)$_{6000}$ causes the decamer to change from a helical to a β conformation in F$_3$EtOH (Fig. 7).

Circular dichroism has been valuable in studies of peptides that exhibit a major preference for the formation of β structures. We observed that both the heptamer and octamer of the Boc-Ile$_n$-OMe series exist as β structures in F$_3$EtOH [49]. Toniolo [14,16,50–52] has carried out CD investigations on peptides containing valine, leucine, phenylalanine, and methionine residues [16]. He reports that most of these have a tendency to assume β-like conformations in solution and that this tendency is often increased by the addition of water. Furthermore, he has given preliminary assessments of the stability of various β structures.

It is widely accepted that water–polymer interactions play a major role in the determination of protein structure. It is tempting to conclude that, when water is added to solutions of aliphatic peptides in organic solvents, the peptide backbone attempts to minimize hydrocarbon–water contacts, which forces the peptide chain to assume a conformation similar to that in the interior of a globular protein. One could thus conclude that water actually encourages the formation of β conformations in solution. Since most aliphatic homooligopeptides are extremely insoluble in water, it is more likely that the addition of water tends to precipitate the peptides from solution. This condition is therefore more representative of the situation in the solid state and may not have much in common with the interior of a globular protein in aqueous solutions.

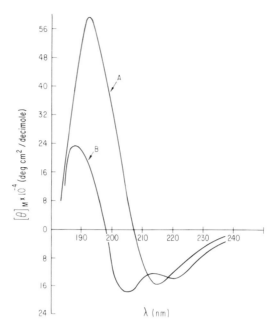

Fig. 7   Circular dichroism spectra in F_3EtOH for (A) the alanine-$(OPEG)_{6000}$ decamer and (B) the *tert*-butoxycarbonylalanine-$(OPEG_{6000})$ decamer [48].

In addition to the CD studies on the above series of homooligopeptides, Lorenzi and co-workers [53,54] have recently begun a systematic investigation of the interaction of the aromatic side chains of amino acids such as tyrosine, tryptophan, and phenylalanine. In particular, they are interested in the dependence of the spectroscopic properties of these residues on the distance that separates them in a peptide chain. This research is still preliminary. The uv and CD spectra of these oligomers are complex and difficult to interpret. Nevertheless, the CD patterns in both the near and far uv depend on the exact sequence of the aromatic residues in model peptides. For tryptophan oligomers, Lorenzi [53,54] believes that a negative band centered at 225 nm indicates some degree of rigidity in the sequence R—CO—Trp—Trp—. Further investigations are necessary before the conclusions from these most interesting studies can be fully understood.

NUCLEAR MAGNETIC RESONANCE

Despite the successful application of molar rotation, uv and CD to the conformational analysis of homooligopeptides, these techniques all have serious limitations. Since these techniques measure the effect of the overall

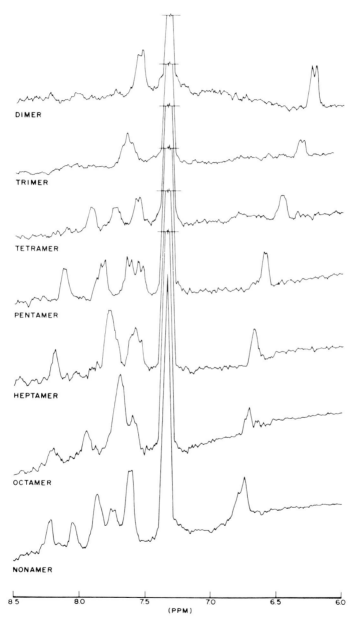

Fig. 8 Partial 220 MHz nmr spectra of the benzyloxycarbonyl γ-ethyl L-glutamate oligomers as 2% solutions in F₃EtOH at 23°C [55].

environment on the peptide chromophore, they can at best give information pertaining to the average conformational state of a molecule. It is impossible, therefore, to obtain any detailed conformational information about individual residues in a peptide chain. Nuclear magnetic resonance techniques can provide such information. In principle, this approach can distinguish between magnetically different atoms whose nuclear spin does not equal zero. The advent of spectrometers with superconducting magnets with pulse Fourier transform attachments permits the analysis of oligopeptides at concentrations similar to those used in uv, CD, and ir investigations. It is therefore possible to gain complementary conformational information from nmr studies and to investigate the environments of individual residues.

Our early successful application of the nmr technique involved γ-ethyl L-glutamate oligopeptides in F_3EtOH [38,39]. Using a 220 MHz spectrometer, individual NH's can be resolved up to the hexamer (Fig. 8). In the higher oligomers of this series, overlap of the NH peaks begins to occur. Careful analysis of model compounds permitted us to assign the N-terminal NH resonance. The change in position of this resonance and the lowest-field NH resonance with chain length fitted S-shaped curves, which were used to determine the onset of secondary structure. The results of this study are consistent with conclusions from the CD investigation of the same compounds.

We continued our nmr studies on the glutamate oligopeptides in $CDCl_3/ F_3AcOH$ mixtures [55]. The data in Fig. 9 show that the variation of τ values of the NH resonances of a glutamate heptamer depends on their exact position in the peptide. The lowest-field NH passes through a minimum value of chemical shift before additional F_3AcOH causes it to shift upfield and merge with other resonances. Similar variations for certain NH resonances are observed in a glutamate tetramer and pentamer. In contrast, we do not find such variations in the analogous dipeptide nor in an aspartate octamer (Fig. 9). The behavior of the lowest-field NH's in the higher glutamate peptides is consistent with the existence of intramolecular hydrogen bonds in these oligopeptides. Since the tetramer is the smallest oligomer that exhibits this characteristic dependence, we believe that the nmr technique has revealed the presence of folded forms in a tetrapeptide.

We have recently applied the nmr technique to the investigation of alanine-[56,57] and methionine-containing peptides [58]. In addition to gaining insights into the "double-peak" phenomenon [56] in polypeptides, we have observed strong indications of a folded conformation for a tetrapeptide derivative of alanine in $CDCl_3$ [57]. By CD, we believed that alanine tetra-, penta-, and hexapeptides were disordered [30]. By nmr, we see definite structures for these peptides. The nmr techniques therefore provide us with a more sensitive procedure to assess the conformational preferences of short peptides in solution.

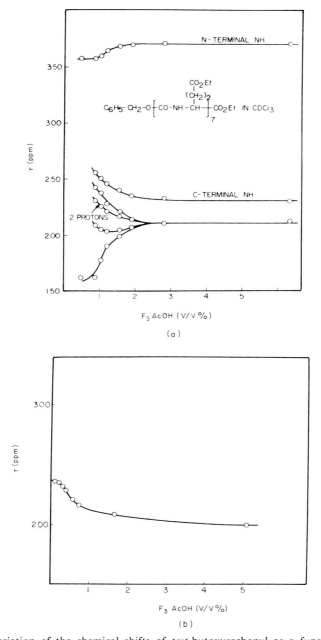

Fig. 9 Variation of the chemical shifts of *tert*-butoxycarbonyl as a function of trifluoroacetic acid (F₃AcOH) concentration in a CDCl₃/F₃AcOH acid solution of (a) a
glutamate heptamer Z—Glu)₇-OEt and (b) an aspartate octamer Z—[Asp]₇-Asp-Oet [55].

Of the 20 naturally occurring amino acids, proline plays a unique role in many biomolecules. It is frequently found in the β bends of globular proteins. It is a major component of fibrous proteins such as collagen and is an essential amino acid in many hormones. The understanding of the effect of proline residues on the conformation of linear peptides is therefore of particular significance. The nmr technique was applied by Deber *et al.* [59] to study *cis–trans* isomerization in linear peptides containing proline. They observed that the α-hydrogen atoms of proline residues give rise to separate resonances for *cis* and *trans* peptide bonds. This finding was used to quantitate the relative amounts of *cis* and *trans* isomers that were present. In the Boc-Pro$_n$-OBzl ($n = 2$–6) series, they observed that various mixtures of *cis* and *trans* peptide bonds existed for $n \leq 4$. At the pentamer the peptide assumed the *all-trans* conformation of polyproline II. Later studies using ^{13}C nmr confirmed many of the earlier results and extended them to cyclic peptides containing proline (see Deber *et al.* [60] for a concise and critical review of this work). The information evolving from studies of this kind is of special importance since it can be applied to the study of hormones containing proline residues.

An interesting and important development is the use of various probes (e.g., 3-oxyl-2,2,4,4-tetramethyloxazolidine) to detect intramolecular hydrogen bonds. Kopple and co-workers [61–63] have applied this procedure to the investigation of a number of peptides believed to form β bends in solution. Differential broadening of the NH resonances is attributable to the presence or absence of peptide NH's which are exposed or shielded from the radical. If found to be generally applicable, this technique should be extremely useful since it will permit us to pinpoint the involvement of individual residues in specific H-bonded structures.

STUDIES OF OLIGOPEPTIDES CONTAINING
MORE THAN ONE AMINO ACID RESIDUE

Most studies emanating from our laboratories have dealt with homooligo-peptides. Few detailed systematic investigations have appeared which attempt to examine synthetic linear oligopeptides containing different amino acids. The spectroscopic basis for the conformational analysis of oligopeptides is now well established, and a natural extension of our work would be to study more complex oligomers.

The most serious attempts to investigate oligopeptides containing different amino acids are those dealing with collagen model compounds. In these studies oligomers with the sequence X-Gly-Pro and X-Pro-Gly were synthesized and examined in both the solid state and solution [64–85]. Many of these investigations employed either solution polymerization or the Merrifield procedure to prepare the oligomers. There is therefore some question as to the homogeneity of these preparations. As has been pointed out recently by Bonora and originally stressed by Blout [77] polydisperity can affect the conformation of collagen models in solution, and it is highly desirable to study monodisperse oligomers. We believe that classic synthetic techniques still afford the best route to the preparation of such sequence-controlled oligopeptides.

One of the earliest studies was reported by Blout and co-workers [77] on oligomers of the sequence Boc-(Ala-Pro-Gly)$_n$-OMe $(n = 1-6)$. They observed that oligomers $(n = 3-6)$ are capable of existing in three independent conformations in the solid state. The exact conformation that an oligomer assumed was related to the method of sample preparation. Different conformations are found when oligomers are dried from aqueous solution, precipitated from somewhat hydrophobic solvents, or dried from F$_3$EtOH. Molecular conformation is, of course, dependent on chain length. Neither the tripeptide $(n = 1)$ nor the hexapeptide $(n = 2)$ assumes the regular solid-state structures. In aqueous solution, F$_3$EtOH, or ethylene glycol/hexafluoro-isopropanol, the CD spectra of all oligomers are consistent with a disordered structure.

In a later study, Engel and co-workers [85] presented results on Z-(Gly-Pro-Pro)$_n$-OBu^t $(n = 3-7)$ oligopeptides. Using both molecular weight determinations and CD spectroscopy, they found that oligopeptides with $n > 3$ assume a collagenlike triple helix in methanol but are in the "random coil" conformation in dilute acetic acid. The number of tripeptide units necessary for the formation of the triple-helical conformation both in solution and in the solid state can be ascertained from such investigations. By carefully studying the relative stabilities of the secondary structures formed, subtle information can also be gained concerning the effect of the positions of given amino acid residues in a sequence on the conformation of these collagenlike model compounds.

Recently, Kataki and Nakayama [86,87] reported the synthesis of several series of sequential oligopeptides having the structure Nps-[L-Glu(OMe)-L-Phe]$_n$-OEt $(n = 1-4)$ and Nps-(L-Ala-L-Leu-Gly)$_n$-OEt $(n = 1-6)$. Few details of the physicochemical characterizations of these compounds are available at present. Further studies on these oligopeptide series should enhance our understanding of factors influencing secondary structure of peptides in solution and in the solid state.

CONCLUSIONS AND PLANS FOR FUTURE STUDIES

Table 1 summarizes information from studies on the conformation of linear homooligopeptides in solution. The table contains details of critical chain lengths for a number of different oligopeptide series under specific conditions. In the few cases where comparison can be made, it appears that CD and nmr are the most sensitive techniques for determining the onset of

TABLE 1
Data on the Conformation of Linear Homooligopeptides in Solution

Oligomer series	Solvent	Preferred conformation[a]	Critical size	Technique	Reference
Z-Ala$_n$-OEt	F$_3$AcOH	Disordered	—	Optical rotation	29
	Cl$_2$AcOH	Ordered	10	Optical rotation	29
	F$_3$EtOH	β	6–7	CD	30
Z-Ala$_n$-NC$_2$H$_4$OC$_2$H$_4$	HFIP	Disordered	—	CD	30
	F$_3$EtOH	β	7–8	CD	30
	F$_3$EtOH–1% H$_2$SO$_4$	α-Helical	6–7	CD	30
MeO(C$_2$H$_4$O)$_2$Ac(Ala)$_n$-NC$_2$H$_4$OC$_2$H$_4$	HFIP	Disordered	—	CD	30
	F$_3$EtOH	β[b]	9	CD	30
	F$_3$EtOH–1% H$_2$SO$_4$	α-Helical	6–7	CD	30
Boc-Ile$_n$-OMe	HFA	Disordered	—	Optical rotation	14
	F$_3$EtOH	β	7	Optical rotation	14
				uv, CD	14, 49
Boc-Leu$_n$-OMe	F$_3$EtOH	α-Helical	7	CD	50
	20% F$_3$EtOH–80% H$_2$O	β	5–6	CD	50

Boc-Met$_n$-OMe	HFIP	Disordered	—	Optical rotation, CD	17
	HFA	Disordered	—	Optical rotation, CD	17, 31, 47
	F$_3$EtOH	Helical	7	CD	17, 47
	EG	β	5	CD	17
	20% F$_3$EtOH–80% H$_2$O	β	5	CD	17
Boc-Phe$_n$-OMe	F$_3$EtOH	β^c	6	uv	16, 52
Boc-Val$_n$-OMe	HFA	Disordered	—	Optical rotation	15
	HFIP	Disordered	—	CD	15, 51
	F$_3$EtOH	β	7	Optical rotation, uv, CD	15, 51
$$Z\!-\!\left[\begin{smallmatrix}\;\;\;\;\text{OMe}\;\;\;\text{OEt}\\\;\;\;\;\mid\;\;\;\;\;\mid\\\text{Glu}\text{---}\text{Glu}\text{---}\text{OEt}\end{smallmatrix}\right]_n$$	Cl$_2$AcOH	Disordered	—	Optical rotation	27
	F$_3$EtOH	α-Helical	9	Optical rotation, uv	36
$$Z\!-\!\left[\begin{smallmatrix}\;\;\;\;\text{OEt}\;\;\;\text{OEt}\\\;\;\;\;\mid\;\;\;\;\;\mid\\\text{Glu}\text{---}\text{Glu}\text{---}\text{OEt}\end{smallmatrix}\right]_n$$	F$_3$EtOH	α-Helical	7	CD, nmr	38
	TMP	α-Helical	7	CD, nmr	38
	DMSO	Disordered	—	nmr	38
$$Z\!-\!\left[\begin{smallmatrix}\;\;\;\;\text{OMe}\;\;\;\text{OEt}\\\;\;\;\;\mid\;\;\;\;\;\mid\\\text{Asp}\end{smallmatrix}\right]_n\!\!-\text{Asp}\text{---}\text{OEt}$$	Cl$_2$AcOH	Disordered	—	Optical rotation	28
	F$_3$EtOH	Helical	14	uv	36

^{a} Denotes conformation of the ordered form. The concentration of the oligomer affects the critical chain length of the β structure.

^{b} There is evidence suggesting that the heptamer and octamer in this series have some helical character.

^{c} This assignment is tentative at present.

197

secondary structure in solution. It must be noted, however, that at present it is impossible to determine oligopeptide conformations using nmr studies alone. Although the NH and α-CH regions of the oligopeptides are well resolved and contain valuable information, there is not as yet a clear relationship between the nature of these resonances and the conformation of the peptide in solution. One must rely on CD or some other procedure to assign the overall conformation.

An interesting, novel approach to the study of oligopeptide conformation in solution was recently reported by Katchalski-Katzir, Steinberg, and their associates [88]. In their approach they attach a donor and an acceptor of electronic excitation to the ends of oligopeptide chains. Energy transfer according to the Förster mechanism provides a basis for end-to-end distances for the oligopeptides in solution. After information is collected by other techniques, nmr can be used to ascertain further details concerning the environment of individual residues.

Until recently, most nmr studies were carried out at an oligomer concentration of at least 1% (w/v), whereas typical CD concentrations are 0.1% (w/v) and lower. Since concentration has a major influence on conformation, comparisons of preferred structure are tenuous at best (see Fig. 6). With the advent of high-field spectrometers and Fourier transform techniques, we have been able to carry out nmr studies at concentrations below 0.1% (w/v). Such studies should permit us to make direct comparison with CD investigations if identical solvents can be used.

The major problem that remains in our studies on linear homooligopeptides is the determination of the local conformation of individual residues. This can best be achieved by assigning individual NH resonances to a specific residue in the oligopeptide. Coupling constants, temperature shifts, and contact shifts of hydrogen-bonding free radicals can then be used to gain information about ϕ, ψ angles and the nature of the hydrogen bonds. Although the proton magnetic resonances of the glutamate and methionine oligomers are well resolved up to the hexamer, we have not been able to assign specific NH resonances to specific residues except for those adjacent to the end groups. The problem can be resolved by synthesizing oligopeptides specifically α-deuterated at certain residues or perhaps by the judicious replacement of a given "host" amino acid by other amino acid residues. We are currently exploring the feasibility of this procedure using methionine oligopeptides:

Boc-Met-Met-Met-Met-Met-Met-OMe ("host peptide")
Boc-Gly-Met-Met-Met-Met-Met-OMe
Boc-Met-Gly-Met-Met-Met-Met-OMe
Boc-Met-Met-Gly-Met-Met-Met-OMe
Boc-Met-Met-Met-Gly-Met-Met-OMe
Boc-Met-Met-Met-Met-Gly-Met-OMe
Boc-Met-Met-Met-Met-Met-Gly-OMe

By substituting the host methionine residue with glycine in six different hexamers, we hope to be able to assign each of the NH resonances in the host compound. This procedure assumes that such replacement will not grossly perturb the conformation of the hexapeptide.

We also intend to synthesize oligopeptides that are multiply labeled with ^{13}C and ^{15}N. These compounds will contain extremely sensitive probes which can yield information about relaxation times and thus specify molecular mobility in various parts of an oligopeptide molecule. It is our belief that this multifaceted approach of (a) α-deuteration, (b) amino acid replacement, and (c) multiple labeling will allow us to learn a great deal about the overall molecular shape of linear peptides in solution. Future studies will be directed toward more complex oligopeptides containing two or perhaps three different amino acid residues in a given sequence. Such model systems should give information concerning the interaction of a residue in a helical section with a residue either in another helical section or in a "β" section. Hopefully, additional information from such investigations will increase our knowledge of the sequence factors that influence the conformations of short sections of proteins and peptide hormones.

ACKNOWLEDGMENTS

This review was supported in part by a grant from the National Institutes of Health, GM 22086 (F. Naider). M. Goodman wishes to acknowledge continuing support by grants from the National Institutes of Health (GM 18694) and from the National Science Foundation (MPS 74-21422). Fred Naider is a Research Career Development awardee of the Public Health Service (GM 00025). We also wish to acknowledge the help of Dr. Michael Verlander, who critically reviewed the manuscript.

REFERENCES

1. R. T. Ingwall and M. Goodman, in "International Review of Science, Amino Acids Peptides and Related Compounds" (H. N. Rydon, ed.), *Org. Chem. Ser.* 2, Vol. 6, pp. 153–182. Butterworth, London, 1976.
2. M. Goodman, E. E. Schmitt, and D. A. Yphantis, *J. Am. Chem. Soc.* **84**, 1283–1288 (1962).
3. M. Goodman, I. G. Rosen, and M. Safdy, *Biopolymers* **2**, 503–517 (1964).
4. M. Goodman and I. G. Rosen, *Biopolymers* **2**, 519–536 (1964).
5. M. Goodman and M. Langsam, *Biopolymers* **4**, 275–303 (1966).
6. M. Goodman and F. Boardman, *J. Am. Chem. Soc.* **85**, 2483–2490 (1963).
7. M. Goodman, R. Rupp, and F. Naider, *Bioorg. Chem.* **1**, 294–309 (1971).
8. M. Goodman and C. Glaser, *Pept.: Chem. Biochem., Proc. Am. Pept. Symp. 1st 1968*, p. 267 (1970).
9. B. Iselin, W. Rittel, P. Sieber, and R. Schwyzer, *Helv. Chim. Acta* **40**, 373–387 (1957).

10. J. Rudinger and J. Honzel, *Collect. Czech. Chem. Commun.* **26**, 2333–2334 (1961).
11. R. Schwyzer, *Angew. Chem.* **71**, 742 (1959).
12. G. W. Anderson, J. E. Zimmerman, and F. M. Callahan, *J. Am. Chem. Soc.* **89**, 5012–5017 (1967).
13. C. Toniolo, *Biopolymers* **10**, 1707–1717 (1971).
14. C. Toniolo, G. M. Bonora, and A. Fontana, *Int. J. Pept. Protein Res.* **6**, 371–380 (1974).
15. C. Toniolo and G. M. Bonora, *Bioorg. Chem.* **3**, 114–124 (1974).
16. G. M. Bonora and C. Toniolo, *Bipolymers* **13**, 2179–2190 (1974).
17. T. Curtius, *Ber. Dtsch. Chem. Ges.* **35**, 3226–3228 (1902).
18. C. Cohen, *Nature (London)* **175**, 129–130 (1955).
19. A. R. Downie, A. Elliot, W. E. Hanby, and B. R. Malcolm, *Proc. R. Soc. London, Ser. A* **242**, 325–340 (1957).
20. P. Doty and R. D. Lundberg, *Proc. Natl. Acad. Sci. U.S.A.* **43**, 213–222 (1957).
21. E. Brand, B. F. Erlanger, and H. Sachs, *J. Am. Chem. Soc.* **73**, 3508–3510 (1951).
22. P. Doty and E. P. Geiduschek *in* "The Proteins" (H. Neurath and K. Bailey, eds.), Vol. 1, Part A, p. 393. Academic Press, New York, 1953.
23. E. R. Blout and C. Djerassi, "Optical Rotatory Dispersion," Chapter 17. McGraw Hill, New York, 1960.
24. R. B. Simpson and W. Kauzmann, *J. Am. Chem. Soc.* **75**, 5139–5152 (1953).
25. I. Tinoco, *J. Am. Chem. Soc.* **81**, 1540–1544 (1959).
26. S. J. Leach, *Rev. Pure Appl. Chem.* **9**, 33–85 (1959).
27. M. Goodman, I. Listowsky, and E. E. Schmitt, *J. Am. Chem. Soc.* **84**, 1296–1303 (1962).
28. M. Goodman, F. Boardman, and I. Listowsky, *J. Am. Chem. Soc.* **85**, 2491–2497 (1963).
29. M. Goodman, M. Langsma, and I. G. Rosen, *Biopolymers* **4**, 305–319 (1966).
30. M. Goodman, F. Naider, and R. Rupp, *Bioorg. Chem.* **1**, 310–328 (1971).
31. F. Naider and J. M. Becker, *Biopolymers* **13**, 1011–1022 (1974).
32. M. Goodman and I. G. Rosen, *Biopolymers* **2**, 537–559 (1964).
33. I. Tinoco, A. Halpern, and W. T. Simpson, *Polyamino Acids, Polypeptides, Proteins, Proc. Int. Symp., 1st 1961* p. 147 (1962).
34. P. Doty and W. B. Gratzer, *Polyamino Acids, Polypeptides, Proteins, Proc. Int. Symp., 1st 1961* p. 111 (1962).
35. W. B. Gratzer, *in* "Poly-α-Amino Acids" (G. D. Fasman, ed.), p. 177. Dekker, New York, 1967.
36. M. Goodman, I. Listowsky, Y. Masuda, and F. Boardman, *Biopolymers* **1**, 33–42 (1963).
37. M. Goodman and I. Listowsky, *J. Am. Chem. Soc.* **84**, 3770–3771 (1962).
38. M. Goodman, A. S. Verdini, C. Toniolo, W. D. Phillips, and F. A. Bovey, *Proc. Natl. Acad. Sci. U.S.A.* **64**, 444–450 (1969).
39. M. Goodman, C. Toniolo, and A. S. Verdini, *in* "Peptides 1969" (E. Scoffone, ed.), pp. 207–221. North-Holland Publ., Amsterdam, 1971.
40. G. Holzwarth and P. Doty, *J. Am. Chem. Soc.* **87**, 218–228 (1965).
41. S. Beychok, *in* "Poly-α-Amino Acids" (G. D. Fasman, ed.), p. 293. Dekker, New York, 1967.
42. W. Moffit, *J. Chem. Phys.* **25**, 467–478 (1956).
43. I. Tinoco, R. W. Woody, and D. F. Bradley, *J. Chem. Phys.* **38**, 1317–1325 (1963).
44. J. A. Schellman and P. Oriel, *J. Chem. Phys.* **37**, 2114–2124 (1962).
45. J. Caspers, W. Hecq, and A. Loffet, *Biopolymers* **14**, 2263–2279 (1975).

46. F. Quadrifaglio and D. W. Urry, *J. Am. Chem. Soc.* **90**, 2760–2765 (1968).
47. J. M. Becker and F. Naider, *Biopolymers* **13**, 1747–1750 (1974).
48. M. Mutter, H. Mutter, R. Uhmann, and E. Bayer, *Biopolymers* **15**, 917–927 (1976).
49. M. Goodman, F. Naider, and C. Toniolo, *Biopolymers* **10**, 1719–1730 (1971).
50. C. Toniolo and G. M. Bonora, *Makromol. Chem.* **175**, 1665–1668 (1974).
51. G. M. Bonora and C. Toniolo, *Makromol. Chem.* **175**, 2203–2207 (1974).
52. E. Peggion, M. Palumbo, G. M. Bonora, and C. Toniolo, *Bioorg. Chem.* **3**, 125–132 (1974).
53. R. Guarnaccia, G. P. Lorenzi, V. Rizzo, and P. L. Luisi, *Biopolymers* **14**, 2329–2346 (1975).
54. P. L. Luisi, V. Rizzo, G. P. Lorenzi, B. Straub, U. Suter, and R. Guarnaccia, *Biopolymers* **14**, 2347–2362 (1975).
55. P. A. Temussi and M. Goodman, *Proc. Natl. Acad. Sci. U.S.A.* **68**, 1767–1772 (1971).
56. M. Goodman, N. Ueyama, and F. Naider, *Biopolymers* **14**, 901–914 (1975).
57. M. Goodman, N. Ueyama, F. Naider, and C. Gilon, *Biopolymers* **14**, 915–925 (1975).
58. F. Naider, J. M. Becker and M. Goodman, unpublished results.
59. C. M. Deber, F. A. Bovey, J. P. Carver, and E. R. Blout, *J. Am. Chem. Soc.* **92**, 6191–6198 (1970).
60. C. M. Deber, V. Madison, and E. R. Blout, *Acc. Chem. Res.* **9**, 106–113 (1970).
61. K. D. Kopple and T. J. Schamper, *J. Am. Chem. Soc.* **94**, 3644–3646 (1972).
62. K. D. Topple, A. Go, T. J. Schamper, and C. Wilcox, *J. Am. Chem. Soc.* **95**, 6090–6096 (1973).
63. K. D. Kopple and A. Go, *Biopolymers* **15**, 1701–1715 (1976).
64. V. N. Rogulenkova, M. I. Millionova, and N. S. Andreeva, *J. Mol. Biol.* **9**, 253–254 (1964).
65. W. Traub and A. Yonath, *J. Mol. Biol.* **16**, 404–414 (1966).
66. J. Engel, J. Kurtz, E. Katchalski, and A. Berger, *J. Mol. Biol.* **17**, 255–272 (1966).
67. P. J. Oriel and E. R. Blout, *J. Am. Chem. Soc.* **88**, 2041–2045 (1966).
68. D. J. Prockop, K. Juva, and J. Engel, *Hoppe-Seyler's Z. Physiol. Chem.* **348**, 553–560 (1967).
69. S. Sakakibara, Y. Kishida, Y. Kikuchi, R. Sakai, and K. Kakiuchi, *Bull. Chem. Soc. Jpn.* **41**, 1273 (1968).
70. W. Traub, *J. Mol. Biol.* **43**, 479–485 (1969).
71. M. Rothe, R. Theysohn, K. D. Steffen, H. J. Schneider, M. Zamami, and M. Kostrzewa, *Angew. Chem., Int. Ed. Engl.* **8**, 919–920 (1969).
72. M. Rothe, R. Theysohn, K. D. Steffen, H. J. Schneider, M. Zamami, M. Kostrzewa, and W. Schindler, *Angew. Chem., Int. Ed. Engl.* **9**, 535 (1970).
73. R. A. Berg, B. R. Olsen, and D. J. Prockop, *J. Biol. Chem.* **245**, 5759–5763 (1970).
74. Y. Kobayashi, R. Sakai, K. Kakiuchi, and T. Isemura, *Biopolymers* **9**, 415–427 (1970).
75. V. A. Shibnev, K. T. Poroshin, and V. S. Grechisko, *Dokl. Chem.* (*Engl. Transl.*) **198**, 446–448 (1971).
76 B. R. Olsen, R. A. Berg, S. Sakakibara, Y. Kishida, and D. J. Prockop, *J. Mol. Biol.* **57**, 589–595 (1971).
77. B. B. Doyle, W. Traub, G. P. Lorenzi, and E. R. *Biochemistry* **10**, 3052–3060 (1971).
78. G. P. Lorenzi, B. B. Doyle, and E. R. Blout, *Biochemistry* **10**, 3046–3051 (1971).
79. S. Sakakibara, Y. Kishida, K. Okuyama, N. Tanaka, T. Ashida, and M. Kakudo, *J. Mol. Biol.* **65**, 371–373 (1972).
80. K. Okuyama, N. Tanaka, T. Ashida, M. Kakudo, Y. Kishda, and S. Sakakibara, *Acta Crystallogr., Sect. A* **28**, Part 54, 538 III 16 (1972).

81. K. Sutch and H. Noda, *Biopolymers* **13**, 2385–2390 (1974).
82. K. Sutch and H. Noda, *Biopolymers* **13**, 2391–2404 (1974).
83. G. M. Bonora and C. Toniolo, *Biopolymers* **13**, 1055–1066 (1974).
84. G. M. Bonora and C. Toniolo, *Biopolymers* **13**, 1067–1078 (1974).
85. P. Bruckner, B. Rutschmann, J. Engel, and M. Rothe, *Helv. Chim. Acta* **58**, 1276–1287 (1975).
86. R. Kataki and Y. Nakayama, *Biopolymers* **15**, 747–755 (1976).
87. R. Katakai, *Biopolymers* **15**, 1815–1824 (1976).
88. E. Haas, M. Wilchek, E. Katchalski-Katzir, and I. Z. Steinberg, *Proc. Natl. Acad. Sci. U.S.A.* **72**, 1807–1811 (1975).

Chemical Synthesis of DNA Fragments: Some Recent Developments

Alexander L. Nussbaum

INTRODUCTION

Why synthesize molecules? One of the traditional functions of this endeavor by the natural-products chemist—to furnish definitive proof of the structure of a substance—threatens to fall by the wayside as increasingly powerful physical probes preempt this need. Nevertheless, synthesis can play a vital role in situations where it is difficult or impossible to isolate the desired substance from nature and where that particular molecule is needed to answer a specific question. A case in point can be found in molecular biology, where nucleic acid-like synthetic molecules of defined sequence have been used in the elucidation of biochemical problems. In this essay, I will review several recent examples.

The literature pertaining to the chemical synthesis of oligo- and poly-nucleotides has been capably evaluated by Kössel and Seliger [1] and by Narang and Wightman [2]; comments relating to methodology will therefore be kept to a minimum. Chemical synthesis in this context is considered to include the use *in vitro* of certain enzymes. A philosophical issue is involved here: As long as it is necessary to employ enzymes as catalysts, a pale remnant of vitalism persists. However, as enzymes themselves are beginning to be synthesized chemically, as their mode of action is better understood in the light of accepted physical principles, and, most satisfactorily, as biochemical catalysis can be mimicked by simpler systems, even though with lower efficiency, this problem will disappear. At present, recourse to enzymatic processing, like photochemistry in the recent past, must be considered to be Clausewitzian continuation of conventional organic chemistry by other means. In what follows, an attempt has been made to select some examples of how certain chemically synthesized molecules are being used to unravel narrowly defined biochemical questions. Other instances might have been chosen; substrates for biophysical studies were intentionally excluded. It is hoped that the interested nonspecialist will become more familiar with the type of thinking currently in vogue in this field of bioorganic chemistry.

REQUIREMENTS FOR REPRESSOR BINDING

When is genetic information only potential, and what is required to make it overt so that it can be observed in the phenotype? A classic example in bacteria is the ability to grow on lactose. *Escherichia coli* harbors in its genome a region of DNA, the lactose operon [3], that contains the structural genes for lactose-metabolizing enzymes. This region is closely associated with another stretch of DNA which is involved in the control of expression (i.e., transcription into messenger RNA and hence synthesis of the pertinent enzymes) of these structural genes. The regulatory region exerts such control in conjunction with certain regulatory proteins, among them the *lac* repressor. The latter prevents transcription by binding to the control region (the *lac* operator) but fails to do so in the presence of certain galactosides, so that ultimately these carbohydrates themselves elicit the overt expression of the bacterium's potential to utilize them for growth. Thus, the system calls for the synthesis of these specialized enzymes only in the presence of the appropriate substrate, a sensible arrangement preventing waste.

What structural features in the regulatory region are required to result in the transcriptional block? This intriguing question has now been largely answered by both an analytical and a synthetic approach. By virtue of the strong affinity of the repressor protein for its cognate DNA, a small fragment of the latter was isolated as a protein–nucleic acid complex, and a short stretch of double-stranded DNA obtainable therefrom (about 27 base pairs) was subjected to sequence analysis [4]. The major portion of this duplex manifests extensive elements of symmetry (Fig. 1), a feature believed to be a

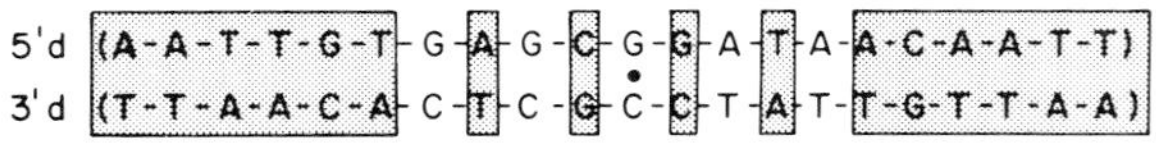

5'd (A-A-T-T-G-T) G -A- G -C- G -G- A -T- A -A-C-A-A-T-T)
3'd (T-T-A-A-C-A) C -T- C -G- C -C- T -A- T -T-G-T-T-A-A)

Fig. 1 A DNA duplex containing 21 base pairs implicated in *lac* repressor binding. Symmetry elements are shown.

significant determinant in the recognition of this region by the repressor protein.

This symmetrical 21-mer duplex has now been synthesized [5]. Oligomers were constructed chemically by a modifier triester method [6], and their sequences were confirmed by the elegant procedures first developed in Sanger's laboratory [7]. Two 21-mers thus obtained were annealed into duplex structure. Two alternate enzymatic procedures, one involving primer-initiated template-guided repair, the other employing the enzyme polynucleotide ligase from coliphage T4, offered shortcuts to the more laborious chemical steps.

The artificial 21-mer duplex actually binds to repressor protein (Fig. 2), as shown in a conventional assay [9]. The absolute extent of this binding amounts to one-quarter of that shown by macromolecular bacterial DNA. It is subject to severe reduction by the addition of a galactoside analog (isopropyl thiogalactoside), a property shared by the natural DNA and understandable in the light of the system's function.

It would thus appear that the information required to exert transcriptional control requires a defined nucleotide sequence. Itakura *et al.* [6] are now in a

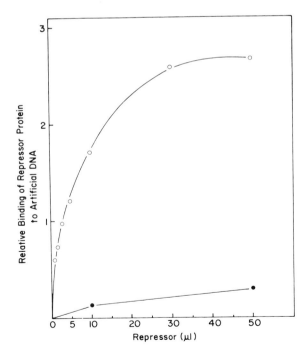

Fig. 2 Repressor protein binding by synthetic DNA. Open circles, no analog; closed circles, galactoside analog present.

position to study fine structure modulation of this protein–nucleic acid interaction by varying the size and nature of the duplex, for instance, by introducing single base pair variations (point mutations). It will be interesting to see what additional information can be extracted from the synthetic approach.

RECOGNITION OF FOREIGN DNA

When certain bacteria are infected with foreign organisms such as bacteriophages, they resist these potentially catastrophic invasions by attacking the invader's genetic instruction, i.e., its DNA [10]. They do this in certain cases by means of a class of endonucleases that recognize specific base sequences in double-stranded DNA and catalyze the hydrolysis of two phosphodiester bonds within such sequences in such a way as to cause a double-stranded break [11]. The invader's DNA, having suffered cleavage and thus a possibly fatal obstacle to its further function, is now said to be restricted; hence, these enzymes are termed restriction endonucleases. In order to prevent attack on their own DNA, with potentially serious consequences, these bacteria also evolved the ability to avoid endonucleolysis at the same specific base sequences in their own genomes by masking the sequence via methylation (modification).

One such system, the *Eco*RI restriction endonuclease and corresponding modification methylase, has been studied intensively [12]. These enzymes occur in *E. coli* carrying certain plasmids (extrachromosomal elements) and act on a defined sequence in the manner shown in Fig. 3. It will not escape the reader's attention that, first, the sequence again demonstrates simple 2-fold symmetry, even without the interruptions seen in the lac repressor binding site discussed above, and, second, that the product of endonucleolytic cleavage gives rise to two DNA termini possessing protruding single-stranded

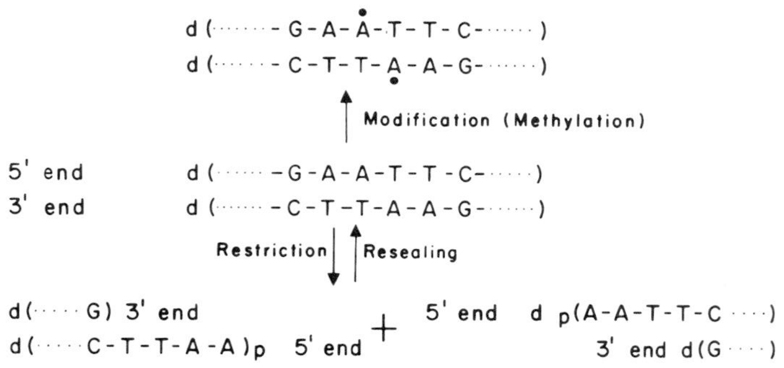

$$
\begin{array}{c}
\text{d} (\cdots - G - A - \overset{\bullet}{A} - T - T - C - \cdots) \\
\text{d} (\cdots - C - T - T - A - A - G - \cdots) \\
\underset{\bullet}{}
\end{array}
$$

↑ Modification (Methylation)

5' end   d (⋯ - G - A - A - T - T - C - ⋯)
3' end   d (⋯ - C - T - T - A - A - G - ⋯)

Restriction ↓ ↑ Resealing

d(⋯ G) 3' end   5' end d p(A - A - T - T - C ⋯)
d(⋯ C - T - T - A - A)p 5' end   +   3' end d(G ⋯)

Fig. 3 Base pair sequence recognized by *Eco*RI restriction endonuclease and modification methylase. Endonuclease cleavages are made at the arrows; methylation occurs at the adenines indicated by dots.

regions that are mutually complementary and can be resealed by the use of the enzyme polynucleotide ligase. It is the latter property that has made this endonuclease, and other sharing the ability of "cutting on a bias," so useful in the propagation of foreign DNA in certain *in vivo* systems [13].

What, in fact, constitutes a necessary and sufficient recognition site for these enzymes? In order to study this question, a self-complementary octanucleotide, dp(T-G-A-A-T-T-C-A), was synthesized [14]. It will be seen that this sequence is identical to that which is observed as the substrate sequence in macromolecular DNA (Fig. 3), except that one additional nucleotide has been added on to each terminus, extending the overall length of self-complementarity, for reasons that will become apparent.

Is this "minimal DNA" recognizable as a substrate by the *Eco*RI system? Exposure of the oligomer to hydrolysis when catalyzed by the endonuclease showed the following reaction [14]:

$$dp(T\text{-}G\text{-}A\text{-}A\text{-}T\text{-}T\text{-}C\text{-}A) \longrightarrow dp(T\text{-}G) + dp(A\text{-}A\text{-}T\text{-}T\text{-}C\text{-}A)$$

Specific cleavage had occurred at the same point as is observed in a natural substrate (*mutatis mutandis*, corresponding results were obtained with the methylation enzyme). The rate at which this reaction occurs, however, is much slower than with macromolecular substrate (Fig. 4). This is not surprising, since the enzyme requires double-stranded DNA, and the oligomer has to anneal before being acted on. At a temperature where the enzyme can function and at practical substrate concentrations, the double-stranded fraction of oligomer becomes severely limiting. This becomes obvious from

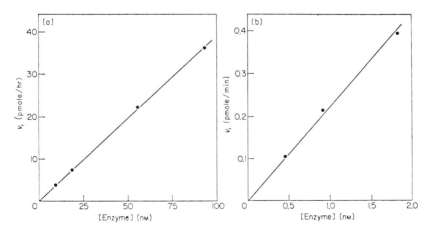

Fig. 4 Rate of cleavage of phosphodiester bonds by *Eco*RI endonuclease as a function of enzyme concentration. (a) Reaction with synthetic oligomer; (b) reaction with macromolecular DNA. (For details, see Greene *et al.* [13].)

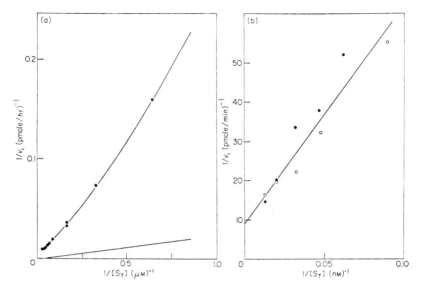

Fig. 5   Double-reciprocal plots of the data in Fig. 4. (a) Synthetic oligomer; (b) macro-molecular DNA.

the shape of a conventional Lineweaver–Burk plot of the kinetic data (Fig. 5). While the natural DNA demonstrates the expected straight line, this is not the case for the oligomer. A theoretical curve can be plotted only when allowance is made for the actual fraction of the molecules present as dimers, i.e., in double-stranded form. This phenomenon also vitiated an attempted control experiment. A second octamer was synthesized in which the lone C had been changed to a U, resulting in dp(T-G-A-A-T-T-U-A). This oligomer was not cleaved by the endonuclease. However, substitution of a GU for a GC pair in the desired double-stranded structure sufficiently displaced the monomer–dimer equilibrium to furnish a population essentially devoid of double-stranded molecules.

Nevertheless, the use of a "minimal DNA" allows one to conclude that it is the nucleotide sequence alone that is recognized by the enzymatic machinery of the *Eco*RI restriction and modification system.

ARTIFICIAL PRIMERS FOR DNA SEQUENCING

A professed aim of molecular biology is to unravel the entire information encoded in the genome of an organism. This requires that one would wish to know, first of all, the full primary structure—the nucleotide sequence—of the

DNA of that organism. One would thus define not only the code-determined concordance of the structural genes with the amino acid sequence of pertinent proteins, when the latter are gene products, but hopefully also the structure–function relationships of those intercistronic regions presumed to be involved in the highly organized control activities necessary for life.

For several reasons, the first nucleic acid sequences discovered were those of the smaller RNA molecules [15], but substantial progress in DNA sequencing has been made in the last several years [8]. One strategy that holds considerable importance involves the use of chemically synthesized oligomers. This approach requires (1) knowledge of a specific DNA sequence from a region of interest in a genome, (2) annealing of a synthetic DNA fragment, complementary to that region in the Watson–Crick sense, of sufficient size to insure specific binding, (3) elongation of the fragment (the primer) by enzymatic means to produce a faithful copy of a neighboring area of the genome under the latter's guidance as a template, and (4) actual sequencing of the enzymatically synthesized copy of the template [16]. The resulting data define the sequence of the area of interest (Fig. 6).

A nice illustration of this procedure deals with the elaboration of a stretch of nucleotides in the genome of bacteriophage ϕX174. It had been known for some time that, under appropriate conditions, ribosomes have the ability to protect specific regions of messenger RNA (ribosome binding sites and contiguous areas) against nuclease degradation and thus permit their isolation and the determination of their nucleotide sequences [17]. Surprisingly, an area of the DNA in ϕX174 was found to be similarly protected [18]. This DNA

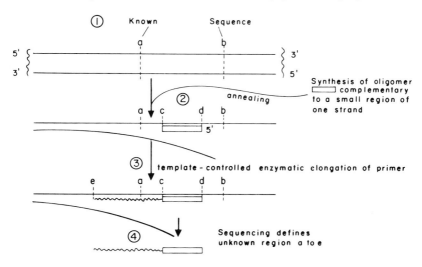

Fig. 6 Sequence by primer elongation. (For details, see text.)

Fig. 7   Relationships of a portion of gene *G* to "spike" protein and control elements.

exists in single-stranded form during most of the phage's life cycle, and the area in question contains the same nucleotide sequence (uracil becoming thymine) as a ϕX174 messenger RNA, so that the protective action appears to be a natural mistake. By application of the sophisticated techniques developed in Sanger's laboratory it was possible to determine a sequence of some 50 nucleotides constituting the protected region [19]. Comparison with a known amino acid sequence revealed that this stretch of DNA constituted a precisely circumscribed region from the genetically defined gene *G*, which codes for the "spike" protein [19]. The sequence data permitted the chemical synthesis [20] of a complementary decamer, an oligomer sufficiently large to allow one to hope for specific binding at a unique site in a genome of the size studied here [21]. Fortunately, this turned out to be the case. With the phage DNA as a template, this decamer was used as a primer in a typical "repair" synthesis catalyzed by *E. coli* polymerase I [22]. Sequence analysis of the resulting product showed that only the proper region of the bacteriophage genome had been copied. In the instance at hand, the results were quite unequivocal since most of the sequence derived in this manner was already known. The data are compiled in Fig. 7. The region defining structural information (i.e., the amino acid sequence of "spike" protein N-terminus) is preceded in the "upstream," or 5'-terminal, direction by certain control information which may include features of secondary structure (a potential hairpin [23]).

A number of synthetic primers have been used in elucidating DNA sequences [8]. It should be pointed out, however, that chemical synthesis is not the only way in which primers for work of this kind can be obtained; restriction enzymes are perhaps a more valuable tool for this purpose [16].

TRANSFER RNA GENES

By far the most ambitious studies in the entire field are being carried out in Khorana's laboratory. This investigator, first introduced to the field in the laboratory of A. R. Todd, developed the methodology used or, at best, modified by other workers in this area of bioorganic chemistry; in fact, most laboratories active in nucleotide chemical synthesis are staffed by linear or lateral descendants of his school [24]. Khorana has set himself the goal of synthesizing biologically functional DNA, that is, genes capable of giving rise to gene products.

It was recognized some dozen years ago that the most straightforward objective would be the gene for a tRNA molecule since the latter bears a defined relationship to the gene that dictates its biogenesis. It is simply a complement, in the Watson–Crick sense, of the codogenic strand. In this respect, it differs from protein (by far the more common gene product), in which this relationship is complicated by the redundancy of the genetic code. Knowledge of the amino acid sequence in a protein does not permit a unique deduction of the nucleotide sequence in the corresponding structural gene. Accordingly, Khorana proceeded to synthesize a DNA duplex defined by the principal yeast alanine-tRNA, the only molecule of its kind for which a sequence had become known [25] at the time the synthetic work was begun. The completion of this formidable effort [26], a duplex of 77 nucleotides per strand, represented a quantum jump in chemical biopolymer synthesis, a feat not even approached in other laboratories until quite recently. It soon became apparent, however, that this kind of a duplex contained insufficient information for the desired *in vitro* transcription into RNA. The heterologous (for yeast) *E. coli* DNA-dependent RNA polymerase catalyzes only improperly initiated and terminated RNA fragments, and those at disappointingly low levels. It was recognized that a potentially much more rewarding homologous system was available: the gene for an *E. coli* tyrosine-tRNA. There were several good reasons for choosing this system. First, the gene in question had been transferred [27] from the relatively huge *E. coli* chromosome to the much smaller genome of a transducing bacteriophage, $\phi80$, which can be easily grown in quantity; this step made the gene available in relatively larger amounts. Second, it was discovered [28] that the primary transcript *in vivo* of this gene was not the mature tRNA at all but a precursor molecule some 40 nucleotides longer. Third, it became clear that additional DNA sequences in the gene were needed to control proper initiation of the act of transcription, as well as the proper course of posttranscriptional events. The relative layout of these several regions is shown in Fig. 8. This realization, of course, makes necessary the synthesis of a much larger molecule than that originally envisaged. However, the availability of the gene under discussion in the relatively small bacteriophage genome permitted Khorana and colleagues to sequence the adjacent purported control regions [29] by the methodology discussed earlier. It turns out that these control regions again show features of symmetry. Finally, it was possible to approach the problem from the other end. Well-circumscribed regions of DNA carrying the gene in

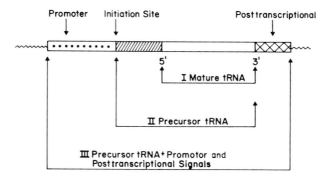

Fig. 8   A tRNA gene (DNA) with presumed control elements.

question were excised from the bacteriophage genome and shown to be appropriately transcribed by the *E. coli* RNA polymerase *in vitro* [30]. The primary transcript, shown to be the expected tRNA precursor, was processed by a bacterial cell extract to give material showing a biological activity (enzymatically mediated esterification with tyrosine) and some of the characteristic modification of the nucleobases observed in the genuine tRNA. The regions of bacteriophage DNA used in these transcription experiments were less than 600 base pairs long, still larger than the expected end product of the synthetic program (closer to 200 base pairs) but not enormously discrepant. How large, in fact, does the synthetic molecule have to be—what information is necessary and sufficient—to result in the controlled transcription of what is destined to become biologically functional tRNA?

The actual nucleotide sequence, deduced either from the primary structure of the precursor RNA itself or from sequence studies carried out on the gene sequestered in the bacteriophage genome, is shown in Fig. 9. Much of this imposing molecule has already been synthesized [31]. The picture that emerged from the work of Altman and Smith [28] indicated that primary transcription is initiated at the base pair numbered 1 in Fig. 9 and that the DNA template is copied into an RNA molecule (the precursor tRNA) which extends to a point a few base pairs to the right of position 126, the 3′ terminus of the mature tRNA. Posttranscriptional trimming removes these latter and the first 41 transcribed ribonucleotides from the precursor molecule.

The methodology for synthesis employs a "leapfrog" approach, in which oligonucleotides of manageable size (some 8–14 nucleotides in length) are chemically synthesized (in Fig. 9 these are delineated by carets) and subsequently joined together enzymatically. The enzymatic reaction requires that any two oligomers to be joined must be held together by a complementary third, overlapping with both so as to produce a nick [32]. Much of the work has been published in outline. The entire DNA duplex defining the precursor

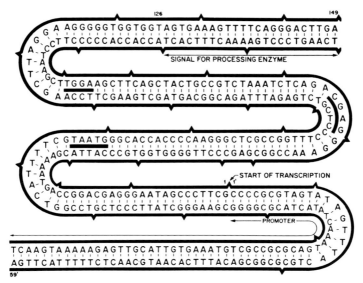

Fig. 9 Synthetic tyrosine-tRNA gene.

molecule plus the 23-base-pair segment at its 3' terminus has been completed. Work on the promoter region, believed to require at most a stretch of 55 base pairs, is in progress and about two-thirds finished at the time of this writing.* Although *in vitro* studies of controlled transcription must await completion and attachment of this control region, it has been possible to achieve a distinctly encouraging preliminary success: When the synthetic molecule constructed so far (i.e., extended partially into what is believed to be the total required promoter region) was exposed to the RNA polymerase in the presence of low concentrations of ribonucleoside triphosphates and a small RNA primer—a tetranucleotide—a primary transcript was obtained (Fig. 10) which could be biochemically processed into what is believed to be bona fide tRNA [33]. What apparently happens is that, in addition to the known [28] specific endonucleolytic fission at the 5' terminus, a second such step occurs some seven base pairs "downstream" from the CCA 3' terminus of the mature tRNA, the final trimming occurring via exonuclease action. The

* So rapidly do matters move in this highly topical field that it becomes necessary to amend this account. Synthesis of the duplex in question has now been completed, and its transcription *in vitro* into a 4 S product has been observed. The molecule has also been transcribed *in vivo* into functional tRNA. See R. Belagaje, E. L. Brown, H.-J. Fritz, M. J. Gait, R. G. Lees, K. E. Norris, T. Seikya, R. Contreras, H. Kupper, and H. G. Khorana, Abstracts of Papers, 172nd American Chemical Society National Meeting, San Francisco, California, August 29–September 3, 1976. Carb 012, and widely disseminated reports in the lay press.

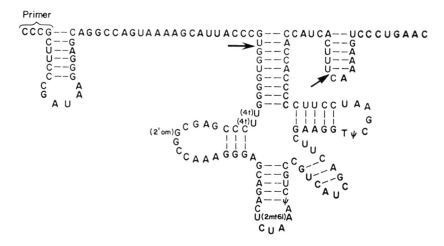

Fig. 10 Synthetic tyrosine-tRNA gene. Primary transcript to tRNA. 2'om, 2'-*O*-methyl, 2mt6I, 2-methylthio-6-isopentenyl.

integrity of this terminus would then be insured by the well-known CCA addition system. It can be seen from Fig. 10 that both of these endonucleolytic steps occur in regions where the RNA precursor is double stranded, a condition requiring symmetry elements in the primary structure of the gene.

This experimental finding represents a most instructive advance in the understanding of tRNA biogenesis and may well change our ideas of what properly constitutes the tRNA precursor molecule (i.e., the primary transcript *in vivo*). While it must be realized that the primer represents a shortcut to correct spontaneous initiation of transcription, it is nevertheless highly gratifying to observe that a synthetic molecule can direct the posttranscriptional processing events observed in nature.

GENES FOR POLYPEPTIDES

An even more ambitious aim of DNA chemistry is the synthesis of genetically meaningful sequences that can be made in some fashion to express themselves in the form of polypeptide structure. Consider some of the problems faced by the designer of such artificial genes. He must select a sequence which codes for a gene product that can be recognized (i.e., scored); he must find a biochemical or biological system that permits the synthesis of such a gene product under the direction of foreign (and artificial) DNA; in view of the currently still meager sequence information developed from DNA, he must derive the actual nucleotide sequence of his synthetic molecule in the face of codon redundancy by making deductions from the amino acid sequence of the intended gene product. Much of the information required for intelligent design simply does not exist yet. Nevertheless, some laboratories are attempting to provide shortcuts to the solution of these problems by synthesizing "minigenes."

One such effort has been undertaken at Hoffmann–La Roche. It had been known for some time [34] that enzymatic cleavage of bovine pancreatic ribonuclease A (Fig. 11) with the bacterial protease subtilisin produces two fragments: an eicosapeptide (*S*-peptide, amino acids 1–20), and the remainder of the protein (*S*-protein). Both the species are separately completely devoid of ribonucleolytic activity, but the latter can be fully restored, without reconstitution of covalency, upon simple mixing of the two fragments. Through the extensive studies of Hofmann, Finn, and their associates, it has been demonstrated [35] that the full *S*-peptide is not required to regenerate activity. A number of truncated peptides, one as small as 12 amino acids in length as compared to the original 20, retain the capacity to reactivate. Since this biochemical activity might provide a handle for scoring the presence of such a peptide, it was decided to synthesize the genetic instruction for such a molecule (the *S*-peptide, amino acids 2–14) with the hope that it might eventually be possible to demonstrate that the peptide in question can be synthesized, in an appropriate biochemical or biological system, under such instruction.

The relationship between the deoxyribonucleotide duplex, a corresponding portion of RNA, and the amino acid sequence for *S*-peptide (amino acids

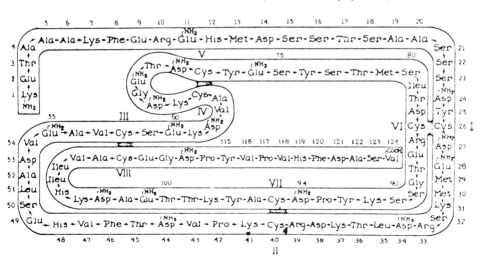

Fig. 11 Bovine ribonuclease A.

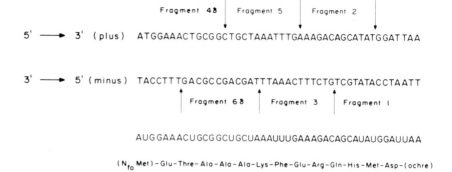

Fig. 12 Relationship of S-Peptide (amino acids 2–14), corresponding "messenger RNA," and "gene" (DNA duplex) sequences. Chemically synthesized fragments are delineated by vertical arrows.

2–14) is shown in Fig. 12. It can be seen that the nucleic acid sequence is provided with translational "punctuation marks": an AUG codon for "start" and a UAA codon for "stop" (marked "ochre"); at present both of these are known to be valid only in prokaryotes. It should be realized that the decision to use that particular sequence was arrived at entirely on chemical grounds [36], the only condition being consistency with the universal genetic code. It should also be realized that several tens of thousands of different sequences can be written for the peptide message (and hence gene) if the full redundancy of the code is taken into consideration. Nevertheless, the validity *in vivo* of the particular codons chosen has been demonstrated, at least in prokaryotes.

The synthesis of this duplex of 45 nucleotides per strand [37] has recently been completed. In Fig. 12, chemically synthesized oligomers are shown between vertical arrows; at these sites, enzymatic joining was performed. Efforts are currently in progress to incorporate the synthetic species into a genome that can be propagated *in vivo*. A first step is the attachment of "sticky ends" (cohesive termini; see above), which would permit the insertion of the "minigene" into foreign DNA in a typical genetic engineering experi-

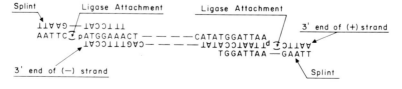

Fig. 13 Scheme for attachment of cohesive termini to the "minigene."

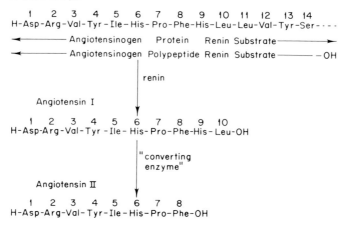

Fig. 14 Formation of angiotensin II from protein precursor.

ment [13]. Current synthesis of "splints" permitting enzymatic attachment of cohesive termini is illustrated in Fig. 13. The synthesis of one of these splints has already been completed.

A similar effort has recently been described in the German literature [38]. The human peptide hormone angiotensin II, an octapeptide, is a pleiotropic tissue hormone with a host of activities, notably vasopressor and antidiuretic [39]. This peptide is cleaved off *in vivo* in several steps from a protein precursor, angiotensinogen protein, as shown in Fig. 14. The protein, obtained from the α_2-globulin fraction of blood plasma, in synthesized on the ribosome.

Köster *et al.* [38] designed a DNA duplex coding for angiotensin II, as shown in Fig. 15. Again, the actual codons used were chosen from among the

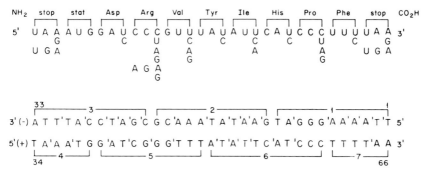

Fig. 15 Design of 33-base-pair duplex chosen from hypothetical messenger for amino acid sequence of angiotensin II [38].

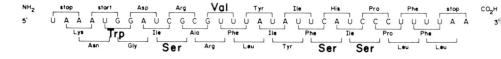

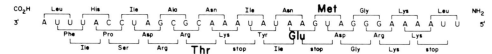

Fig. 16 Peptide sequences resulting from symmetrical transcription or out-of-phase translation [38].

large number of possibilities permitted by the redundancy of the genetic code (shown on top). This resulted in the sequence of the duplex shown in the lower part of Fig. 15. The actual oligomers chemically synthesized are indicated by brackets; again, these were then enzymatically joined in the fashion described earlier. There is, however, an ingenious additional feature in this design. Any translation that would occur from either a transcript of the "wrong" (i.e., the noncodogenic) strand of the duplex or by out-of-phase translation from the transcript of either strand was designed to be detected by virtue of unique amino acids, shown in boldface in Fig. 16. Only the desired sequence has the proper "punctuation."

ADDENDUM

It is well known that a respectable portion of the evidence leading to the nexus of concepts known as molecular biology has come from chemistry. Outstanding examples are Chargaff's painstaking analysis of the base composition of DNA [40], an important component of the data leading to the structure of the genetic material, and Khorana's incisive synthetic studies vital in establishing the genetic code [41]. Much current work continues in this vein. Synthetic DNA fragments, both single and double stranded, have been constructed in order to shed light on a variety of topics of interest to molecular biology. In this essay, I have discussed a stretch of DNA involved in the control of transcription by association with the *lac* repressor, a target site for a bacterial restriction endonuclease and modification enzyme, a primer for copying an area of the DNA of bacteriophage ϕX174, and the synthesis of genes for tRNA molecules and for amino acid sequences occurring in mammalian proteins. This cursory survey is intended to demonstrate that chemistry continues to be a useful tool in the field.

REFERENCES

1. H. Kössel and H. Seliger, *Prog. Chem. Org. Nat. Prod.* **32**, 298–508 (1975).
2. S. A. Narang and R. H. Wightman, *in* "The Total Synthesis of Natural Products" (J. ApSimon, ed.), Vol. 1, pp. 280–330. Wiley (Interscience), New York, 1973.
3. J. R. Beckwith and D. Zipser, eds., "The Lactose Operon." Cold Spring Harbor Lab., Cold Spring Harbor, New York, 1970.
4. W. Gilbert and A. Maxam, *Proc. Natl. Acad. Sci. U.S.A.* **70**, 3581 (1973).
5. C. P. Bahl, R. Wu, K. Itakura, N. Katagiri, and S. A. Narang, *Proc. Natl. Acad. Sci., U.S.A.* **73**, 91 (1976).
6. K. Itakura, N. Katagiri, S. A. Narang, C. P. Bahl, K. J. Marians, and R. Wu, *J. Biol. Chem.* **250**, 4592 (1975).
7. G. G. Brownlee and F. Sanger, *Eur. J. Biochem.* **11**, 395 (1969). These procedures have now been significantly improved: see C.-P. D. Tu, E. Jay, C. P. Bahl, and R. Wu, *Anal. Biochem.* **74**, 73 (1976); see also Wu *et al.* [8].
8. R. Wu, R. Bambara, and E. Jay, *Crit. Rev. Biochem.* **2**, 455 (1974).
9. A. D. Riggs, H. Suzuki, and S. Bourgeois, *J. Mol. Biol.* **48**, 67 (1970).
10. W. Arber and S. Linn, *Annu. Rev. Biochem.* **38**, 467 (1969).
11. H. W. Boyer, *Fed. Proc., Fed. Am. Soc. Exp. Biol.* **38**, 1125 (1974).
12. A. Dugaiczyk, J. Hedgpeth, H. W. Boyer, and H. M. Goodman, *Biochemistry* **13**, 503 (1974).
13. See, for instance, J. F. Morrow, S. N. Cohen, A. C. Y. Chang, H. M. Goodman, and R. B. Helling, *Proc. Natl. Acad. Sci. U.S.A.* **71**, 1743 (1974).
14. P. J. Greene, M. S. Poonian, A. L. Nussbaum, L. Tobias, D. E. Garfin, H. W. Boyer, and H. M. Goodman, *J. Mol. Biol.* **99**, 237 (1975).
15. G. G. Brownlee, *Lab. Tech. Biochem. Mol. Biol.* 3, Part 1, 1–265 (1972).
16. For greater detail, see F. Sanger's Croonian Lectures, *Proc. R. Soc. London, Ser. B* **191**, 317 (1975).
17. See, for instance, J. A. Steitz, *Nature (London)* **224**, 957 (1969).
18. H. D. Robertson, B. G. Barrell, H. L. Weith, and J. E. Donelson, *Nature (London), New Biol.* **241**, 38 (1973).
19. B. G. Barrell, H. L. Weith, J. E. Donelson, and H. D. Robertson, *J. Mol. Biol.* **92**, 377 (1975).
20. H. Schott, *Makromol. Chem.* **175**, 1683 (1974).
21. C. A. Thomas, Jr., *Prog. Nucleic Acid Res. Mol. Biol.* **5**, 315. (1966).
22. J. E. Donelson, B. G. Barrell, H. L. Weith, H. Kössel, and H. Schott, *Eur. J. Biochem.* **58**, 383 (1975).
23. H. M. Sobell, *Adv. Genet.* **17**, 411–476 (1973).
24. The author of this essay, while spending a sabbatical at Stanford University, was himself guided in his first steps in this field by John G. Moffatt, a student of Khorana.
25. R. W. Holley, J. Apgar, G. A. Everett, J. T. Madison, M. Marquisee, S. H. Merrill. J. R. Penswick, and Z. Zamir, *Science* **147**, 1462–1465 (1965).
26. K. L. Agarwal, H. Büchi, M. H. Caruthers, N. Gupta, H. G. Khorana, K. Kleppe, A. Kumar, E. Ohtsuka, U. L. RajBhandara, J. H. van de Sande, V. Sgaramella, H. Weber, and T. Yamada, *Nature, (London)* **227**, 27 (1970).

27. R. L. Russell, J. N. Abelson, A. Landy, M. L. Gefter, S. Brenner, and J. D. Smith, *J. Mol. Biol.* **47**, 1 (1970).
28. S. Altman and J. D. Smith, *Nature (London), New Biol.* **233**, 35 (1971).
29. H. G. Khorana, *in* "Proceedings of the International Symposium on Macromolecules" (E. B. Mano, ed.), pp. 371–395. Elsevier, Amsterdam, 1975. A more recent leading reference is T. Sekiya, H. van Ormondt, and H. G. Khorana, *J. Biol. Chem.* **250**, 1087 (1975).
30. H. Küpper, R. Contreras, A. Landy, and H. G. Khorana, *Proc. Natl. Acad. Sci. U.S.A.* **72**, 4754 (1975).
31. H. G. Khorana, *Int. Congr. 24th, Pure Appl. Chem., 1973* Vol. 2, pp. 19–43 (1974). Full details have appeared during this writing; see H. G. Khorana, K. L. Agarwal, P. Besmer, H. Büchi, M. H. Caruthers, P. J. Cashion, M. Fridkin, E. Jay, K. Kleppe, R. Kleppe, A. Kumar, P. C. Loewen, R. C. Miller, K. Minamoto, A. Panet, U. L. RajBhandary, B. Ramamoorthy, T. Sekiya, T. Takeya, and J. H. van de Sande, *J. Biol. Chem.* **251**, 565 (1976), and following papers.
32. See, for instance, B. Weiss and C. C. Richardson, *Proc. Natl. Acad. Sci. U.S.A.* **57**, 1021 (1967).
33. H. G. Khorana, personal communication (1976).
34. F. M. Richards, *Proc. Natl. Acad. Sci. U.S.A.* **44**, 162 (1958).
35. F. M. Finn and K. Hofmann, *Acc. Chem. Res.* **6**, 169 (1973).
36. G. J. Powers, R. L. Jones, G. A. Randall, M. H. Caruthers, J. H. van de Sande, and H. G. Khorana [*J. Am. Chem. Soc.* **97**, 875 (1975)] describe an algorithm for arriving at a sequence of this type. Similar considerations led to the structure here discussed. See also T. Lewinson, *J. Theor. Biol.* **31**, 557 (1971).
37. C. L. Harvey, K. Olson, A. deCzekala, and A. L. Nussbaum, *Nucleic Acids Res.* **2**, 2007 (1975), and earlier references cited therein.
38. H. Köster, H. Blöcker, R. Frank, S. Geussenhainer, and W. Kaiser, *Hoppe-Seyler's Z. Physiol. Chem.* **356**, 1585 (1975).
39. See Angiotensin, *Handb. Exp. Pharmakol.*, **37**, (1974).
40. E. Chargaff, *Experientia* **6**, 201 (1950).
41. H. G. Khorana, *Harvey Lect.* **62**, 79 (1968).

CHAPTER

11

Synthetic Transformations of Naturally Occurring Nucleosides and Nucleoside Antibiotics

Morris J. Robins

A wide variety of "minor component" nucleosides with modified carbohydrate and/or heterocyclic base structures have been isolated from ribonucleic acids (tRNA, rRNA, mRNA, nRNA)* [1]. A number of "nucleoside antibiotics" and other naturally occurring nucleosides exist in the free state in microbiological systems, plants, sponges, etc. [2]. Owing to the diverse biological responses associated with metabolic processes and specific enzyme control mechanisms involving these potent biomolecules and synthetic analogs, extensive efforts have been directed toward their chemical and biochemical syntheses [3].

Three general approaches have been employed in syntheses of modified nucleosides: (1) coupling of suitably constructed and protected sugar and base derivatives followed by appropriate transformations and deblocking, (2) elaboration of the desired heterocyclic base at the "anomeric carbon" of an appropriately functionalized carbohydrate derivative, and (3) transformations of intact naturally occurring nucleosides [3]. Base–sugar coupling

---

* Abbreviations used in this chapter are as follows: tRNA, transfer ribonucleic acid; rRNA, ribosomal ribonucleic acid; mRNA, messenger ribonucleic acid; nRNA, nuclear ribonucleic acid; DMF, N,N-dimethylformamide; DME, 1,2-dimethoxyethane; DMPP, 4,4-dimethyl-3-pivaloxypent-2-enoyl; AIBN, α,α'-azobisisobutyronitrile; BSA, N,O-bistrimethylsilylacetamide; DBN, 1,5-diazabicyclo[4.3.0]non-5-ene; THF, tetrahydrofuran; tlc, thin-layer chromatography.

procedures have been developed and employed extensively. This represents the most widely used approach to base-analog nucleosides, and it has also been used frequently to prepare modified sugar nucleosides. Heterocyclic base elaboration beginning with a functionalized sugar precursor is central to most approaches to C-nucleosides [3a] and has also been used in syntheses of "natural" N-nucleosides. The third general approach, natural product trans-formation, has been explored widely only with pyrimidine nucleosides [3,4]. Easily accessible $O^2 \rightarrow$ sugar cyclonucleoside (anhydronucleoside) syntheses, the possible nucleophilic opening of these anhydronucleosides at either the sugar or base terminus, and the stability of pyrimidine nucleosides toward glycosyl bond cleavage make these compounds readily amenable to sugar (and certain base) transformations. Analogous purine-$X^8 \rightarrow$ sugar cyclonucleoside chemistry has been explored, primarily by Ikehara and co-workers [5]. However, additional problems are involved in the purine series since the naturally occurring purine nucleosides are unsubstituted at C-8. Therefore, appropriate functionalization at C-8 requires a minimum of two steps, and (except in the case of desulfurization) an "unnatural" 8-substituent remains after cyclonucleoside formation and ring opening. Purine deoxynucleosides are also very susceptible to glycosyl bond cleavage. A further limitation arises with nucleoside antibiotics or chemical species in which the base structure precludes functionalization at the position corresponding to C-8 of purines.

We have been interested primarily in this third general area, transformation of natural products, for several years. We have been concentrating on the development of general reactions for carbohydrate transformations whose success is not dependent on base structural features. Obvious advantages arising (especially in cases of component precursors that are themselves not easily accessible chemically) include (1) retention of biologically determined stereochemical and positional isomeric purity at unmodified loci, (2) omission of otherwise specifically demanded blocking–condensation–de-blocking steps, (3) elimination of preparation of specific precursor base and/or sugar structures, and (4) availability of *practical* quantities of biologically derived structures via fermentation techniques. Examples of base transfor-mations while maintaining constant sugar structures and of sugar transfor-mations in a given base series have been investigated.

Chloromercuri salt [6] and fusion [7] coupling procedures were employed originally to give 6-chloro-9-(2-deoxy-β-D-*erythro*-pentofuranosyl)purine (**3**) (6-chloropurine deoxyriboside) in about 6% overall yield (plus the corres-ponding α-anomer). Direct transformation of the purin-6-one system of tri-O-acetylinosine to give 6-chloropurine riboside triacetate had been effected under Vilsmeier–Haack conditions (thionyl chloride/DMF in chloro-form solution) [8]. However, attempted application of this procedure to di-*O*-acyl derivatives of 2'-deoxyinosine led to glycosyl bond cleavage, and hypoxanthine was isolated in greater than 90% yields (also observed previously by Dr. A. D. Broom and by Dr. J. Žemlička, personal communications). A 3'-*O*-*p*-toluenesulfonyl group had been observed [9] to markedly stabilize 2'-deoxyadenosine against acid-catalyzed glycosyl hydrolysis. This is consistent with the postulated hydrolysis mechanism involving unimolecular dissociation of the protonated base, leaving the ring-oxygen-stabilized glycosyl cation [10]. Strongly electron withdrawing groups on the sugar moiety would be expected to destabilize developing positive character and thus retard glycosyl cleavage. Treatment of 2'-deoxyinosine (2) (Scheme 1) [obtained in high yield by enzymatic deamination of commercially available 2'-deoxyadenosine (1)] with trifluoroacetic anhydride gave the bistrifluoroacetate derivative quantitatively [11]. Careful treatment of this product with SOCl$_2$/DMF in refluxing methylene chloride and methanolysis of blocking groups during column chromatography of the crude product mixture on alumina gave 6-chloropurine deoxyriboside (3) in 81% yield from (2). Approximately 5–10% glycosyl cleavage (hypoxanthine) was observed in contrast to ∼90% when the di-*O*-acetyl derivative was subjected to identical conditions.

The chlorine atom of (3) is easily displaced nucleophilically, and mercapto (4a), methylmercapto (4b), *p*-nitrobenzylmercapto (4c), benzylamino (4d), and hydroxylamino (4e) derivatives were prepared in the usual manner [11]. Amination of (3) in liquid ammonia gave 2'-deoxyadenosine (1). This represents a two-stage [(2) → (3) → (1)] chemical reversal of the enzymatic deamination [(1) → (2)] and constitutes the first noted conversion of deoxyinosine to deoxyadenosine. Trimethylamine converted (3) to the quaternary ammonium chloride salt (4f), which was treated with fluoride anion to give the previously unknown 6-fluoropurine deoxyriboside (4g). The 6-chloro (3), 6-hydroxylamino (4e), and 6-fluoro (4g) [as well as 6-amino (1)] deoxynucleosides were effective substrates for the enzyme adenosine deaminase [adenosine aminohydrolase (EC 3.5.4.4)].

The potent cytotoxic 5-fluorouracil and 5-fluorocytosine bases, nucleosides, and nucleotides had always been synthesized by chemically *de novo* pathways beginning with the highly toxic ethyl fluoroacetate [12]. Chemical (or enzymatic) coupling of appropriately blocked and/or derivatized bases and sugars gave nucleosides that were sometimes obtained in less than desirable yields after tedious separations and deblocking. Our first attempted direct entry into 5-fluoropyrimidines involved treatment of 1-methyl-5-bromo-6-methoxy-5,6-dihydrouracil with fluoride nucleophiles. However, elimination of the elements of methanol occurred to give 1-methyl-5-bromouracil in good yield [13]. Since the dihydrouracil adduct starting material was prepared by addition of methyl hypobromite to 1-methyluracil [14, 14a], an analogous

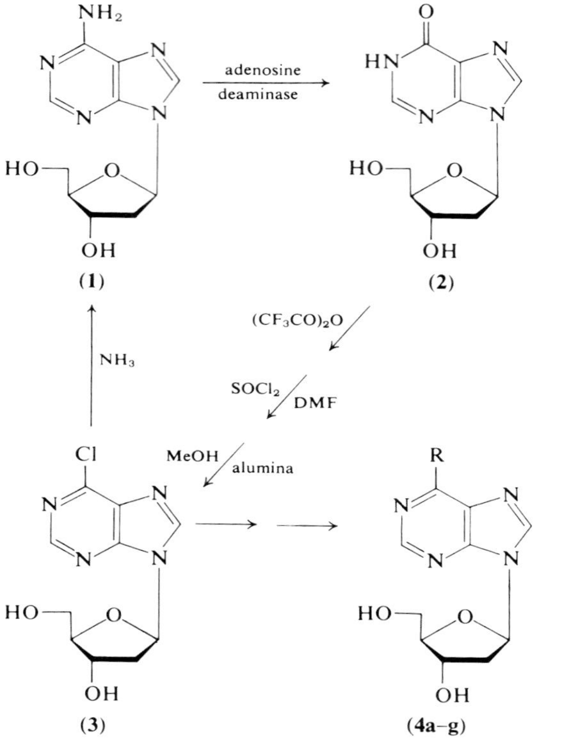

	R
a	SH
b	SCH_3
c	$SCH_2C_6H_4NO_2(p)$
d	$NHCH_2Ph$
e	NHOH
f	$\overset{+}{N}(CH_3)_3Cl^-$
g	F

Scheme 1

addition–elimination approach was begun [13] employing trifluoromethyl hypofluorite. Barton and co-workers had explored this reagent extensively in the terpenoid area and had observed formation of predominantly *cis*-fluoro trifluoromethoxy adducts [15]. However, treatment of a cold (−78°C)*

* **Caution:** CF_3OF is a strongly oxidizing pseudohalogen of fluorine and *should not be added directly* to alcohols or other easily oxidized solutions *especially at room temperature.*

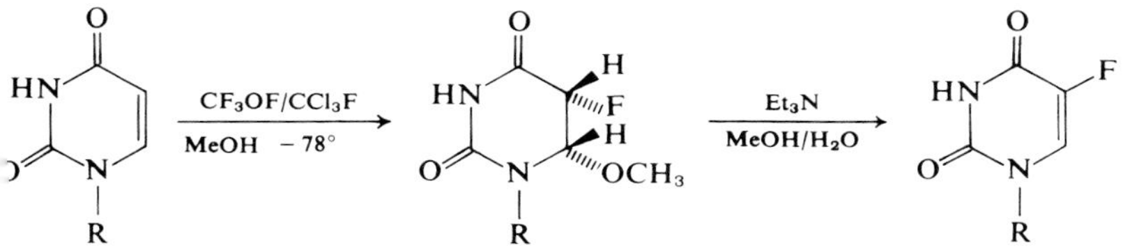

(5a,b) **(6a,b)** **(7a,b)**

a R = H **b** R = CH_3

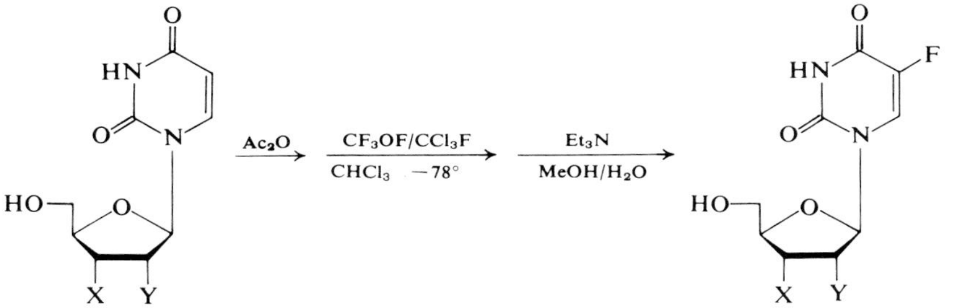

(8a–e) **(9a–e)**

	X	Y
a	OH	OH
b	OH	H
c	OH	OCH_3
d	OCH_3	OH
e	OCH_3	OCH_3

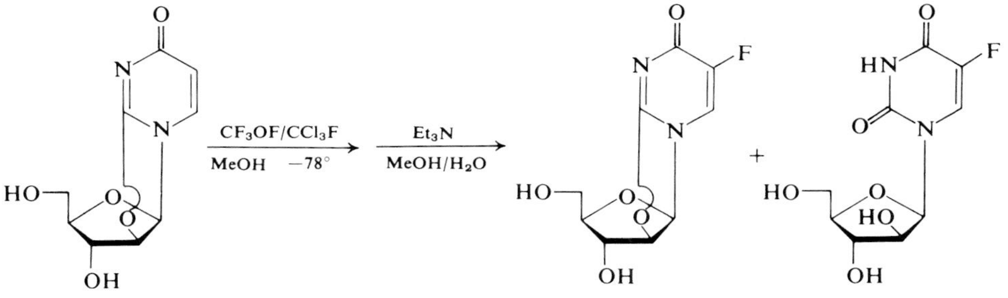

(10) **(11)** **(12)**

Scheme 2

methanolic solution of uracil (**5a**) or 1-methyluracil (**5b**) with a cold (−78° or −98°C) solution of trifluoromethyl hypofluorite (CF_3OF) in trichlorofluoromethane (CCl_3F, Freon 11) gave high yields of the *cis*-5-fluoro-6-methoxy-5,6-dihydrouracil adducts (**6a,b**) (Scheme 2). The exclusive incorporation of solvent methoxyl at C-6 and the complete *cis* stereoselectivity of fluorine and methoxy orientation (by cationic or anionic processes) were indicated spectroscopically and confirmed by single-crystal X-ray analysis [16]. The previously assigned *trans* stereochemistry [14] of the 5-fluorouracil photo-hydrate (5-fluoro-6-hydroxy-5,6-dihydrouracil) was also corrected to *cis*. Synthetic procedures, spectroscopic details, and discussion of the stereoelectronically selective mechanisms are fully detailed [17]. Treatment of the adducts with base (triethylamine in aqueous methanol is a convenient system) gave 5-fluorouracil (7a) or its 1-methyl derivative (7b) in 90% overall crystalline yields [13,17]. Uridine (8a) and its tri-*O*-acetyl derivative were treated ($-78°C$) with CF_3OF/CCl_3F in methanol and chloroform solutions, respectively. The nucleoside adducts were less stable than the simple heterocyclic structures and were treated with base directly after evaporation of the reaction mixtures. The reactions proceeded in higher yields with the acetylated nucleosides in chloroform solution to give 82% of pure 5-fluorouridine (9a) vs. 47% from unprotected uridine in methanol. Analogous treatment of 2'-deoxyuridine (8b), 2'-*O*-methyluridine (8c), 3'-*O*-methyluridine (8d), and 2',3'-di-*O*-methyluridine (8e) gave the corresponding 5-fluorouracil nucleosides (9b–e) in moderate to high yields.

It was found [13,17] that acetylation of nucleosides using a catalytic amount of 4-*N,N*-dimethylaminopyridine in neat acetic anhydride gave high yields of pure peracetylated products. Certain technical difficulties involved with the usual pyridine/acetic anhydride procedures are thereby circumvented (e.g., colored solutions and residual pyridine by-products) and 3',5'-di-*O*-acetyl-2'-deoxyuridine was obtained in over 90% yield by direct crystallization. Since the base treatment necessary for "aromatization" of the 5,6-dihydro adducts concomitantly removes the acetyl blocking groups, an experimentally convenient three-stage procedure (acetylation, addition, elimination–deblocking) is realized.

Direct treatment of a methanolic solution of $O^2 \to 2'$-anhydroarabino-furanosyluracil (10) followed by base treatment under the usual conditions gave 5-fluoro-$O^2 \to 2'$-anhydroarabinofuranosyluracil (11) (60%) and 5-fluoro-1-β-D-arabinofuranosyluracil (12) (20%). In view of the facile acid- and base-catalyzed hydrolyses of uracil-O^2-anhydronucleosides, the mildness of this two-stage *in situ* procedure is accentuated [17].

Previous syntheses of 5-fluorocytosine and its nucleoside derivatives have been effected by various heterocyclic transformations from the corresponding 5-fluorouracil series. Cytosine (13) was rapidly converted to a 270 nm transparent adduct with $CF_3OF/CCl_3F/MeOH$ at $-78°C$. Evaporation of the reaction mixture and treatment of the residual adduct with base gave 5-fluorocytosine (14) in over 80% yield [18] plus a small quantity of 5-fluorouracil (7a) by deamination (Scheme 3). Cytidine (15a) and its per-*N,O*-acetylated derivative were treated analogously in methanol and chloroform solutions,

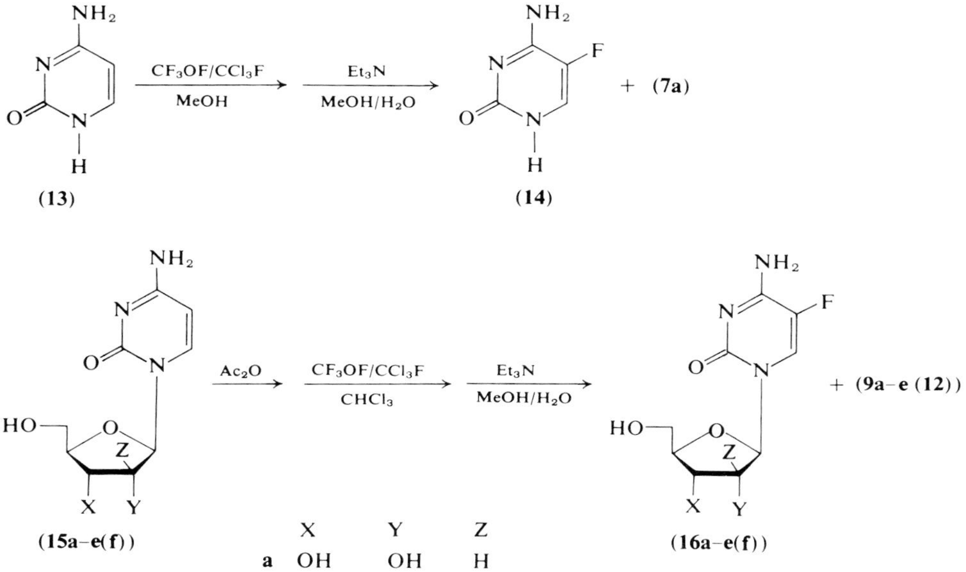

	X	Y	Z
a	OH	OH	H
b	OH	H	H
c	OH	OCH_3	H
d	OCH_3	OH	H
e	OCH_3	OCH_3	H
f	OH	H	OH

Scheme 3

respectively, followed by treatment with base to give 5-fluorocytidine (**16a**) plus 5-fluorouridine (**9a**). The sequential procedure involving peracetylation, addition, elimination–deblocking was applied to 2′-deoxycytidine (**15b**), 2′-*O*-methylcytidine (**15c**), 3′-*O*-methylcytidine (**15d**), 2′,3′-di-*O*-methylcytidine (**15e**), and 1-β-D-arabinofuranosylcytosine (**15f**) (araC, Cytarabine). The corresponding 5-fluorocytosine nucleosides (**16a–f**) plus 5-fluorouridine nucleosides (**9a–e, 12**) (by deamination) were obtained in over 80% (combined) yields [17]. Deamination of the 5-fluorocytosine to uracil products proceeded continuously in the basic medium, and after an extended period of time (~48 hr) only the 5-fluorouracil products were observed. Closely related isomeric sugar nucleosides in the cytosine series are readily separated by anion-exchange chromatography (Dowex 1-X2 [OH^-], the Dekker column) [19], whereas the corresponding uracil products (pK_a ~ 9.2) are not amenable to this strongly basic column separation unless *N*-3 derivatized [20]. Therefore, the *O*′-methylcytidines actually provide more convenient access to the corresponding 5-fluoro-*O*′-methyluridines. The 5-fluorouracil and cytosine

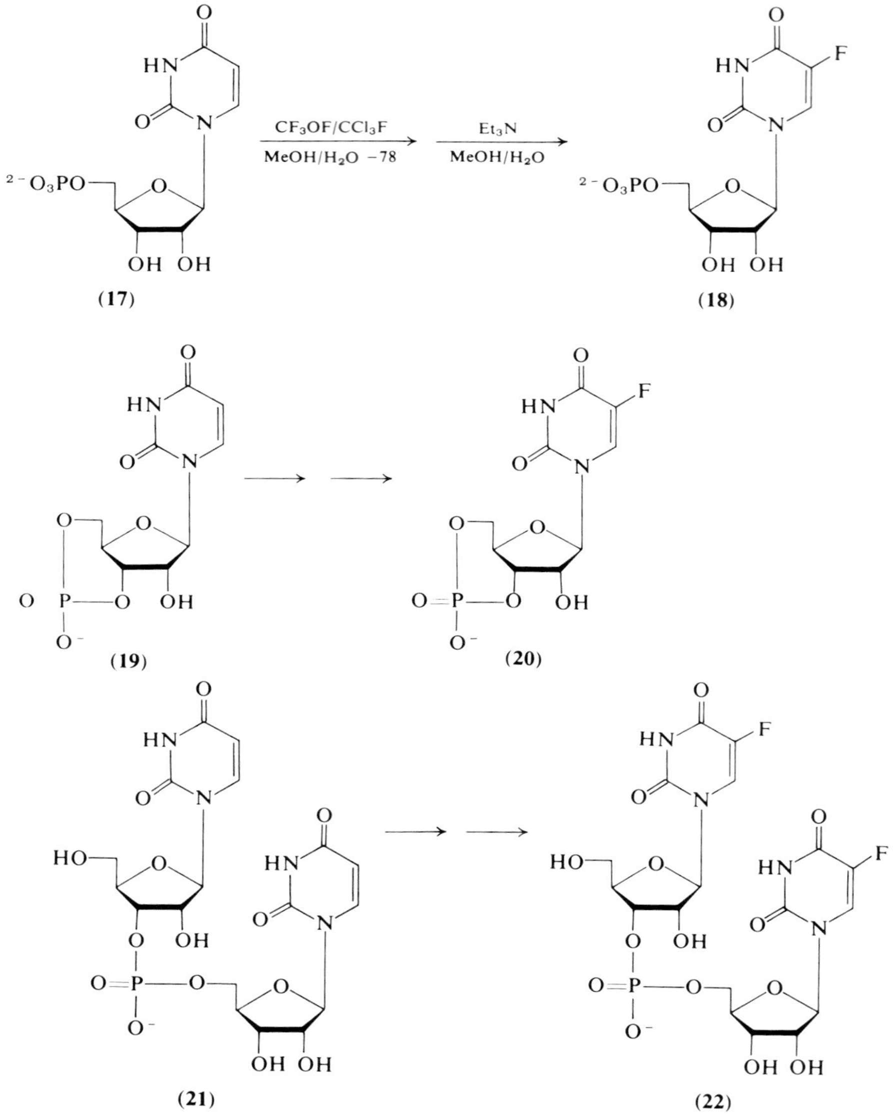

Scheme 4

products have markedly different chromatographic mobilities and can be readily separated. Thus, convenient access into both series can be gained from the cytosine precursor. The relative yields of the two pyrimidine products can be controlled by quenching the basic reaction medium at the desired point as monitored by tlc. Exclusive formation of the 5-fluorocytosine product can be realized by treatment of the 5,6-dihydro adduct with anhydrous dimethyl-amine in absolute ethanol. Such processing gave ~90% yields of 5-fluorocyt-idine (16a) and 5-fluoro-2',3'-di-O-methylcytidine (16e) with none of the corresponding 5-fluorouracil products [(9a) and (9e)] detected [17].

Extension of this general procedure to nucleotides using aqueous methanol solutions for fluorination gave moderate (~60%) yields of 5-fluorouridine-5'-phosphate (18) and 5-fluorouridine 3',5'-cyclic phosphate (20) from the corresponding uridylic acids (17) and (19), respectively [21] (Scheme 4). Application to dinucleoside monophosphates was explored briefly, and a low yield of 5-fluorouridylyl-(3' → 5')-5-fluorouridylate (22) (5FUp5FU) from UpU (21) was obtained and characterized by enzymatic and spectroscopic methods [21]. However, the aqueous solutions required for nucleotide solubility give rise to adducts (presumably 5-fluoro-6-hydroxy-5,6-dihydro structures) which do not undergo clean elimination to the 5-fluorouracil product in basic media [14].

Isolation of the ubiquitous 2'-O-methyl nucleosides from tRNA's and rRNA's [1] has been followed recently by their discovery in mRNA's and nRNA's [22]. Syntheses of these sugar–ether derivatives have involved base coupling with the methylated sugar [23], methylation of intact nucleosides using diazomethane in hot aqueous DME [24], and alkylation in strong aqueous base [25]. However, these procedures give rise to variable yields of the various possible isomers of the multifunctional structures. We have found that stannous chloride [26] and a number of other inorganic Lewis acids [27] markedly catalyze the monomethylation of vicinal cis-diol group-ings using diazomethane. The $O^{2'}$- to $O^{3'}$-methyl isomeric ratio with adenosine can be altered from ~1:1 to ~4:1 by the choice of catalyst [27]. However, the generality and structural basis of these observations have been explored in only a preliminary fashion at present.

Stannous chloride dihydrate is a colorless, innocuous, crystalline salt, which allows observation of the persistence of yellow diazomethane when reactions are complete. Treatment of methanolic solutions of a variety of nucleosides that have no "acidic" heterocyclic protons, including adenosine (23a), 6-chloropurine riboside (23b), the antibiotic tubercidin (23c) (4-amino-7-β-D-ribofuranosylpyrrolo[2,3-d]pyrimidine),   4-methoxy-1-β-D-ribofuranosyl-2-pyrimidinone (23d), and cytidine (15a) (Scheme 5), with diazomethane in the presence of stannous chloride gave quantitative conversions to the corre-sponding 2'-O-methyl [(24a–d),(15c)] and 3'-O-methyl [(25a–d),(15d)] nucleo-sides [28]. The 6-chloropurine derivatives (24b) and (25b) were treated with hydrosulfide to give the analogous 6-mercaptopurine (6-thioinosine) products. These were converted to the 6-p-nitrobenzylthio derivatives (using p-nitro-benzyl bromide) [28] for evaluation of sugar hydroxyl binding specificity in cell-wall-mediated nucleoside transport systems [29]. The 4-methoxy group

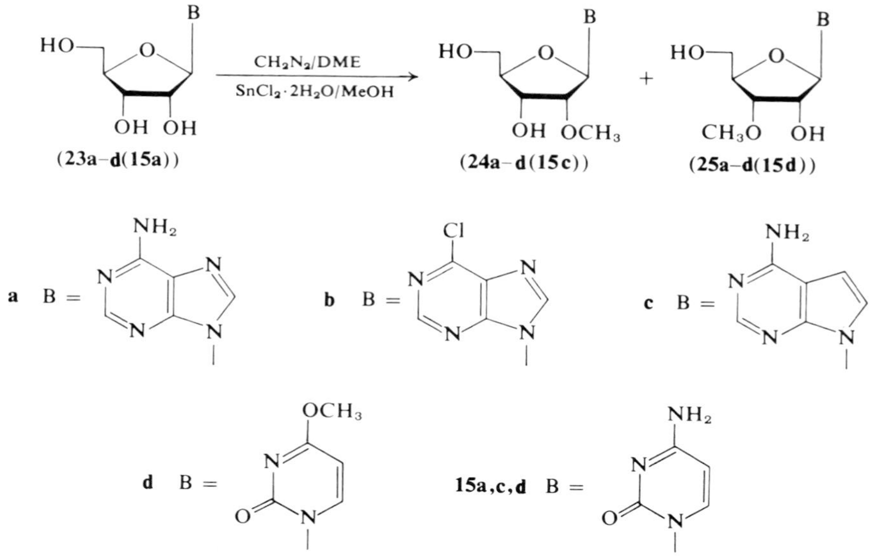

Scheme 5

of (**24d**) was nucleophilically displaced by hydrosulfide to give 2'-*O*-methyl-4-thiouridine [30], a 4-thiouridine derivative that would be expected to be resistant to nucleoside phosphorylase and hydrolase enzymes [31], and by methylamine to give $N^4,O^{2'}$-di-methylcytidine [30], a base–sugar doubly modified nucleoside isolated from *Escherichia coli* RNA [32]. The anion-exchange chromatographic procedure devised by Dekker [19] readily separates 2'-*O*-methyl and 3'-*O*-methyl isomers of nucleosides which are stable in the strongly basic resin column.

   A second series of nucleosides that have at least one acidic heterocyclic ring proton was also investigated [28] (Scheme 6). The antibiotic formycin (**26a**) (7-amino-3-β-D-ribofuranosylpyrazolo[4,3-*d*]pyrimidine) was selectively monomethylated at O-2' and O-3' in over 80% combined yield in the presence of the acidic ($pK_a \sim 9.5$) pyrazole proton (N-1—H). Guanosine (**26b**) and inosine (**26c**) are very readily methylated at N-7 by diazomethane. Slow addition of diazomethane to warm solutions of (**26b**) and (**26c**) in DMF containing stannous chloride gave $\sim 50\%$ combined yields of (**27b**) plus (**28b**) and (**27c**) plus (**28c**), respectively, plus other products and traces of starting materials. Uridine (**8a**) ($pK_a \sim 9.2$), which is rapidly methylated with diazomethane in methanolic solution to give $\sim 90\%$ of 3-methyluridine, was carefully treated under the catalytic conditions to give $\sim 85\%$ of O-2' (**8c**)

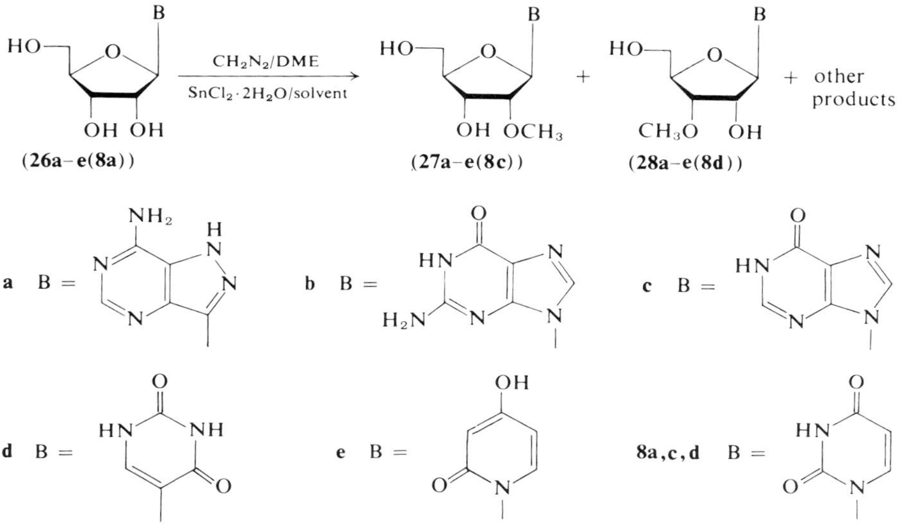

Scheme 6

and O-3' (**8d**) monomethylated products. Pseudouridine (**26d**) (pK_a ~ 8.9) has two acidic ring protons but was again selectively O-2' (**27d**) and O-3' (**28d**) monomethylated in ~80% combined yield in the presence of stannous chloride [28]. Comparably convenient syntheses of the base–sugar doubly modified natural product (**27d**) are difficult to envisage.

Finally, the antibacterial and antileukemia active synthetic nucleoside 3-deazauridine (**26e**) (4-hydroxy-1-β-D-ribofuranosyl-2-pyridinone) [33] was subjected to this reaction [34]. This ketopyridinol nucleoside is seen to correspond in acidity (pK_a ~ 6.3) to an electronegatively substituted phenol, and it is rapidly and quantitatively converted to a mixture of 4(2)-methoxy-2(4)-pyridinone ribosides using diazomethane in methanol [34]. However, in the presence of stannous chloride a 67% combined yield of the O-2' (**27e**) and O-3' (**28e**) monomethylated products was isolated, again emphatically demonstrating the catalytic efficiency of this presumably metal—diol complexed system. Various anion-exchange and adsorption chromatographic techniques were employed to resolve the sugar-methylated nucleoside isomers of acidic bases [28,30,34].

Alkylation of 2'-O-methyladenosine (**24a**) with methyl iodide gave $N^1,O^{2'}$-dimethyladenosine (**29**) (Scheme 7), which underwent smooth Dimroth rearrangement in methanolic dimethylamine to give $N^6,O^{2'}$-dimethyladenosine (**30**) [35]. Treatment of 6-chloropurine riboside (**23b**) with diazomethane in the presence of stannous chloride followed by stirring of the crude reaction

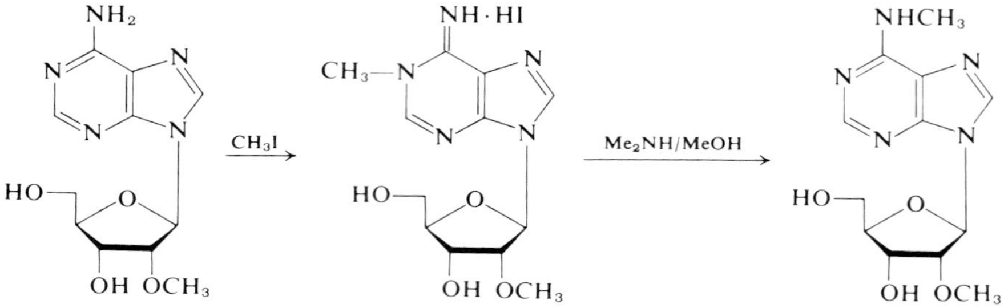

Scheme 7

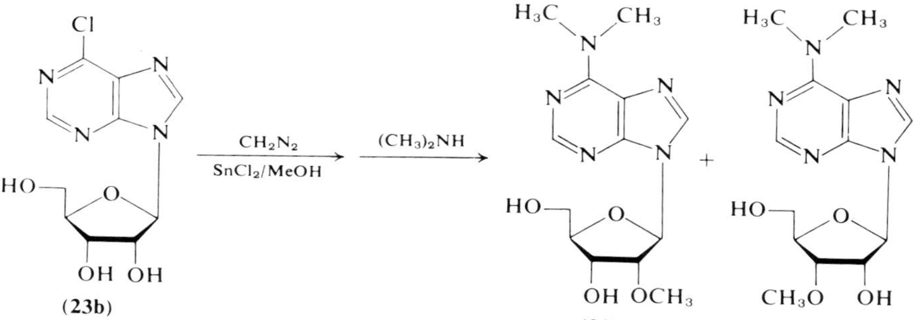

product with dimethylamine gave $N^6,N^6,O^{2'}$-trimethyladenosine (31) and its 3'-O-methyl isomer (32). Compound (30) was recently reported to be the penultimate nucleoside in the highly modified 5'-terminal "cap" of a number of viral and mammalian mRNA's [36] and to occur at an internal position in nRNA [36a]. On the basis of ratios of radioactive methyl incorporation into the base and sugar moieties of modified "cap" nucleosides of adenovirus mRNA's, the presence of trimethyl derivative (31) has been suggested [37]. Since these doubly modified nucleosides occur in mRNA's in the range of one per thousand nucleoside units, the usual spectroscopic techniques are effectively precluded, and radioactive tracer label was employed for detection on paper chromatography and electrophoresis [36,37]. However, compounds (30) and (31) migrate at essentially equal mobilities in a number of paper chromatographic systems. The use of adenosine deaminase cleanly differentiates between these two methylated derivatives since the 6-methylamino (30) product is completely converted to the inosine analog (27c), whereas the 6-dimethylamino (31) compound shows no substrate activity [35].

A final area of interest involves structural transformations of the sugar portion of intact nucleosides. The antibiotics tubercidin (23c) and formycin (26a) closely resemble adenosine (23a) chemically and have been observed to undergo many of the corresponding enzymatic transformations [2]. These antibiotics are amenable to large-scale production by fermentation, whereas the multistep routes devised for their total syntheses [38] make analogous approaches with modified sugars rather uninviting. The purine-8-cyclonucleoside approach of Ikehara [5] is not readily applicable in the tubercidin (23c) series and is precluded in the case of formycin (26a), which has N-2 at the position corresponding to C-8 of adenine. A general procedure, independent of specific base structures, was devised [39] on the basis of nucleophilic functionalization via sugar acyloxonium ions generated from 2',3'-O-orthoesters.

Adenosine (23a) was quantitatively converted to 2',3'-O-methoxyethyl-ideneadenosine (33) (Scheme 8) by acid-catalyzed orthoester exchange with methyl orthoacetate [40]. Treatment of (33) with inorganic Lewis acids in the presence of a large excess of sodium iodide or with pyridine hydrohalides resulted in significant quantities of 2'(3')-O-acetyl products formed by overall hydrolysis of the orthoacetate function (after workup). Acetyl halides in hot pyridine led to dark intractable mixtures. The α,α,α-trisubstituted acyl halide pivalyl chloride (in which ketene formation is precluded) in refluxing pyridine effected conversion of (33) to a mixture containing the expected 6-N-pival-amido-9-(3-chloro-3-deoxy-2-O-acetyl-5-O-pivalyl-β-D-xylofuranosyl)purine (34a) and its 2'-chloro-3'-O-acetylarabino isomer (35a) [40]. A second pair of products identified as 6-N-pivalamido-9-(3-chloro-3-deoxy-2-O-[4,4-dimethyl-3-pivaloxypent-2-enoyl]-5-O-pivalyl-β-D-xylofuranosyl)purine (34b) and its 2'-chloro-3'-O-DMPP isomer (35b) were formed in minor amount. Treatment of this mixture with tri-n-butyltin hydride in the presence of AIBN followed by deblocking gave 3'-deoxyadenosine (37) (the antibiotic cordycepin) and 2'-deoxyadenosine (1). Methanolic sodium methoxide treatment of the mixture effected deblocking and concomitant epoxide closure of the resulting chloro-hydrins to give 2',3'-anhydroadenosine (36) in ~60% yields [40]. This epoxide product (36) suffers intramolecular cyclization readily in warm water to give the aminoimidazolecarboxamidine cyclonucleoside (38) by initial attack of N-3 at C-3' followed by attack of water at C-2 of the positively charged pyrimidine ring and opening of the heterocycle [41]. Benzoylation of (36) gave the $N^6,N^6,O^{5'}$-tribenzoyl derivative (39). This easily crystallized product is stabilized [42] against N-3 $\rightarrow$ C-3' nucleophilic attack and is also soluble in organic solvents. Nucleophilic opening of the epoxide ring with benzoate, methanol/borohydride, azide, and fluoride occurred with high regioselectivity at C-3' to give (40a,b,c,e), respectively, after deblocking [41]. Reduction of (40c) gave (40d), which is the 3'-amino epimer of the antibiotic 3'-amino-3'-deoxyadenosine [2].

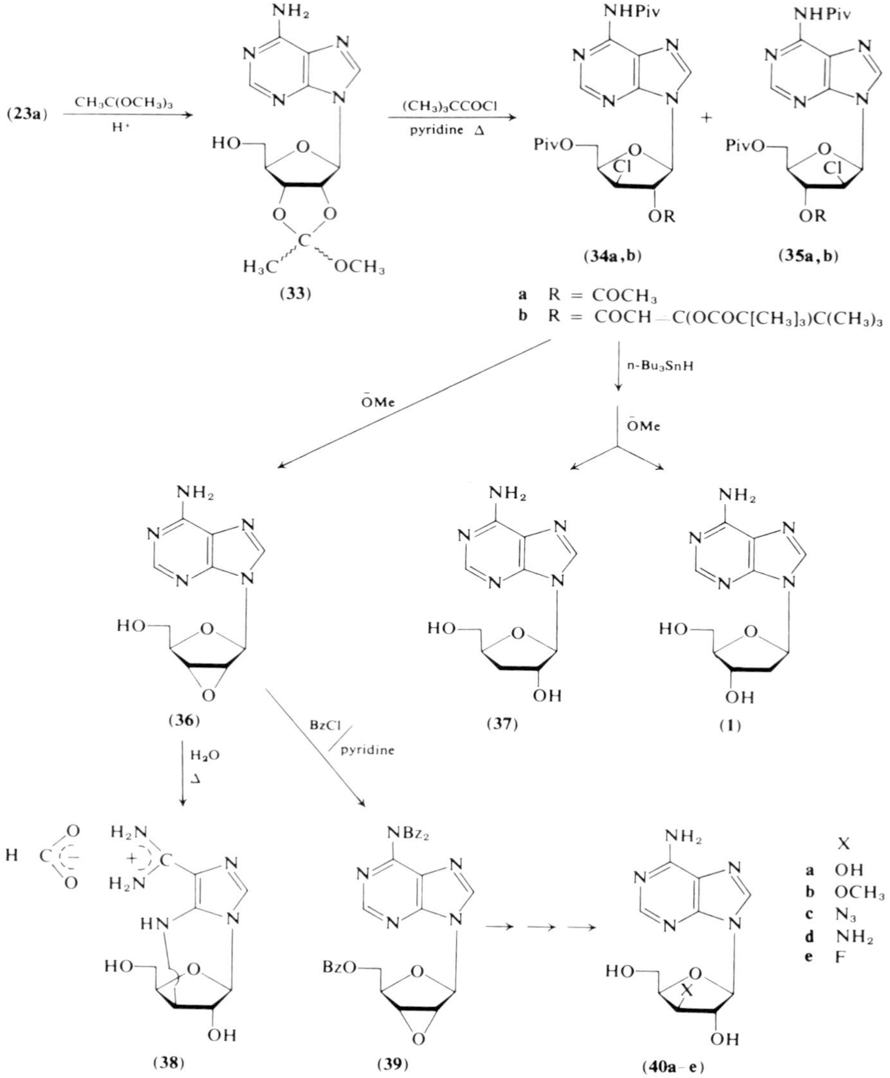

Scheme 8

Treatment of orthoester (33) with pivalyl chloride and excess sodium iodide in refluxing pyridine (Scheme 9) effected rapid conversion to a mixture containing 6-N-pivalamido-9-(2-iodo-2-deoxy-3-O-[4,4-dimethyl-3-pivaloxy-pent-2-enoyl]-5-O-pivalyl-β-D-arabinofuranosyl)purine (41); the corresponding 3'-iodo-2'-O-DMPP derivative (42); the 3'-ene product (43), derived from (42)

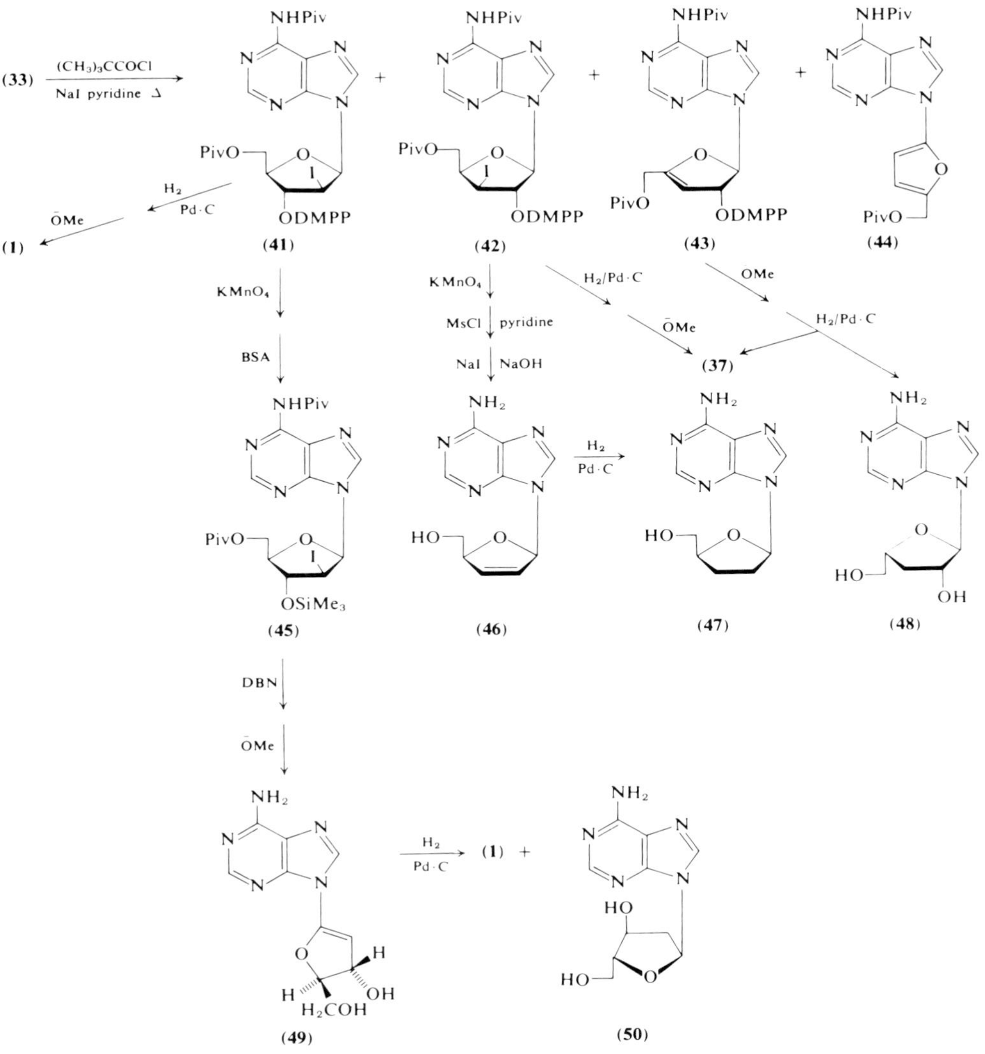

Scheme 9

by elimination of hydrogen iodide; and the pivaloxymethylfuran derivative (**44**) [40,43]. Catalytic hydrogenolysis of the iodo derivative (**41**) followed by deblocking gave 2′-deoxyadenosine (**1**) in over 90% yield. Analogous treatment of the major 3′-iodo component (**42**) gave a similar excellent yield of cordycepin (**37**) [40]. Silver acetate effected selective elimination of hydrogen iodide from (42) to yield 3'-ene (43) quantitatively. The deblocked crystalline 3'-ene product, 9-(3-deoxy-β-D-*glycero*-pent-3-enofuranosyl)adenine, was hydrogenated to give cordycepin (37) and its 4'-epimer, 9-(3-deoxy-α-L-*threo*-pentofuranosyl)adenine (48), in excellent combined yield after anion-exchange column chromatographic separation [43]. Selective oxidative removal of the DMPP group was effected quantitatively using potassium permanganate in cold aqueous pyridine. Such treatment of (42) followed by mesylation of the freed hydroxyl group gave a somewhat unstable *trans*-iodomesylate. Clean elimination of iodine and mesylate from this intermediate was effected by stirring with sodium iodide in aqueous sodium hydroxide solution. The initial purple (iodine) color faded to yellow (hypoiodite), and side reactions including glycosyl cleavage which occurred under usual conditions for this nucleophilic elimination process were avoided. Concomitant deblocking occurred to give the 2'-ene, 9-(2,3-dideoxy-β-D-*glycero*-pent-2-enofuranosyl)adenine (46), in high yield [43]. Hydrogenation of (46) gave the DNA chain terminator 2',3'-dideoxyadenosine (47) in over 90% yield. Oxidative removal of the DMPP function from (41) followed by protection of the hydroxyl group using BSA gave the 2'-iodo-3'-trimethylsilyloxy product (45). Treatment of (45) with DBN followed by deblocking gave the first [44] authentic 1',2'-unsaturated nucleoside, 9-(2-deoxy-D-*erythro*-pent-1-enofuranosyl)adenine (49), in 90% overall crystalline yield from (45). Hydrogenation of (49) gave 2'-deoxy-adenosine (1) plus its α-anomer, 9-(2-deoxy-α-D-*erythro*-pentofuranosyl)-adenine (50) in a ratio of ~5:1. Although the overall yield of (50) from adenosine is prohibitively low as described [43,44] [owing primarily to the minor formation of (41) in the original reaction and the unusual stereoselectivity in the hydrogenation of (49)], this sequence represents the novel conversion of a β-ribonucleoside to its α-2'-deoxynucleoside without rupture and reforming of glycosyl or C-1'—O-4' bonds.

Published data on a previously described [45] product purported to be a 1',2'-unsaturated uridine derivative led us to suspect the validity of that claim. Treatment of $O^2 \rightarrow$ 2'-anhydro-1-β-D-arabinofuranosyluracil (10) with 2-methoxypropene/acid or *tert*-butyldimethylsilyl chloride/imidazole gave the 3',5'-O-bis(2-methoxyprop-2-yl) (51a) or 3',5'-O-bis-(*tert*-butyldimethylsilyl) (51b) derivatives, respectively (Scheme 10). Treatment of these blocked $O^2 \rightarrow$ 2'-cyclonucleosides with potassium *tert*-butoxide in dimethylformamide gave the corresponding blocked 1'-ene derivatives, (52a) or (52b), by elimination of H-1' and uracil 2-oxide [46]. The extremely acid labile N,O-ketene acetal compound (52b) was treated with tetraethylammonium fluoride to give the free 1'-ene, 1-(2-deoxy-D-*erythro*-pent-1-enofuranosyl)uracil (54). Crystalline compound (54) was completely characterized and was found to exhibit spectroscopic and chromatographic properties different from the previously reported product of reaction of reduced hydroxycobalamin (vitamin $B_{12}s$)

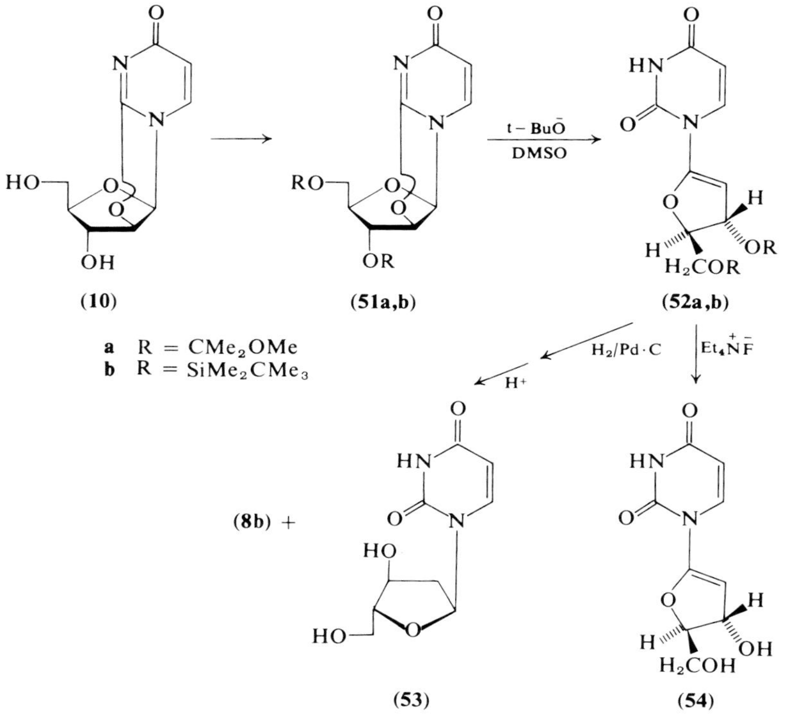

Scheme 10

with 2'-bromo-2'-deoxyuridine, which had been assigned structure (54) [45]. Thus, the presently described [46] transformation represents the first example of a pyrimidine 1',2'-unsaturated nucleoside. Hydrogenation of the more readily accessible 1'-ene derivative (52a) followed by acidic deblocking gave the acid-stable 2'-deoxyuridine (8b) and its α-anomer, 1-(2-deoxy-α-D-*erythro*-pentofuranosyl)uracil (53), in approximately equal amounts. In this sequence, the high-yield exclusive conversions to (52a) and the absence of β-anomer stereoselectivity in its hydrogenation (observed in the adenine counterpart) result in a preparatively viable route to α-2'-deoxyuridine (53) from uridine (8a) [46].

The 2'- and 3'-deoxy derivatives of formycin (26a) and 3-'deoxytubercidin (59) were prepared [47] using the "abnormal Mattocks reaction" [48], which was first investigated systematically and applied to nucleoside chemistry by Moffatt and co-workers [49,49a]. However, epimeric sugar products of these antibiotics and the biochemically interesting 2'-deoxyturbecidin remained unknown. Analogous repetitions of the total synthesis routes

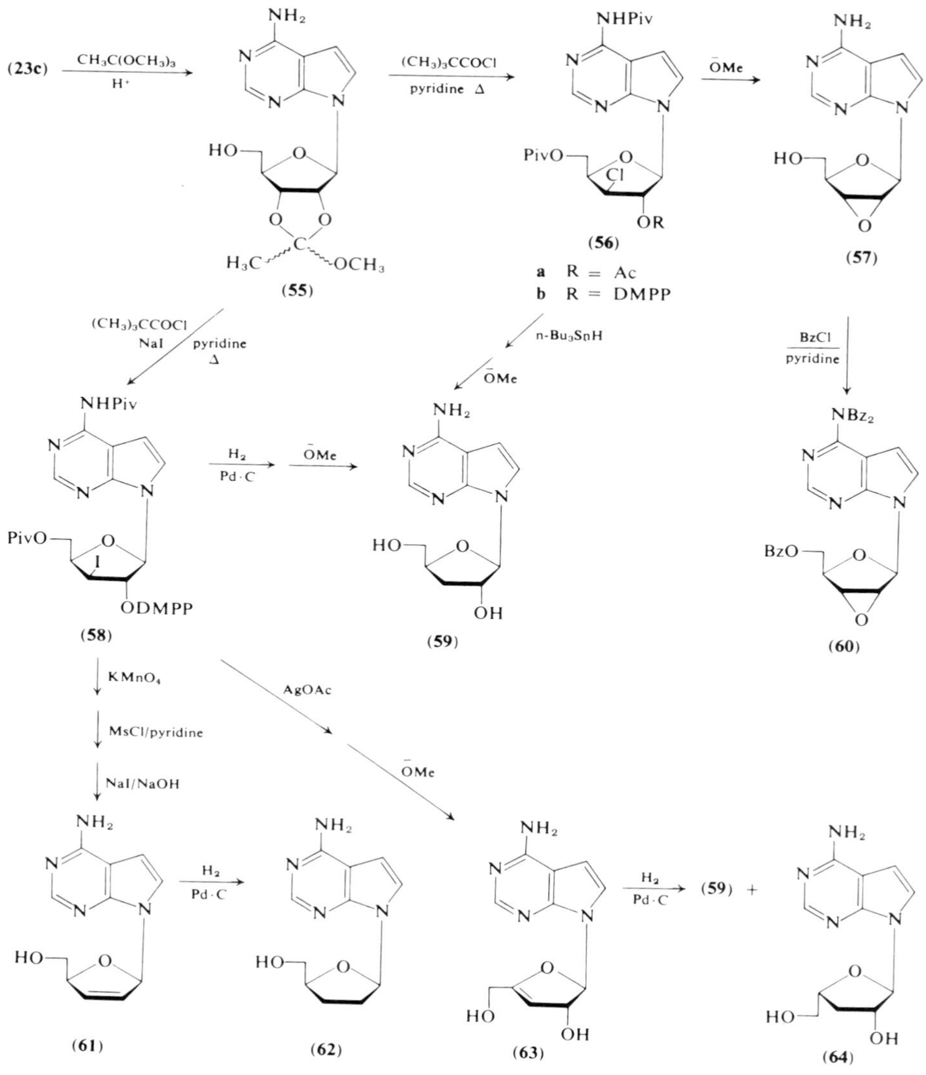

Scheme 11

[38] with other sugar derivatives were not appealing. Treatment of tuber-cidin (**23c**) with methyl orthoacetate gave 2',3'-O-methoxyethylidene-tubercidin (**55**) in high yields [50] (Scheme 11). This product was treated with pivalyl chloride in hot pyridine to give 4-N-pivalamido-7-(3-chloro-3-deoxy-2-O-acetyl-5-O-pivalyl-β-D-xylofuranosyl)pyrrolo[2,3-d]pyrimidine (**56a**) plus a

minor amount of the corresponding DMPP derivative analogously to the reaction with adenosine. However, no 2'-chloro isomers were detected [50] in harmony with our previous results [47] and those of Moffatt [49a]. Treatment of this mixture with tri-*n*-butyltin hydride/AIBN and deblocking [50] gave the known 3'-deoxytubercidin (**59**) [47]. The original reaction mixture (**56a,b**) was stirred with methanolic sodium methoxide to effect deblocking and concomitant epoxide closure, and 2',3'-anhydrotubercidin (**57**) was obtained in 50–70% yields [50]. Epoxide (**57**) is extremely susceptible to presumed N-1 → C-3' cyclization with attendant decomposition to purple materials. Benzoylation of (**57**) gave the $N^4,N^4,O^{5'}$-tribenzoyl derivative (**60**), which is a colorless stable crystalline product. Treatment of orthoester (**55**) with pivalyl chloride and excess sodium iodide in hot pyridine gave the 3'-iodo-2'-*O*-DMPP product (**58**) plus unsaturated derivatives [50]. Hydrogenolysis of the iodo function of (**58**) followed by deblocking gave (**59**) in good yield. Oxidative removal of the 2'-*O*-DMPP group from (**58**) was followed by mesylation of the freed hydroxyl group and then iodide-induced elimination of the *trans*-iodomesylate functions. Concomitant deblocking occurred in the basic aqueous solution to give the 2'-ene, 4-amino-7-(2,3-dideoxy-β-D-*glycero*-pent-2-enofuranosyl)pyrrolo[2,3-*d*]pyrimidine (**61**). Hydrogenation of (**61**) gave 2',3'-dideoxytubercidin (**62**). Treatment of (**58**) with silver acetate effected selective elimination of hydrogen iodide and the resulting blocked 3'-ene was deprotected to give 4-amino-7-(3-deoxy-β-D-*glycero*-pent-3-enofuranosyl)pyrrolo[2,3-*d*]pyrimidine (**63**). Hydrogenation of (**63**) gave 3'-deoxytubercidin (**59**) plus its 4'-epimer, 4-amino-7-(3-deoxy-α-L-*threo*-pentofuranosyl)pyrrolo[2,3-*d*]pyrimidine (**64**), in high combined yield after anion-exchange column chromatographic separation [50].

Owing to the marked susceptibility of 2',3'-anhydrotubercidin (**57**) to undergo transient $N^1 → C^{3'}$-cyclonucleoside formation and then degradation, milder reaction conditions were investigated. Treatment of 4-amino-7-(3-iodo-3-deoxy-2-*O*-acetyl-5-*O*-[2,5,5-trimethyl-1,3-dioxolan-4-on-2-yl]-β-D-xylofuranosyl)pyrrolo[2,3-*d*]pyrimidine, which was prepared quantitatively in one step from tubercidin (**23c**) [47], with alcoholic ammonia and rapid silica gel column chromatographic purification gave (**57**) in 96% overall yield from (**23c**) [51]. The tribenzoyl derivative (**60**) was treated with boron trifluoride etherate to effect 3' opening of the epoxide ring with 3',5'-benzoxonium ion participation. Deblocking of the intermediate(s) gave 4-amino-7-β-D-xylofuranosylpyrrolo[2,3-*d*]pyrimidine (xyloTu) (**65**) in 91% overall yield from the parent antibiotic (**23c**) [51] (Scheme 12). Formation of the 3',5'-*O*-isopropylidine derivative, mesylation of the 2'-hydroxyl group, acidic removal of the acetonide group, and epoxide closure in methanolic sodium methoxide solution gave 4-amino-7-(2,3-anhydro-β-D-lyxofuranosyl)pyrrolo[2,3-*d*]pyrimidine (**66**). Treatment of the lyxo epoxide (**66**) with sodium benzoate in hot

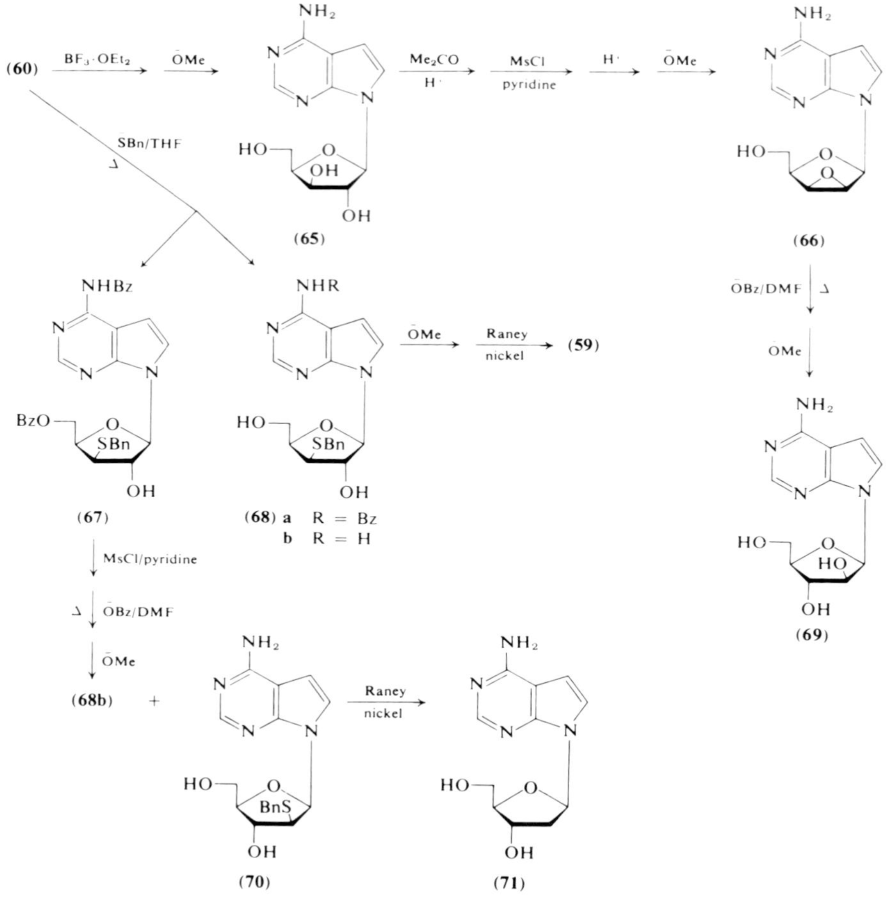

Scheme 12

moist DMF followed by deblocking gave 4-amino-7-β-D-arabinofuranosyl-pyrrolo[2,3-d]pyrimidine (araTu) (**69**) plus a small quantity of (**65**). The araTu (**69**) and xyloTu (**65**) are seen to be antibiotic analogs of araA (9-β-D-arabinofuranosyladenine) and xyloA (9-β-D-xylofuranosyladenine). The latter two adenine nucleosides are active antiviral and antitumor agents and have been studied extensively [52]. However, they are efficient substrates for adenosine deaminase, which occurs in reasonable amounts in mammalian blood and digestive tracts, and are thereby rapidly converted to the corresponding hypoxanthine products [53]. In contrast, tubercidin (**23c**), araTu (**69**), and xyloTu (**65**) exhibit no substrate activity with this enzyme over extended periods of time [2,51]. The ara and xylo products (**69**) and (**65**) have significant human tumor inhibitory activity, and biochemical and biological studies of these diastereomers of (23c) are underway [54].

Careful treatment of (60) with sodium benzylthiolate in dry THF gave 4-N-benzamido-7-(3-S-benzylthio-3-deoxy-5-O- benzoyl -β- D - xylofuranosyl)-pyrrolo[2,3-d]pyrimidine (67) plus its 5'-deblocked product (68a) in a ratio of ∼3:1 and 90% combined yield [55]. Again, no product of C-2' attack was observed. Mesylation of (67) proceeded quantitatively, and the amorphous trans-benzylthiomesylate was treated with sodium benzoate in hot DMF. Deblocking in the usual manner gave 4-amino-7-(2-S-benzylthio-2-deoxy-β-D-arabinofuranosyl)pyrrolo[2,3-d]pyrimidine (70) and the corresponding 3'-S-benzylthioxylo isomer (68b) in a ratio of ∼3:2 and 90% combined yield after separation by anion-exchange column chromatography. Desulfurization of (70) with Raney nickel in hot DMF gave the elusive 2'-deoxytubercidin (71) in 77% yield [55]. This corresponds to an overall yield of 27% for the eight-stage conversion of tubercidin (23c) to 2'-deoxytubercidin (71). Analogous desulfurization of (68b) gave 3'-deoxytubercidin (59) in equivalent yield. Thus, the combined overall yield of ∼60% of (71) and (59) [from assumed pooling of (68a) and (68b)] demonstrates the viability of the high-yield transformations in this sequence, which relied on intramolecular migration of the benzylthio group via a thiiranium intermediate [56] to place a hydrogenolizable functionality at C-2'.

The successful development of these various direct transformations of naturally occurring nucleosides makes readily available a variety of new biologically interesting molecules with modified base and/or sugar structures which would not be easily accessible by coupling or ring elaboration routes. In addition, a number of known compounds that elicit established biological responses have been synthesized in a stereochemically pure state and in high yields from commercially available nucleosides. High-yield multistage conversions of tubercidin to its ara and xylo epimers and deoxynucleosides vindicate our hopes that such transformations of antibiotics derived from fermentation would provide these biomolecule analogs in practical, accessible quantities for desired biochemical and biological evaluation. Mechanistic discussions concerning the new reactions described and reference bibliographies relative to the known reactions and compounds covered in this chapter can be found in the primary sources quoted.

ACKNOWLEDGMENTS

We thank the National Research Council of Canada (A5890), the National Cancer Institute of Canada, and The University of Alberta for generous financial support. Stimulating discussions and pleasant associations with, as well as dedicated and expert experimental efforts by, my coworkers indicated in the primary publications quoted are gratefully and enthusiastically acknowledged.

REFERENCES

1. R. H. Hall, "The Modified Nucleosides in Nucleic Acids." Columbia Univ. Press, New York, 1971; S. Nishimura, *Prog. Nucleic Acid Res. Mol. Biol.* **12**, 49 (1972).
2. R. J. Suhadolnik, "Nucleoside Antibiotics." Wiley (Interscience), New York, 1970.
3. L. Goodman, *Basic Princ. Nucleic Acid Chem.* **1**, 93–208 (1974).
3a. S. Hanessian and A. G. Pernet, *Adv. Carbohydr. Chem. Biochem.* **33**, 111 (1976).
4. J. J. Fox, *Pure Appl. Chem.* **18**, 233 (1969).
5. M. Ikehara, *Acc. Chem. Res.* **2**, 47 (1969).
6. R. H. Iwamoto, E. M. Acton, and L. Goodman, *J. Org. Chem.* **27**, 3949 (1962).
7. M. J. Robins and R. K. Robins, *J. Am. Chem. Soc.* **87**, 4934 (1965).
8. M. Ikehara and H. Uno, *Chem. Pharm. Bull.* **13**, 221 (1965); J. Žemlička and F. Šorm, *Collect. Czech. Chem. Commun.* **30**, 1880 (1965).
9. M. J. Robins and R. K. Robins, *J. Am. Chem. Soc.* **86**, 3585 (1964).
10. J. A. Zoltewicz, D. F. Clark, T. W. Sharpless, and G. Grahe, *J. Am. Chem. Soc.* **92**, 1741 (1970).
11. M. J. Robins and G. L. Basom, *Can. J. Chem.* **51**, 3161 (1973).
12. C. Heidelberger, *Prog. Nucleic Acid Res. Mol. Biol.* **4**, 1 (1965).
13. M. J. Robins and S. R. Naik, *J. Am. Chem. Soc.* **93**, 5277 (1971); **94**, 2158 (1972).
14. R. Duschinsky, T. Gabriel, W. Tautz, A. Nussbaum, M. Hoffer, E. Grunberg, J. H. Burchenal, and J. J. Fox, *J. Med. Chem.* **10**, 47 (1967).
14a. L. Szabo, T. I. Kalman, and T. J. Bardos, *J. Org. Chem.* **35**, 1434 (1970).
15. D. H. R. Barton, *Pure Appl. Chem.* **21**, 285 (1970).
16. M. N. G. James and M. Matsushima, *Acta Crystallogr., Sect. B*, **32**, 957 (1976).
17. M. J. Robins, M. MacCoss, S. R. Naik, and G. Ramani, *J. Am. Chem. Soc.* **98**, 7381 (1976).
18. M. J. Robins and S. R. Naik, *J. Chem. Soc., Chem. Commun.* p. 18 (1972).
19. C. A. Dekker, *J. Am. Chem. Soc.* **87**, 4027 (1965).
20. K. D. Philips and J. P. Horwitz, *J. Org. Chem.* **40**, 1856 (1975).
21. M. J. Robins, G. Ramani, and M. MacCoss, *Can. J. Chem.* **53**, 1302 (1975).
22. R. C. Desrosiers, K. H. Friderici, and F. M. Rottman, *Biochemistry* **14**, 4367 (1975) and references therein.
23. A. H. Haines, *Tetrahedron* **29**, 2807 (1973).
24. A. D. Broom and R. K. Robins, *J. Am. Chem. Soc.* **87**, 1145 (1965).
25. J. T. Kusmierek, J. Giziewicz, and D. Shugar, *Biochemistry* **12**, 194 (1973).
26. M. J. Robins and S. R. Naik, *Biochim. Biophys. Acta* **246**, 341 (1971).
27. M. J. Robins, A. S. K. Lee, and F. A. Norris, *Carbohydr. Res.* **41**, 304 (1975).
28. M. J. Robins, S. R. Naik, and A. S. K. Lee, *J. Org. Chem.* **39**, 1891 (1974).
29. C. E. Cass and A. R. P. Paterson, *Biochim. Biophys. Acta* **419**, 285 (1976).
30. M. J. Robins and S. R. Naik, *Biochemistry* **10**, 3591 (1971).
31. M. Honjo, Y. Kanai, Y. Furukawa, Y. Mizuno, and Y. Sanno, *Biochim. Biophys. Acta* **87**, 696 (1964).
32. J. L. Nichols and B. G. Lane, *Biochim. Biophys. Acta* **119**, 649 (1966).
33. M. J. Robins and B. L. Currie, *Chem. Commun.* p. 1547 (1968).
34. M. J. Robins and A. S. K. Lee, *J. Med. Chem.* **18**, 1070 (1975).
35. M. J. Robins, M. MacCoss, and A. S. K. Lee, *Biochem. Biophys. Res. Commun.* **70**, 356 (1976).
36. C.-M. Wei, A. Gershowitz, and B. Moss, *Nature (London)* **257**, 251 (1975).
36a. H. Shibata, T. S. Ro-Choi, R. Reddy, Y. C. Choi, D. Henning, and H. Busch, *J. Biol. Chem.* **250**, 3909 (1975).

37. S. Sommer, M. Salditt-Georgieff, S. Bachenheimer, J. E. Darnell, Y. Furuichi, M. Morgan, and A. J. Shatkin, *Nucleic Acids Res.* **3**, 749 (1976).
38. R. L. Tolman, R. K. Robins, and L. B. Townsend, *J. Am. Chem. Soc.* **90**, 524 (1968); **91**, 2102 (1969); E. M. Acton, K. J. Ryan, D. W. Henry, and L. Goodman, *J. Chem. Soc., Chem. Commun.* p. 896 (1971); R. A. Long, A. F. Lewis, R. K. Robins, and L. B. Townsend, *J. Chem. Soc. C* p. 2443 (1971).
39. M. J. Robins, R. Mengel, and R. A. Jones, *J. Am. Chem. Soc.* **95**, 4074 (1973).
40. M. J. Robins, R. Mengel, R. A. Jones, and Y. Fouron, *J. Am. Chem. Soc.* **98**, 8204 (1976).
41. M. J. Robins, Y. Fouron, and R. Mengel, *J. Org. Chem.* **39**, 1564 (1974).
42. W. Jahn, *Chem. Ber.* **98**, 1705 (1965).
43. M. J. Robins, R. A. Jones, and R. Mengel, *J. Am. Chem. Soc.* **98**, 8213 (1976).
44. M. J. Robins and R. A. Jones, *J. Org. Chem.* **39**, 113 (1974).
45. V. I. Borodulina-Shvets, I. P. Rudakova, and A. M. Yurkevich, *Zh. Obshch. Khim.* **41**, 2801 (1971).
46. M. J. Robins and E. M. Trip, *Tetrahedron Lett.* p. 3369 (1974).
47. M. J. Robins, J. R. McCarthy, Jr., R. A. Jones, and R. Mengel, *Can. J. Chem.* **51**, 1313 (1973).
48. A. R. Mattocks, *J. Chem. Soc.* pp. 1918 and 4840 (1964).
49. S. Greenberg and J. G. Moffatt, *Abstr. 155th Natl. Meet., Am. Chem. Soc., San Francisco, Calif.*, Article C 054 (1968).
49a. T. C. Jain, A. F. Russell, and J. G. Moffatt, *J. Org. Chem.* **38**, 3179 (1973).
50. M. J. Robins, R. A. Jones, and R. Mengel, *Can. J. Chem.* **55**, 1251 (1977).
51. M. J. Robins, Y. Fouron, and W. H. Muhs, *Can. J. Chem.* **55**, 1260 (1977).
52. D. Pavan-Langston, R. A. Buchanan, and C. A. Alford, Jr., eds., "Adenine Arabinoside: An Antiviral Agent."Raven, New York, 1975; D. B. Ellis and G. A. LePage, *Mol. Pharmacol.* **1**, 231 (1965).
53. A. Bloch, M. J. Robins, and J. R. McCarthy, Jr., *J. Med. Chem.* **10**, 908 (1967); G. A. LePage, *Can. J. Biochem.* **48**, 75 (1970).
54. C. E. Cass and M. J. Robins, *Proc. Am. Assoc. Cancer Res.* **18**, 92 (1977).
55. M. J. Robins and W. H. Muhs, *J. Chem. Soc., Chem. Commun.* p. 269 (1976).
56. C. D. Anderson, L. Goodman, and B. R. Baker, *J. Am. Chem. Soc.* **81**, 3967 (1959).

12

Chemical Reactions of Nucleic Acids

Robert Shapiro

INTRODUCTION

The nucleic acids and their component nucleotides and nucleosides undoubtedly fall within the realm of organic chemistry. They contain carbon, of course, and hydrogen, oxygen, nitrogen, and phosphorus for good measure. They pose numerous mechanistic, structural, and synthetic problems, as engrossing and worthy of attack as those found in more conventional areas of organic chemistry such as alkaloids and terpenes. Yet the nucleic acids get little space or attention in most organic texts or courses, they have no section of their own in *Chemical Abstracts*, and relatively few academic chemists do research in the area.

One reason for the situation may be that their physical properties are inconvenient and differ from those of the majority of organic compounds. They are not volatile, and they are insoluble in most organic solvents but soluble in water. They melt poorly, do not distill, and yield to mass spectrometry and gas chromatography with difficulty. A conventionally trained chemist must master new techniques, such as gel filtration, ion exchange, and electrophoresis, to work in the area. Most of the techniques have become readily available only within the last two decades. Furthermore, it is only within that same time period that the central importance of nucleic acids in the process of life has been recognized.

Whatever the reason for the historical lack of attention given to nucleic acids by organic chemists, it is unfortunate. The area has as much intrinsic chemical interest as any other, for all types of chemical study. In addition, the nucleic acids are the hereditary and key control elements of all living systems. For this reason, any discovery made in the course of a chemical study may have found profound implications in other disciplines such as biochemistry, genetics, and public health.

In this review, I first discuss the types of applications that may arise from studies on the chemical reactivity of nucleic acids. The points are then illustrated by examples of specific reagents and transformations that have received attention from our research group in the past years.

CHEMICAL REACTIONS OF NUCLEIC ACIDS AS MODELS FOR EVENTS IN LIVING CELLS

Many important consequences may result from the reaction of DNA or RNA with a chemical in its cellular environment.

Inactivation

A number of chemical alterations of DNA may lead to its loss of function [1]. The list includes such obvious changes as backbone breakage or cross-linking of the two chains, which prevents them from separating for purposes of replication. Less severe chemical changes, such as base alterations, may also block the action of the enzymes that copy DNA sequences into new DNA or RNA molecules. The significance of these alterations also depends on the ability of the cell to repair the inflicted damage [2]. Permanent loss of function of a vital gene in DNA would lead to cell death. Because of the multiplicity of RNA molecules, damage to a single RNA molecule would be less significant unless the reactions were so extensive that a large proportion of RNA molecules of one type were affected. Reagents that inactivate nucleic acids are toxic in multicellular organisms. If however, they show some selectivity between cell types or species, they may be valuable as antibiotics or antitumor agents [3].

Mutations

A chemical reaction of a base in DNA may alter its hydrogen-binding pattern so that it pairs with a substance other than its normal Watson–Crick hydrogen-binding partner (see, for example, the discussions on nitrous acid and sodium bisulfite below). Such a change produces a point mutation and leads to the alteration of a single amino acid in a protein. Other types of change may lead to frameshift mutations, with more drastic alterations in protein sequence, or to larger deletions, with loss of genetic information [1]. Some mutations, such as those produced by ultraviolet light, arise through defective repair mechanisms rather than as a direct consequence of the chemical change produced [4]. Chemical mutagens have been of great value as research tools in both pure and applied genetics. Recently, there has been much concern that environmental chemicals may be inflicting unfavorable mutations on human beings, resulting in enhanced birth defects and ill health [4].

Carcinogenesis

It is firmly established that the reaction of chemicals with DNA can lead to mutations. A large number of scientists believe that the reaction of a chemical carcinogen with DNA may be the significant step in initiating the process that leads to cancer, but this has not been proved [5]. Great progress has been made recently in exploring the pathways by which unreactive polycyclic aromatic hydrocarbons and amines are converted to metabolites that react readily with DNA. From the study of these reactions, detailed theories have arisen suggesting how the bound carcinogen may strongly influence the functioning of DNA. One example is the base displacement theory [6].

Aging Theories

This area, unlike those discussed above, is entirely speculative. One theory holds that aging is due to a cumulative loss of information from DNA resulting from the formation of cross-links or other types of damage [7].

CHEMICAL REACTIONS OF NUCLEIC ACIDS IN STUDIES OF THEIR STRUCTURE AND FUNCTION

Chemical modification of nucleic acids has been widely used as an aid to the study of their structure and function [8]. It is necessary, of course, that the modification not severely distort the property that is under study. In studies of conformation or function, reagents must be applied in aqueous (or occasionally strongly polar organic) solution in the approximate pH range 4–10 and at temperatures not usually exceeding about 50°C. Some of the principal classes of use are described below.

Study of Primary Structure (Sequence)

The most important tools used in sequencing nucleic acids are specific enzymes. One use for chemicals is in further restricting the specificity of enzymatic cleavage. Pancreatic ribonuclease, for example, cleaves RNA at the phosphodiester bond linking the 3′ position of uridine or cytidine to the 5′ position of another nucleoside. Modification of RNA with Gilham's reagent (a substituted carbodiimide) converts the uridine residues to a derivative that is not recognized by the enzyme [9]. Cleavage of the modified RNA by pancreatic ribonuclease is then limited to the bonds following the cytidine residues. One direct chemical method of cleavage has come into common use: the depurination of DNA by acid. This technique is discussed below. A chemical technique involving periodate oxidation and elimination has also been useful in sequencing short RNA segments by stepwise degradation from one end [10].

Secondary and Tertiary Structure of Nucleic Acids

The term "secondary structure" refers to antiparallel hydrogen bonding between stretches of bases in a nucleic acid. This leads to the formation of large double helices in DNA and smaller helical segments in the "cloverleaf" structure of transfer RNA. Additional, isolated hydrogen bonds that determine a particular three-dimensional structure for a molecule such as a tRNA [11] are classified as tertiary interactions. The most definite information about such structures has been provided by X-ray diffraction studies. Chemical modification studies have been valuable in the great majority of cases where rigorous X-ray work has not been done and have closely supported the X-ray results in the case of yeast phenylalanine-tRNA [12]. Chemical studies are useful because a number of reagents such as Gilham's reagent [9] and O-methylhydroxylamine are single-strand-specific. They do not react with bases that are involved in secondary or tertiary interactions. An additional single-strand-specific reagent, sodium bisulfite, is discussed below.

Quaternary Structure

Nucleic acids may interact with proteins to form more complex particles, such as ribosomes and viruses. Examples of the use of chemical reactions to obtain information about the fine structure of such particles are given in the sections dealing with nitrous acid and α-ketoaldehydes.

Function of Nucleic Acids

Chemical modifications have been used to explore the manner in which nucleic acids function biologically. For example, O-methylhydroxylamine modification of f2 virus RNA increases the number of sites active in initiating protein synthesis [13]. Various modifications impair the amino acid acceptor ability of tRNA species (see below).

EXAMPLES OF CHEMICAL MODIFICATION

We have studied a variety of nucleic acid reactions in our research at New York University in the past 15 years. The chemistry of a number of the more significant findings is discussed here, with examples of subsequent applications in other areas.

Nitrous Acid

The principal reactions of nitrous acid with nucleic acid components involve deamination [14]. Adenine, cytosine, and guanine are converted to hypoxanthine, uracil, and xanthine, respectively (Fig. 1). The same reactions can occur within living organisms. As a result, nitrous acid is an effective mutagen in a variety of microorganisms [15]. The formation of uracil and hypoxanthine are believed to be mutagenic, as they hydrogen bond to adenine and cytosine in DNA. The evidence with respect to xanthine is somewhat contradictory, but the consensus is that its formation is inactivating [1,14,15]. In addition, nitrous acid causes a number of puzzling and unusual chemical and genetic events, which have been summarized [14]. These include an unknown cross-linking reaction, a reaction that consumes guanine, the formation of deletion mutants, and a weak mutagenic transformation of thymine to cytosine in certain bacteriophages.

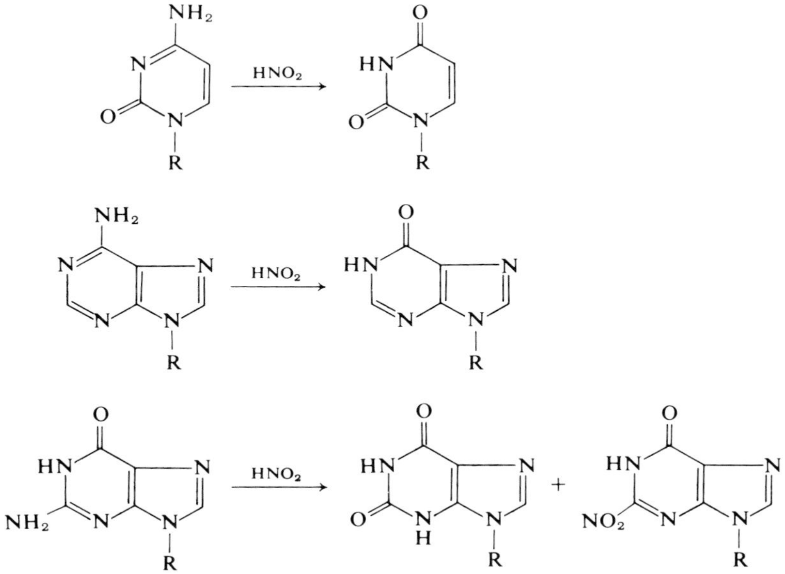

Fig. 1 Reactions of nucleic acid components with nitrous acid.

In order to gain some insight into the undiscovered side reactions, we thoroughly reexamined the reaction of nitrous acid with adenosine, cytidine and guanosine [14,16]. We found that a nitro compound, 2-nitroinosine (Fig. 1), was a side product of guanosine deamination. This was of chemical interest as the first known 2-nitropurine. No clue was obtained as to the nature of the cross-linking reaction or the unusual mutagenic transformations. In more recent work, however (S. Dubelman and R. Shapiro, unpublished), we have isolated cross-linked nucleosides from nitrous acid-treated DNA. Structural studies are in progress.

As the guanosine reaction was the least desirable one for mutagenic purposes, we performed kinetic studies to explore whether varying the conditions would improve the specificity of nitrous acid for the other bases [14]. Under all conditions studied, however, the relative order of reactivity was guanine > adenine > cytosine. Our data lent emphasis to a remarkable observation reported elsewhere in the literature [17,18]. The order of reactivity in the RNA of tobacco mosaic virus was the same as that observed at the nucleoside level. When intact tobacco mosaic virus particles were treated with nitrous acid, however, adenine and cytosine were deaminated, while the reaction of guanine was negligible. Apparently the coat protein protects the amino group of guanine from deamination within the virus in a highly specific manner.

Another question of reactivity to nitrous acid also claimed our attention. Because the amino groups of adenine, cytosine, and guanine are directly involved in Watson–Crick hydrogen bonding, it had been assumed that incorporation of one of these bases within a double helix would serve to protect it from deamination by nitrous acid [19,20]. We measured the reactivity of cytosine as a nucleotide, in a single strand, and in a double helix [21]. No strict single-strand specificity was observed, although the double helix was less than half as reactive as the single strand. Nitrous acid was therefore not an effective agent for studying the secondary structure of polynucleotides, which contrasted with its dramatic effectiveness as a probe for protein–nucleic acid interactions in tobacco mosaic virus.

Hydrolytic Deamination

One of our long-term goals had been to devise an improved mutagen, one that would deaminate cytosine to uracil, as nitrous acid does, without the undesirable deamination of guanine. As the manipulation of the reaction conditions used with nitrous acid did not bring us closer to that goal, other means were sought. It was known that adenine, cytosine, and guanine could all be deaminated by heating in strong acid. When we explored the reactivity of the three bases in weaker acids, we received a happy surprise. Cytosine

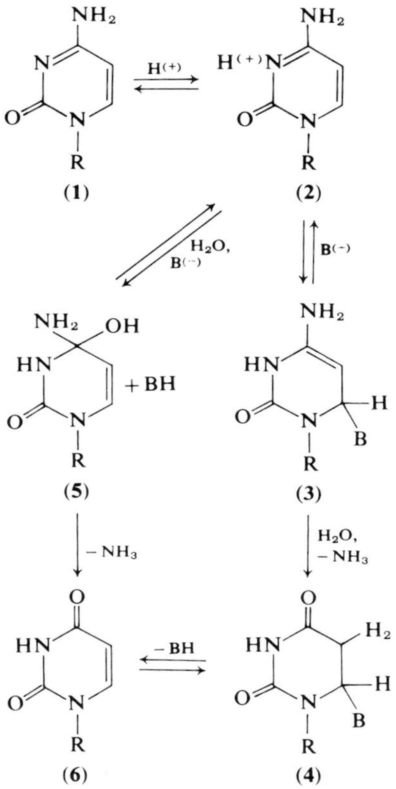

Fig. 2 Mechanisms for deamination of cytosine and cytidine; R = H or β-D-ribofuranosyl, $B^{(-)}$ = buffer anion. (Reprinted with permission from R. Shapiro and R. S. Klein, *Biochemistry* **5**, 2358 (1966). Copyright by the American Chemical Society.)

deaminated at pH 3–5, while adenine and guanine were unaffected [22]. The nature and concentration of the buffer employed had a marked effect on the rate of deamination of cytosine. In moderately concentrated citrate buffers, a rate maximum was observed at pH 3.5. On the basis of our kinetic results and our intuition, we preferred an indirect mechanism for this transformation [(2) → (3) → (4) → (6), Fig. 2] involving a transient saturation of the 5–6 double bond of cytosine. The alternative was the direct route [(2) → (5) →(6)] analogous to the hydrolysis of an amide. Subsequently, more extensive studies by others have confirmed the addition–elimination path proposed by us [23,24].

Upon examination of the mechanism, we realized that the pathways involved could be applied to other purposes as well. When the reaction was

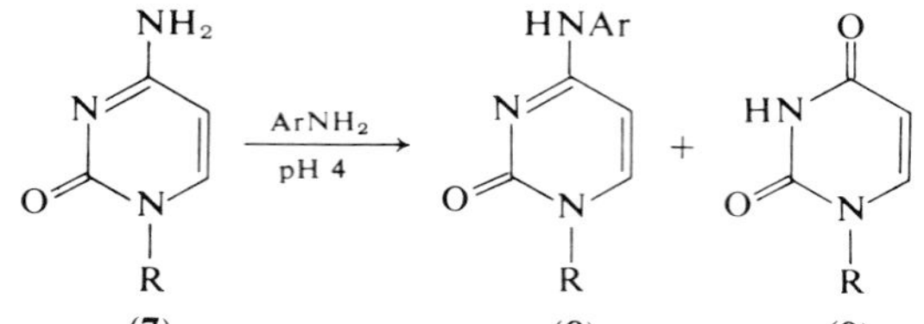

Fig. 3 Transamination and deamination of cytosine derivatives. (Reprinted with permission from R. Shapiro and R. S. Klein, *Biochemistry* **6**, 3576 (1967). Copyright by the American Chemical Society.)

run in D_2O, deuterium exchange preceded deamination, and a 5-deuterocytidine could be prepared by this method [25]. When an aromatic amine was used as the buffer, rather than a carboxylate, the amine displaced the amino group of the saturated cytidine intermediate, and an N^4-arylcytidine [(**8**, Fig. 3] was the principal product.

The hydrolytic deamination reaction had the specificity that we were seeking in a potential mutagen. In fact, we suggested that this reaction was responsible for a weak mutagenic effect, with $GC \rightarrow AT$ specificity, that was observed when T4 phage was warmed in weakly acidic carboxylate buffers [22,26]. However, the slowness of the reaction at physiological temperatures made it unsuitable for general use as a mutagen or as a nucleic acid modification reagent. We felt that the reaction would be improved considerably if a more effective buffer catalyst could be found.

Deamination by Sodium Bisulfite

Sodium bisulfite, a powerful nucleophile in many organic reactions, proved to be the best catalyst for hydrolytic deamination. The reaction went smoothly and rapidly in 1 M NaHSO$_3$, pH 5–6, at room temperature (Fig. 4) [27], and the saturated intermediates were sufficiently stable to be observed by nuclear magnetic resonance spectroscopy. The uracil–bisulfite adduct [(**12**), Fig. 4] was, in fact, stable indefinitely at acid pH, and the solution had to be made slightly alkaline to complete the conversion to uracil. The uracil–bisulfite adduct could readily be formed from uracil and sodium bisulfite at neutral pH. Shortly after our initial publication, these results were reported independently by Dr. H. Hayatsu and his co-workers [28] in Tokyo. It appeared to us that this reagent would be valuable for synthetic purposes, as a tool for the modification of nucleic acids, and as a specific chemical mutagen. Questions of public health were also raised by this discovery, as sodium bisulfite is a common food additive and, in the form of sulfur dioxide, a widespread urban air pollutant.

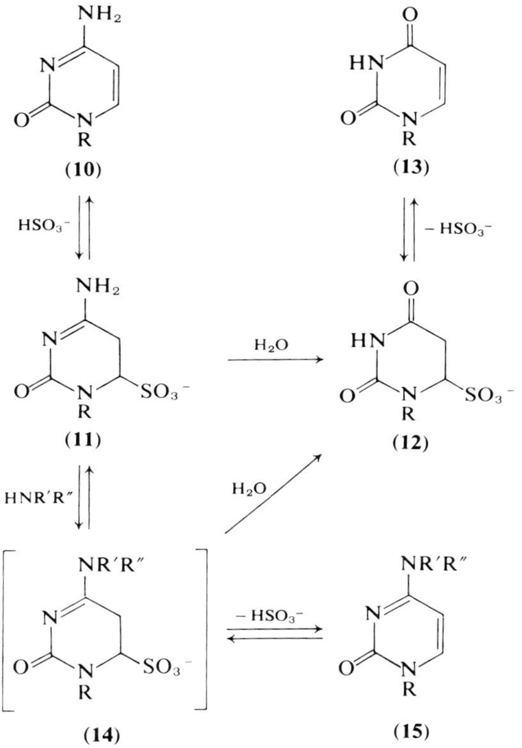

Fig. 4 Deamination and transamination of cytosine derivatives by bisulfite. (From Shapiro and Weisgras [38].)

The synthetic applications were readily tested. Deoxycytidine was converted to deoxyuridine in 94% yield by treatment with sodium bisulfite, pH 5, followed by a brief alkali treatment. Similarly, the dinucleoside phosphate, CpA, was converted by specific deamination to CpU in quantitative yield [27]. This conversion, which could not have been achieved with nitrous acid, was possible with bisulfite because of its unreactivity to adenine and guanine.

Yeast RNA was treated with sodium bisulfite to test the effectiveness of the reaction at the polynucleotide level [29]. Specific deamination of cytosine was observed once again. Preliminary studies with DNA also indicated that sodium bisulfite was single-strand-specific. Denatured DNA reacted readily, but double-helical DNA completely resisted deamination [29]. In later studies, we showed that it was the initial step of the reaction, addition by bisulfite to the 5–6 double bond of cytosine, that was inhibited. We also found that the addition of bisulfite to uracil was a single-strand-specific reaction [30]. It has been one of the paradoxes of nucleic acid chemistry that deamination of cytosine by bisulfite is single-strand-specific, while deamination by nitrous acid is not (see above). Surprisingly, double-helix formation does not prevent attack by nitrous acid at the amino group of cytosine, a site directly involved in hydrogen bonding, but it does block bisulfite addition to the 5–6 double bond, which is remote from the hydrogen-bonding sites.

A number of workers have applied the deamination of cytosine by bisulfite as a probe of the structure and function of nucleic acids. In a typical study, treatment of *Escherichia coli* formylmethionine-tRNA with sodium bisulfite led to the conversion of six cytosines to uracils [31]. The reactive residues were all in single-stranded areas. Cytosines in double-stranded areas did not react. The cytosines that reacted were the same ones that showed considerable reactivity in the bisulfite–methylamine procedure (see next section). They are indicated by the lined circles in Fig. 5. Not every cytosine in a single-stranded portion of the cloverleaf structure was deaminated. The unreacted ones were presumably involved in tertiary interactions.

It was subsequently shown, by partial deamination with bisulfite, that reaction of the cytosines in the anticodon (position 35) or adjacent to the 3′-terminal adenosine (position 76) led to complete loss of amino acid acceptor ability by this tRNA [32]. The other deaminations did not affect acceptor ability. On the other hand, deamination at the 5′ terminus (position 1) endowed the molecule with the ability to form a complex with T factor (a bacterial protein synthesis factor) and GTP. The unmodified tRNA does not form this complex [33].

The application of bisulfite deamination to tobacco mosaic virus was of interest [34]. Cytidine residues in isolated tobacco mosaic virus RNA deaminated normally. However, no deamination was observed in the whole virus particles. This result should be contrasted with the case of nitrous acid (above), in which the deamination of cytosine was obtained in the virus, but that of guanine was blocked.

The discovery of the bisulfite reactions came as the end result of our attempt to devise a specific mutagen. At my request, a colleague, Dr. Frank Mukai, treated a suitable strain of the bacterium *E. coli* with bisulfite. As we had hoped, the treatment induced mutations at genetic sites that contained guanine–cytosine base pairs, but not at sites with adenine–thymine base pairs [35]. The mutagenicity of bisulfite has now been demonstrated in a number of other cases, which are listed in an article by Shapiro *et al.* [30]. The mutagenicity of bisulfite in microorganisms raises the possibility that it may also be mutagenic in man [36]. If so, the hazard would be serious because of the widespread public exposure to this reagent. Although the optimal pH for this reaction is between pH 5 and 6, there is a measurable rate at neutral pH [37]. The resolution of this question must await suitable testing.

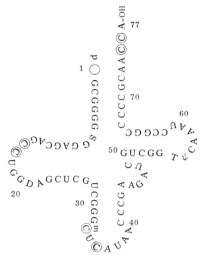

Fig. 5 Sites of N^4-methylcytidine residues in tRNA^{fMet} treated with [^{14}C] methylamine and sodium bisulfite. The symbol ◯ indicates heavily modified residues; ◎ indicates residues modified to an intermediate extent; and Ⓒ indicates residues that have undergone only limited reaction (From Schulman *et al.* [40].)

Transamination by Bisulfite

As incorporation of aromatic amines in a hydrolytic deamination reaction mixture had led to the formation of the N^4-arylcytosine as the principal product, we expected that the same reaction would be catalyzed more effectively by bisulfite. In fact, the reaction of cytosine derivatives with bisulfite and a number of aliphatic and aromatic amines went smoothly in high yield, at temperatures below 50°C [38]. The path is given in Fig. 4. The reaction was best applied at pH 7, because deamination was minimized or eliminated at that pH. It was possible to introduce methyl or phenyl groups specifically onto the cytosines of yeast RNA [39]. The reaction showed strict single-strand specificity, as did deamination by bisulfite.

We felt that the conversion of cytosine to N^4-methylcytosine within a nucleic acid was particularly suitable as a probe for secondary and tertiary structure. Other reagents with single-strand specificity exist, of course [12], including deamination by bisulfite, and several of them, like the transamination reaction, are specific for a single base. The unique advantage of the transamination is that methylation of cytosine would not appreciably alter the position of an oligonucleotide on a chromatographic column or plate. With other reagents, elution volumes or R_f values are altered by modification. After the reaction, the modified RNA is digested by specific enzymes, and each new peak or spot must laboriously be sequenced and identified. This is quite feasible with a tRNA but would be exhausting with a viral or ribosomal RNA. An RNA subjected to transamination with methylamine and digestion would afford the same chromatographic pattern as unmodified RNA. The identity of each peak or spot would be known from the original sequencing work. The reactivity of the cytosines in each oligonucleotide could be determined directly by using radioactive methylamine and counting the incorporated counts.

We tested this procedure with *E. coli* formylmethionine-tRNA, in collaboration with Professor L. H. Schulman [40]. Oligonucleotide peaks from the hydrolyzates by pancreatic ribonuclease or ribonuclease T_1 were separated by anion-exchange chromatography. The position of the eluted anion-exchange peaks was determined by following the ultraviolet absorbance and the reactivity of the cytosines in each peak by determining the counts of $^{14}CH_3$-NH_2 incorporated. The results, illustrated in Fig. 5, were consistent with those obtained by *O*-methylhydroxylamine modification or deamination by bisulfite. We expect that the transamination procedure will be of considerable value in exploring the structure of more complex RNA's.

Among the amines used successfully in the bisulfite-catalyzed transamination reaction was glycine [38]. We suggested that bisulfite, in a similar manner, could catalyze the cross-linking of proteins and nucleic acids. Budowsky and co-workers have subsequently verified our prediction [41,42]. The RNA-containing bacteriophage MS2 was treated with 1 M bisulfite at pH 7. After this treatment, the phage RNA was separated from protein by either gel filtration or phenol treatment. Between 1 and 2% of the protein remained with the RNA during the treatment, indicating RNA–protein cross-links. After acidic hydrolysis, the expected transamination product of cytosine and lysine was detected [42].

Cross-linking of proteins and nucleic acids by transamination represents an additional mechanism by which bisulfite can inflict biological damage, with possible relevance to human health. We have suggested that the chromosomes, where DNA exists closely bound to lysine-rich histones, may be a particularly vulnerable target for cross-linking [43]. Such cross-links would be expected to lead to inactivation, deletion mutations, and other genetic defects.

Addition by Bisulfite to Uracil

As mentioned above, an additional reaction of bisulfite with nucleic acids is its reversible addition to uracil derivatives [27,28,44]. This reaction [(13) → (12), Fig. 4] occurred most rapidly at neutral pH, with an equilibrium constant of 1×10^3 liters mole^{-1} at 25°C. The possibility that it might be a cause of the biological damage inflicted by sulfur dioxide and bisulfite caught our attention (sulfur dioxide is in rapid equilibrium with bisulfite in aqueous solution). Uracil, of course, occurs in RNA, not DNA, while thymine, the related component in DNA, has a much lower equilibrium constant for reaction with bisulfite [44]. The uracil reaction, unlike deamination and transamination, could cause immediate metabolic disruption, but not long-range genetic damage. The microbial growth-inhibiting effects of bisulfite have long been appreciated [45], while statistical evidence about the effects of atmospheric sulfur dioxide on human health [46] have led to legal restrictions on this pollutant. While a number of reactions of bisulfite with cellular constituents have been recorded [47], no clear indication exists as to which is primarily responsible for the biological effects.

We set out to explore the effects of uracil saturation by bisulfite in model biochemical systems. We found that reaction of polyuridylic acid with bisulfite progressively eliminated its ability to form a double-helical complex with polyadenylic acid [48]. The uracil–bisulfite adduct had no ability to form Watson–Crick bonds with adenine. When polyuridylic acid, which had reacted with bisulfite to the extent of 2.6% was used as a messenger in the cell-free protein synthesis system derived from *E. coli*, synthesis of polyphenyl-alanine was reduced by about one-half [48]. As oligopeptides containing as few as four or five amino acids can be detected by the assay used, the results indicated that each uracil–bisulfite adduct terminates the synthesis of a polypeptide chain.

In further studies [49,50], we found that bisulfite inhibited various functions of polyuridylic acid, of the RNA of phage MS2, and of the ribosomes in protein synthesis. In recent work [50a], we have found that uracil–bisulfite adducts are formed in the RNA of *E. coli* under conditions of growth inhibition by bisulfite. Synthesis of the enzyme β-galactosidase is also blocked under these conditions. These results support the possibility that the reaction of bisulfite with uracil in RNA is responsible for the antimicrobial properties of bisulfite. The possibility that this reaction occurs in human beings is unproved but must be considered seriously.

Acidic Hydrolysis of N-Glycosyl Bonds in Nucleosides

In considering the hydrolytic deamination of pyrimidine nucleosides, I wondered whether the dihydropyrimidine intermediates such as (3) and (4) (Figure 2) might play some role in other transformations of pyrimidine nucleosides as well. In particular, I had never been happy with the generally accepted Kenner–Dekker theory [51,52] of nucleoside hydrolysis. According to this theory, the important events in acidic hydrolysis of the N-glycosyl bond of nucleosides by acid were protonation on the sugar ring oxygen [(16) → (20), Fig. 6] followed by opening of the sugar ring [(21) → (22)]. Although this theory was plausible, the facts assembled to support it were inconsistent and

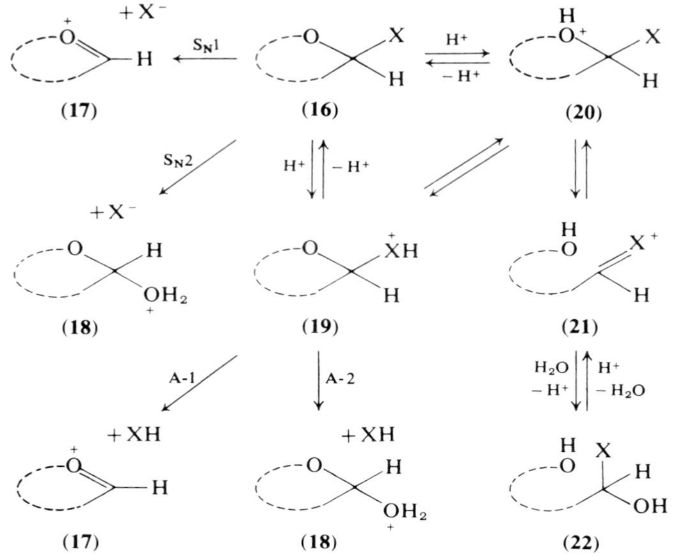

Fig. 6 Possible mechanisms for acidic hydrolysis of nucleosides. (Reprinted with permission from R. Shapiro and S. Kang, *Biochemistry* **8**, 1806 (1969). Copyright by the American Chemical Society.)

unconvincing. In fact, both this theory and our idea of dihydropyrimidines as intermediates in this reaction were incorrect.

Our first surprise came when we reinvestigated the hydrolysis of deoxyuridine derivatives in weakly acidic solution. Hydrolysis rates were constant in the pH range 2–8 and independent of the concentration of the buffer [53]. The relative rates of hydrolysis of various substituted deoxyuridines were proportional to the stability of the bases as leaving groups, with 5-bromo-deoxyuridine the fastest to cleave. Our data were most consistent with an S_N1 or S_N2 pathway for the reaction (**16**) → (**17**) or (**18**) (Fig. 6). By analogy, we expected that hydrolysis in acidic solution would follow an A-1 or A-2 pathway [(**16**) → (**19**) → (**17**) or (**18**)]. In all of these processes, the central feature is direct ionic cleavage of the N-glycosyl bond, without sugar ring opening. These expectations were confirmed by additional studies with pyrimidine deoxynucleosides [54] and deoxynucleotides [55]. Important factors were the pH–rate profiles, substituent effects, and absence of isomerization in the sugar ring. In the interim, Zoltewicz and co-workers had arrived at similar mechanistic conclusions concerning the purine nucleosides [56]. An additional feature of the studies was that glycosyl cleavage could occur via diprotonation on the base as well as from the monoprotonated form (Fig. 7). In the case of uracil nucleosides, the neutral molecule could cleave

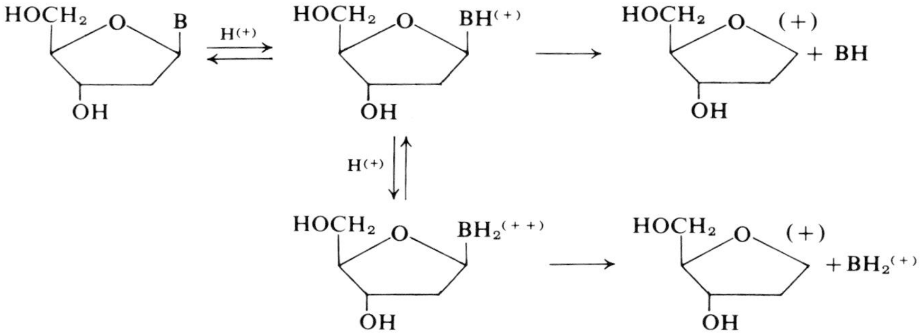

Fig. 7 Mechanism of hydrolysis of pyrimidine deoxyribonucleosides. (From Shapiro and Danzig [55].)

as well. Purine nucleosides form dications more readily than do pyrimidine nucleosides, which explains in part their greater lability to acidic hydrolysis [57].

We suggested then that protonation on the base followed by ionization of the N-glycosyl bond was the most generally operating mechanism of nucleoside hydrolysis [54]. This view has since been supported by a number of other mechanistic and theoretical papers. In the specific case of deoxyuridine and thymidine (but not with 5-bromodeoxyuridine or nucleosides of the other bases) a small percentage of pyranosides was found during hydrolysis by 2 M HClO$_4$ at 90°C [58]. For these conditions, the Kenner–Dekker mechanism may function as a minor, second pathway to hydrolysis.

Applications of the acidic hydrolysis of nucleosides did not, of course, await the elaboration of the mechanism. Some time ago, Chargaff and co-workers used selective hydrolysis of DNA by acid to prepare apurinic acid, a molecule in which only the pyrimidines remained attached to the backbone [57]. This procedure has been used to obtain specific cleavage of DNA for sequencing purposes. Well-defined pyrimidine tracts are obtained after degradation of apurinic acid [59].

The loss of bases from DNA appears to occur *in vivo*. From kinetic data, it has been estimated that a mammalian cell could lose from 2000 to 10,000 purines and a few hundred pyrimidines in a 20 hr cell generation time [60,61]. These events have been anticipated in nature, as cellular repair systems exist that are capable of repairing apurinic sites in DNA [62].

Reactions with α-Ketoaldehydes

It was reported by Staehelin that glyoxal and Kethoxal (a registered trademark of the Upjohn Co., Kalamazoo, Michigan, for α-keto-β-ethoxy-butyraldehyde) reacted readily and specifically with guanine nucleotides [63].

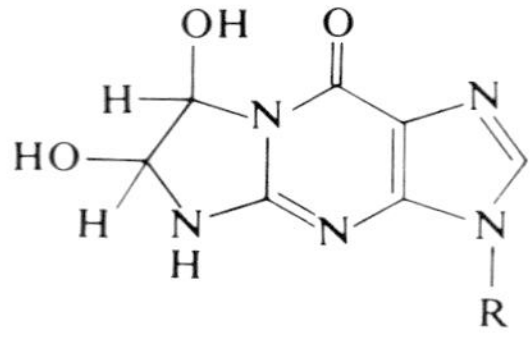

(23a), R = β-D-ribofuranosyl
(23b), R = H

Fig. 8 Adduct of glyoxal with guanine and guanosine. (Reprinted with permission from R. Shapiro *et al.*, *Biochemistry* **8**, 238 (1969). Copyright by the American Chemical Society.)

Binding to tobacco mosaic virus RNA also took place readily, and this binding was presumably related to the potent antiviral activity of these compounds [64].

We set out to explore the chemistry and scope of this reaction. The isolation of the reaction product of guanosine with glyoxal was possible but tedious because of the instability of the product, free of glyoxal, at neutral or alkaline pH [65]. We were more fortunate with the glyoxal–guanine adduct, as it precipitated directly from the reaction medium [66]. The structures of the products, as established by analysis and spectra, are given in Fig. 8. Pyruvaldehyde and Kethoxal reacted similarly, but there was a structural ambiguity about the orientation of the ketoaldehyde with respect to guanine. This problem was solved by periodate cleavage of the adducts, which smoothly

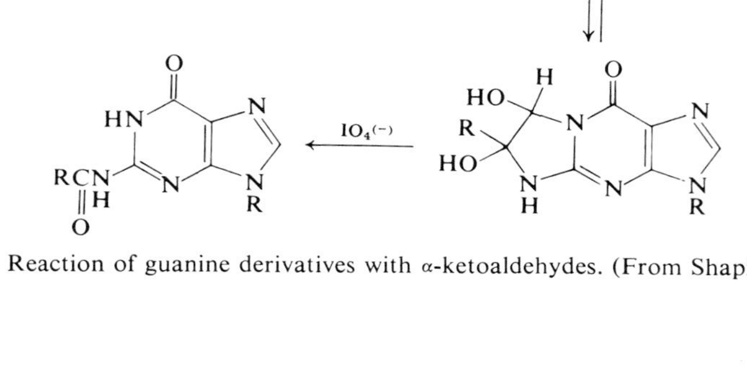

Fig. 9 Reaction of guanine derivatives with α-ketoaldehydes. (From Shapiro [67].)

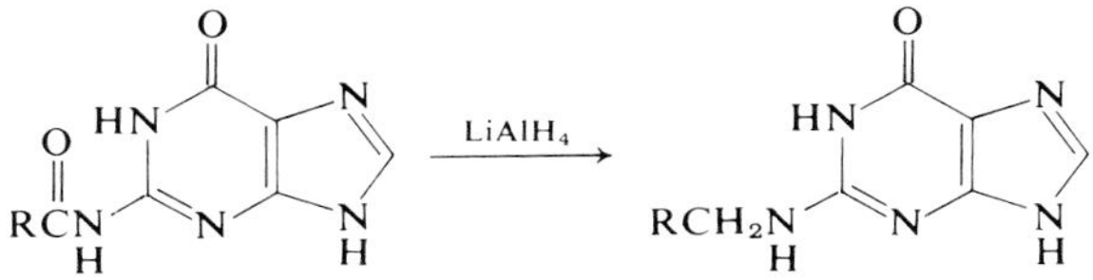

Fig. 10   Reduction of acylguanines (From Shapiro [67].)

yielded acylguanines (Fig. 9) [66,67]. The acylguanines were identified by lithium aluminum hydride reduction to known N^2-alkylguanines, which also provided a new synthetic path to these substances (Fig. 10) [66].

The specificity for guanine of glyoxal was not absolute. At higher concentrations, labile adducts are formed with cytidine and adenosine, presumably through addition of the amino groups across one carbonyl group of glyoxal [68]. Addition of periodate to an adenine–glyoxal mixture led to the isolation of N^6-formyladenine [69]. The adenine and cytosine adducts are quite labile and suffer from a poorer equilibrium constant so that, at lower concentrations of glyoxal, specificity for guanine is regained [68].

A number of applications of these reactions have been made. We have shown that reaction of Kethoxal followed by periodate can be used to achieve specific acylation of the guanine amino group in RNA [69]. Kethoxal represents another example of a single-strand-specific reagent and has been used to explore the structure and function of tRNA [70]. Glyoxal and Kethoxal have also been used to study the conformation of 5 S ribosomal RNA. Two sites that react readily in the free 5 S RNA are also reactive in the intact ribosomal 50 S subunit, which indicates that the ribosomal proteins do not render it inaccessible [71]. In the DNA field, glyoxal has been used to prevent renaturation of partly denatured DNA in a technique called denaturation mapping [72].

The medicinal properties of this series of compounds are also noteworthy. Their antiviral activity was mentioned above. In addition, they are claimed to be regulators of cell division [73,74] and to be effective against cancer [75]. The extent of these properties and their relation to the nucleic acid reactions remain unexplored.

I do not wish to imply, in this chapter, that every novel finding in the chemistry of nucleic acids immediately finds application. One of the most unique and surprising reactions encountered in our studies was that of ninhydrin with cytosine derivatives [76] (Fig. 11). A number of complex transformations were necessary in order to establish the structure of the product. Guanine derivatives also react readily with ninhydrin, in the fashion of glyoxal [65]. The guanine reaction, unlike that of cytosine, is reversible, so specificity for cytosine can be achieved. As the cytosine modification can be carried out without heat at neutral pH, many applications were expected. In fact, no

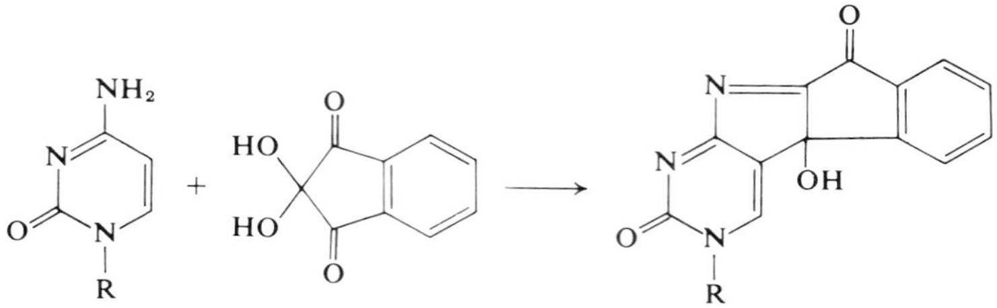

Fig. 11 Reaction of cytosine derivatives with ninhydrin. (Reprinted with permission from R. Shapiro and S. C. Agarwal, *J. Am. Chem. Soc.* **90**, 474 (1968). Copyright by the American Chemical Society.)

important ones have arisen. We still feel, however, that this reaction will eventually find appropriate uses.

CONCLUSION

In many areas of organic chemistry, interest in a reaction diminishes or terminates when the fundamentals of its chemistry have been explored. In nucleic chemistry, there is a continual stream of new developments as the reagent is applied to more complex biochemical situations or as implications develop for areas such as toxicology or genetics. These features make the study of nucleic acids particularly rewarding and well worth the effort required to learn additional techniques. Very much remains to be done. I recommend the area heartily to any enterprising chemist in search of unlimited horizons.

ACKNOWLEDGMENTS

The research in our laboratory could not have been done without the valuable contributions of my co-workers cited in the references. I wish to thank the Division of Environmental Health Sciences, NIH, the National Science Foundation, and the Damon Runyon Cancer Fund for support of this work. I am indebted to the Division of General Medical Sciences, NIH, both for support of this work and for a Career Development Award.

REFERENCES

1. E. Freese, *in* "Chemical Mutagens" (A. Hollaender, ed.), Vol. 1, pp. 1–56. Plenum, New York, 1971.
2. L. Grossman, A. Braun, R. Feldberg, and I. Mahler, *Annu. Rev. Biochem.* **44**, 19 (1975).
3. F. E. Hahn, *Prog. Mol. Subcell. Biol.* **3**, 1 (1971).
4. M. S. Legator and W. G. Flamm, *Annu. Rev. Biochem.* **42**, 683 (1973).

5. C. C. Irving, *Methods Cancer Res.* **7**, 189 (1973).
6. A. F. Levine, L. M. Fink, I. B. Weinstein, and D. Grunberger, *Cancer Res.* **34**, 319 (1976).
7. A. Comfort, *Mech. Ageing Dev.* **3**, 1 (1974).
8. N. K. Kochetkov and E. I. Budowsky, *Prog. Nucleic Acid Res. Mol. Biol.* **9**, 403 (1969).
9. N. W. Y. Ho and P. T. Gilham, *Biochemistry* **6**, 3262 (1967).
10. G. Keith and P. T. Gilham, *Biochemistry* **13**, 3601 (1974).
11. S. H. Kim, J. L. Sussman, F. L. Suddath, G. J. Quigley, A. McPherson, A. O. J. Wang, N. C. Seeman, and A. Rich, *Proc. Natl. Acad. Sci. U.S.A.* **71**, 4970 (1974).
12. J. D. Robertus, J. E. Ladner, J. T. Finch, D. Rhodes, R. S. Brown, B. F. C. Clark, and A. Klug, *Nucleic Acids Res.* **1**, 927 (1974).
13. W. Filipowitcz, A. Wodnar, L. Zagorsky, and P. Szafranski, *Biochem. Biophys. Res. Commun.* **49**, 1272 (1972).
14. R. Shapiro and S. H. Pohl, *Biochemistry* **7**, 448 (1968).
15. L. Fishbein, H. L. Falk, and W. G. Flamm, "Chemical Mutagens: Environmental Effects on Biological Systems," pp. 25–26 and 252–253. Academic Press, New York, 1970.
16. R. Shapiro, *J. Am. Chem. Soc.* **86**, 29 (1964).
17. H. Schuster and R. C. Wilheim, *Biochim. Biophys. Acta* **68**, 554 (1963).
18. O. P. Sehgal and M. M. Soong, *Virology* **47**, 239 (1972).
19. B. Singer and H. Fraenkel-Conrat, *Prog. Nucleic Acid Res. Mol. Biol.* **9**, 1 (1969).
20. A. M. Michelson and F. Pochon, *Biochim. Biophys. Acta* **174**, 604 (1969).
21. R. Shapiro and H. Yamaguchi, *Biochim. Biophys. Acta* **281**, 501 (1972).
22. R. Shapiro and R. S. Klein, *Biochemistry* **5**, 2358 (1966).
23. R. E. Notari, M. L. Chin, and A. Girdoni, *J. Pharm. Sci.* **59**, 28 (1970).
24. E. R. Garrett and J. Tsau, *J. Pharm. Sci.* **61**, 1052 (1972).
25. R. Shapiro and R. S. Klein, *Biochemistry* **6**, 3576 (1967).
26. E. B. Freese, *Brookhaven Symp. Biol.* **12**, 63 (1959).
27. R. Shapiro, R. E. Servis, and M. Welcher, *J. Am. Chem. Soc.* **92**, 420 (1970).
28. H. Hayatsu, Y. Wataya, and K. Kai, *J. Am. Chem. Soc.* **92**, 724 (1970).
29. R. Shapiro, B. I. Cohen, and R. E. Servis, *Nature (London)* **227**, 1047 (1970).
30. R. Shapiro, B. Braverman, J. B. Louis, and R. E. Servis, *J. Biol. Chem.* **248**, 4060 (1973).
31. J. P. Goddard and L. H. Schulman, *J. Biol. Chem.* **247**, 3864 (1972).
32. L. H. Schulman and J. P. Goddard, *J. Biol. Chem.* **248**, 1341 (1973).
33. L. H. Schulman and M. O. Her, *Biochem. Biophys. Res. Commun.* **51**, 275 (1973).
34. S. Kajita and C. Matsui, *Virus* **24**, 170 (1974).
35. F. Mukai, I. Hawryluk, and R. Shapiro, *Biochem. Biophys. Res. Commun.* **39**, 983 (1970).
36. R. Shapiro and J. B. Louis, *Mutat. Res.* **31**, 327 (1974).
37. R. Shapiro, V. Difate and M. Welcher, *J. Am. Chem. Soc.* **96**, 906 (1974).
38. R. Shapiro and J. M. Weisgras, *Biochem. Biophys. Res. Commun.* **40**, 839 (1970).
39. R. Shapiro, D. C. F. Law, and J. M. Weisgras, *Biochem. Biophys. Res. Commun.* **49**, 358 (1973).
40. L. H. Schulman, R. Shapiro, D. C. F. Law, and J. B. Louis, *Nucleic Acids Res.* **1**, 1305 (1974).
41. M. F. Turchinsky, K. S. Kusova, and E. I. Budowsky, *FEBS Lett.* **38**, 304 (1974).
42. E. I. Budowsky, M. F. Turchinsky, I. V. Boni, and Yu. M. Skoblov, *Nucleic Acids Res.* **3**, 261 (1976).

43. R. Shapiro, *in* "Aging, Carcinogenesis, and Radiation Biology" (K. C. Smith, ed.), pp. 225–242. Plenum, New York, 1976.
44. R. Shapiro, M. Welcher, V. Nelson, and V. Difate, *Biochim. Biophys. Acta* **425**, 115, (1976).
45. D. F. Chichester and F. W. Tanner, Jr., *in* "Handbook of Food Additives" (T. E. Furia, ed.), 2nd ed., pp. 142–147. CRC Press, Cleveland, 1972.
46. National Air Pollution Control Administration, "Air Quality Criteria for Sulfur Oxides," Publ. No. AP-50. NAPCA, Washington, D.C., 1969.
47. D. H. Petering and N. T. Shih, *Environ. Res.* **9**, 55 (1975).
48. R. Shapiro and B. Braverman, *Biochem. Biophys. Res. Commun.* **47**, 544 (1972).
49. R. Shapiro, B. Braverman, and W. Szer, *Mol. Biol. Rep.* **1**, 123 (1973).
50. B. Braverman, R. Shapiro, and W. Szer, *Nucleic Acids Res.* **2**, 501 (1975).
50a. B. Braverman, R. Shapiro, T. Rossman, and W. Troll, submitted for publication.
51. G. W. Kenner, *Chem. Biol. Purines, Ciba Found. Symp., 1956* pp. 312–313 (1957).
52. C. A. Dekker, *Annu. Rev. Biochem.* **29**, 453 (1960).
53. R. Shapiro and S. Kang, *Biochemistry* **8**, 1806 (1969).
54. R. Shapiro and M. Danzig, *Biochemistry* **11**, 23 (1972).
55. R. Shapiro and M. Danzig, *Biochim. Biophys. Acta* **319**, 5 (1973).
56. J. A. Zoltewicz, D. F. Clark, J. W. Sharpless, and G. Grahe, *J. Am. Chem. Soc.* **92**, 1741 (1970).
57. C. Tamm, M. E. Hodes, and E. Chargaff, *J. Biol. Chem.* **195**, 49 (1952).
58. J. Cadet and R. Teoule, *J. Am. Chem. Soc.* **96**, 6517 (1974).
59. K. Murray and R. W. Old, *Prog. Nucleic Acid Res. Mol. Biol.* **14**, 117 (1974).
60. T. Lindahl and B. Nyberg, *Biochemistry* **11**, 3610 (1972).
61. T. Lindahl and O. Karlstrom, *Biochemistry* **12**, 5151 (1973).
62. W. G. Verly, F. Gossard, and P. Crine, *Proc. Natl. Acad. Sci. U.S.A.* **71**, 2273 (1974).
63. M. Staehelin, *Biochim. Biophys. Acta* **31**, 448 (1959).
64. B. J. Tiffany, J. B. Wright, R. B. Moffett, R. V. Heinzeman, R. E. Strube, B. D. Aspergren, E. H. Lincoln, and J. L. White, *J. Am. Chem. Soc.* **79**, 1682 (1957).
65. R. Shapiro and J. Hachmann, *Biochemistry* **5**, 2974 (1966).
66. R. Shapiro, B. I. Cohen, S.-J. Shiuey, and H. Maurer, *Biochemistry* **8**, 238 (1969).
67. R. Shapiro, *Ann. N.Y. Acad. Sci.* **163**, 624 (1969).
68. N. E. Braude and E. I. Budowsky, *Biochim. Biophys. Acta* **254**, 380 (1971).
69. R. Shapiro, B. I. Cohen, and D. C. Claggett, *J. Biol. Chem.* **245**, 2633 (1970).
70. C. M. Greenspan and M. Litt, *FEBS Lett.* **41**, 297 (1974).
71. N. F. Noller and W. Herr, *J. Mol. Biol.* **90**, 181 (1974).
72. D. Johnson, *Nucleic Acids Res.* **2**, 2049 (1975).
73. A. Szent-Gylörgi, L. G. Egyud, and J. A. McLaughlin, *Science* **155**, 539 (1967).
74. J. A. Edgar, *Nature (London)* **227**, 24 (1970).
75. L. G. Egyud and A. Szent-Györgi, *Science* **160**, 1140 (1968).
76. R. Shapiro and S. C. Agarwal, *J. Am. Chem. Soc.* **90**, 474 (1968).

Bioorganic Stereochemistry:
Unusual Peptide Structures*

Claudio Toniolo

INTRODUCTION

This chapter gives an account of the recent developments in the field of some relatively unusual structures adopted by polypeptide molecules both in the solid state and in solution. Even in this restricted area only those contributions that appeared to be particularly pertinent to the scope of this chapter are examined.

For the description of the conformation of polypeptide chains the convention proposed by the IUPAC–IUB Commission on Biochemical Nomenclature in 1969 [1] is followed (Fig. 1). The hard-sphere-model conformational map [2] showing the ϕ, ψ torsional angles of the most common polypeptide structures, i.e., those *not* examined in this chapter is illustrated in Fig. 2. For general discussion of this subject, readers are referred to recent review articles and books [3–12].

Additional structural elements of peptides are the intramolecularly H-bonded local forms. An H bond between N—H of an amino acid residue of sequence number m and C=O of a residue of the sequence number n is designated as $m \to n$. Therefore, the possible structures in the system of three linked peptide units [13] considered throughout this chapter are the

---

* This work is part 35 of the "Linear Oligopeptides" series. For part 34, see C. Toniolo and M. Palumbo, *Biopolymers* **16**, 219 (1977).

266 CLAUDIO TONIOLO

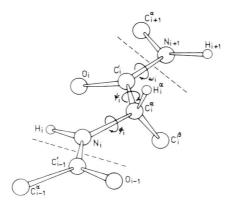

Fig. 1 Perspective drawing of a section of polypeptide chain representing two peptide units. The limits of a residue are indicated by dashed lines, and recommended notations for atoms and torsion angles are indicated. The chain is shown in a fully extended conformation ($\phi_i = \psi_i = \omega_i = +180°$), and the residue illustrated is in the L configuration.

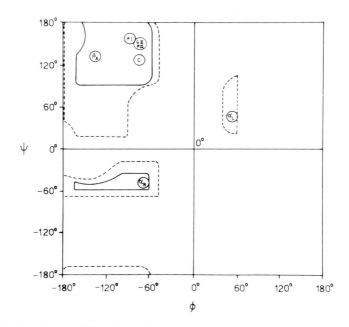

Fig. 2 Hard-sphere-model conformational map showing the ϕ, ψ torsional angles of the most common polypeptide structures (from L-amino acids): α_R, right-handed α helix; α_L, left-handed α helix; β_A, antiparallel-chain pleated-sheet β form; P I, polyproline I, P II, polyproline II; G II, polyglycine II, C, collagen triple helix.

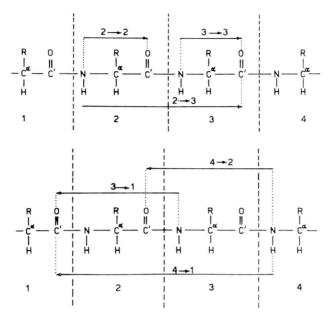

Fig. 3 The possible intramolecularly H-bonded structures in a system of three linked peptide units.

$2 \rightarrow 2$ (or $3 \rightarrow 3$), the $2 \rightarrow 3$, the $3 \rightarrow 1$ (or $4 \rightarrow 2$), and the $4 \rightarrow 1$ intramolecularly H-bonded conformations (Fig. 3.)

THE INTERMOLECULARLY H-BONDED PARALLEL-CHAIN PLEATED-SHEET β STRUCTURE

The extended or near-extended β structure was proposed many years ago for the silk produced by the larvae of *Bombix mori* [14] and for hair keratin that had been steam-stretched [15]. In this form, H bonding takes place between the N—H groups of one chain and the C=O groups of the chains on either side, so making a sheet held together by H bonds. Neighboring sheets are then held together by van der Waals forces. This interpretation of the β form has never been seriously questioned, and later work has concentrated on whether the chains in a sheet are all parallel, with the —CO—NH—CHR— sequence always in the same direction, or whether there is an alteration of this sequence (antiparallel chains) [16]. The possibility of small contractions of the polypeptide chains has also been examined, and precise

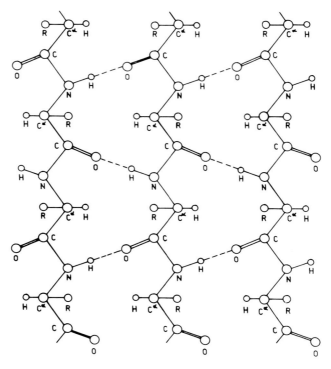

Fig. 4 The parallel-chain pleated-sheet β structure ($\phi = -119°$, $\psi = +113°$; see IUPAC–IUB Commission on Biochemical Nomenclature [1]).

conformations for parallel (Fig. 4) and antiparallel (Fig. 5) pleated-sheet β forms have been proposed [16].

Experimental verification of the occurrence of intermolecularly and intramolecularly H-bonded antiparallel-chain and intramolecularly H-bonded parallel-chain pleated-sheet β structures has been obtained in several cases [11,17]. However, although the existence of the intermolecularly H-bonded parallel-chain pleated-sheet β structure was proposed in synthetic high molecular weight polypeptides and *fibrous* proteins (poly-β-n-propyl-L-aspartate, poly-γ-methyl-L-glutamate, β-keratin, silk I), this was later shown not to be so [11].

Only recently, in our laboratory, has the first example of intermolecularly H-bonded parallel-chain β form been authenticated in synthetic low molecular weight polypeptides [18,19]. In fact, in contrast to the homoheptapeptides derived from L-alanine, L-norvaline (Nva), L-leucine, S-methyl-L-cysteine, and L-methionine, in those derived from L-valine, L-isoleucine, and L-phenylala-

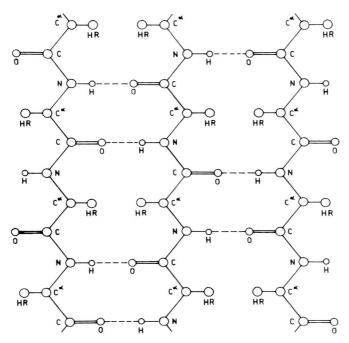

Fig. 5 The antiparallel-chain pleated-sheet β structure ($\phi = -139°$, $\psi = +135°$; see Fig. 2 and IUPAC–IUB Commission on Biochemical Nomenclature [1]).

nine this chain arrangement has been found to predominate in the solid state. It is our contention that the parallel-chain pleated-sheet β structure of these oligopeptides should be of the intermolecular type. Support for the exclusion of folded forms is given by the limited chain length of these peptides, the position of the amide A bands in the absorption spectra below 3300 cm^{-1} [11,20], and the scattered examples of conformational analyses of homo-oligopeptides carried out in the solid state by infrared (ir) absorption and X-ray diffraction techniques [18].

The occurrence of this uncommon structure was verified as follows:

1. The high-frequency component of the split amide I band near 1690 cm^{-1}, indicative of polypeptides in the antiparallel β form [21,22], is not visible in the ir absorption spectra of the three aforementioned homohepta-peptides [18,19]. Also, Chirgadze and Nevskaya [22], on the basis of the perturbation theory in a dipole–dipole approximation, predicted a $\simeq 10$ cm^{-1} shift to higher frequencies of the strong low-frequency component of the amide I band near 1630 cm^{-1} for the parallel-chain pleated-sheet β structure

if compared to the antiparallel-chain pleated-sheet β structure. This theoretical finding is in very good agreement with our experimental results [18,19].

2. The sharp amide III band at 1228 cm^{-1} in the Raman spectrum of L-valine heptamer [23] is different from other peptide and protein frequencies ever seen [24]. Consequently, valine heptamer is not in the α-helical, antiparallel-chain β or unordered form; certainly, the parallel-chain β form seems the most likely.

3. The circular dichroism (CD) spectrum of a film of L-valine heptamer [25] exhibits in the far-ultraviolet region, i.e., in the 210–160 nm region, where the various components of the amide $\pi \rightarrow \pi^*$ transition of polypeptides are found, a splitting pattern very near to that calculated for a parallel-chain pleated-sheet β conformation [26–29]. In particular, the CD is negative in the region of maximum absorption (190 nm).

4. The X-ray diffraction pattern of L-valine homoheptamer is not only unequivocally different from those of α-helical and unordered forms, but also clearly distinct from that of the antiparallel-chain pleated-sheet β form of the L-alanine homoheptamer [30].

Further experimental and theoretical studies are required to shed light on the factors (nature of amino acid side chains, length of peptide chain, etc.) directing chain arrangement in β-forming polypeptides. In this context, it is intriguing that the relative stabilities of the intermolecularly H-bonded β structures of the L-homoheptamers in solution decrease in the order Ile > Val > Cys(Me) > Ala > Nva > Leu > Met [31].

In addition, a statistical analysis of *globular* proteins, the three-dimensional structure of which has been clarified by X-ray diffraction technique, could be of some interest in order to obtain the intramolecularly H-bonded parallel-chain β-conformation parameters [32] for the various amino acid residues.

THE $2 \rightarrow 2$ AND $2 \rightarrow 3$ INTRAMOLECULARLY H-BONDED PEPTIDE CONFORMATIONS

The $2 \rightarrow 2$ intramolecularly H-bonded peptide conformation is the smallest possible conformation among those generated by an H bond linking an N—H of a preceding residue and a C=O of a succeeding residue, i.e., those of the type $2 \rightarrow j$, $j \geq 2$ (Fig. 3). It is quite similar to the fully extended form ($\phi = \psi = +180°$) (Fig. 1). The relative disposition of the two dipoles, N(2)—H(2) and C'(2)=O(2), is such that there is obviously some interaction between them (Fig. 6). These two sites, together with the C$^\alpha$(2) atom, are involved in a pentagonal ring, and it is for this reason that this conformation is also called the C$_5$ structure [33–38]. This conformation has been experimentally identified in model peptides by using ir absorption,

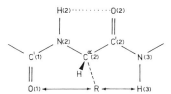

Fig. 6   Internal steric repulsions that induce the warping of the $2 \rightarrow 2$ intramolecularly H-bonded peptide conformations.

proton magnetic resonance (pmr), and dipole moment measurements and an apolar noninteracting solvent such as carbon tetrachloride (at a very low concentration in order to avoid any self-association of the solute) [33–35,37]. Glycyl derivatives show the highest content of $2 \rightarrow 2$ intramolecularly H-bonded peptide structure in the conformational equilibrium mixtures if compared to the derivatives of the other amino acid residues. The influence of the bulkiness of the side substituent can easily be explained by considering the intramolecular nonbonded interactions between the group R and the atoms H(3) and O(1) as shown in Fig. 6. Due to its peculiar symmetry the $2 \rightarrow 2$ intramolecularly H-bonded conformation of the glycyl derivative is very near the fully stretched structure; on the contrary, the dissymmetry introduced by increasing the size of the substituent in the derivatives of the other amino acid residues induces consistently increasing warping of these molecules. However, when the side chain is not completely "inactive," anomalies are observed in the abnormal stabilization of the $2 \rightarrow 2$ intramolecularly H-bonded conformation [34,35].

A recent unequivocal verification of the existence of this intramolecularly H-bonded peptide conformation in the solid state has been obtained in our laboratory in the case of t-BOC-Gly-L-Pro-OH (BOC denotes butyloxycarbonyl) using ir absorption and X-ray diffraction techniques [39]. The positions of the urethane N—H and amide C=O bands have been found at 3417 cm^{-1} and 1644 cm^{-1} respectively, indicating that both groups are H bonded, although not strongly. The X-ray diffraction analysis definitely demonstrated the presence of an intramolecular H bond between the urethane N(2)—H(2) and the amide carbonyl groups (Fig. 7). The H(2)$\cdots$O(2) distance is 2.13 Å. The torsional angles observed for the glycyl residue are $\phi_2 = +172°$ and $\psi_2 = +177°$, very near to the optimal values for a five-membered-ring $2 \rightarrow 2$ intramolecularly H-bonded peptide conformation [33–38]. Not surprisingly, this clear example of five-membered-ring intramolecularly H-bonded peptide conformation in the solid state has been found in the most favorable case, i.e., in a glycyl derivative. In this context, a solid-state investigation of the existence of the $2 \rightarrow 2$ intramolecularly H-bonded peptide conformation stabilized by a further intramolecular

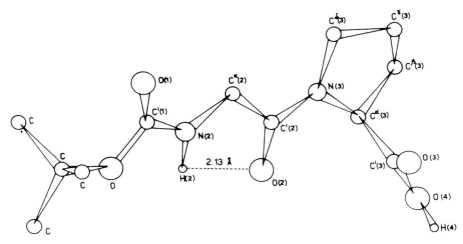

Fig. 7   Molecular structure of t-BOC-Gly-L-Pro-OH.

interaction [34,35] would be of special interest. Finally, it should be mentioned that the presence of this five-membered-ring peptide structure has been considered in theoretical calculations [36,38,40] and proposed on the basis of experimental evidence also in more complex conformations involving the simultaneous occurrence of multiple H bonds [41–43] (for instance, see the section below dealing with intramolecularly bifurcated H-bonded peptide conformations).

To my knowledge, no studies on the $2 \rightarrow 3$ intramolecularly H-bonded peptide conformation (Fig. 3) have yet appeared in the literature. This type of intramolecular H bond has been considered not possible for allowed conformations by Venkatachalam [13] and Ramachandran and Sasisekharan [4]; also, the possibility of occurrence of this folded conformation has been discarded by Urry $et\ al.$ [44] for the pentapeptide model of elastin having the sequence L-Val-L-Pro-Gly-L-Val-Gly.

However, from examination of molecular models and an analysis of data from N-acetyl-N'-methyl dipeptides [45], it appears that the existence of an eight-membered-ring peptide form, although unlikely, is not impossible. In fact, if the configuration of the central amide group in Fig. 8 is taken as cis and those of the first and third amide groups as $trans$, it is possible to build this bent H bond with the $N \cdots O$ distance within the ideal values ($\simeq 3.0$ Å). In the case of the sequence -Gly-L-Pro- the most favorable dihedral angles are $\phi_2 = +173°$, $\psi_2 = +85°$, $\phi_3 = -75°$, $\psi_3 = +167°$. In addition, if the ω value of the central amide group is allowed to vary, a consistent stabilization of the structure is achieved when $|\Delta\omega|$ is in the range of $10°$–$15°$.

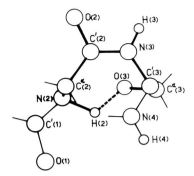

Fig. 8 The 2 → 3 intramolecularly H-bonded peptide conformation (the central amide group is in the *cis* configuration).

Nonplanar deformation of the amide unit (in the *trans* configuration) with $|\Delta\omega|$ up to 15° has been recently shown to be quite probable, since the energy increase for such deviations from planarity are only of the order of 0.5 kcal/mole [46]; also, nonplanar distortions at the nitrogen atom of the peptide unit have been observed in the solid state in cyclic peptides having *cis*-amide bonds [47]. Permitting some bending of bond angles would distribute the energy of distortion value over several smaller values. Crystallographic studies on medium-ring lactams carried out by Dunitz and co-workers showed that in the case of enantholactam the eight-membered ring is not large enough to accommodate the torsion angle of approximately 180° of the *trans*-amide configuration. The constraints of the geometry of the eight-membered ring force the amide in the *cis* configuration with both N—H and C=O groups pointing outward [48].

In principle, a relatively high probability of occurrence of this new type of reverse turn exists in the case of cyclotetrapeptides of the sequence c-(X-Y₃) (where X = α-amino acid residue, and Y = N-alkyl-α-amino acid residue). From the low-temperature pmr spectrum of cyclotetrasarcosyl it was concluded that there is only one conformer in solution and that by symmetric arguments the amide configuration sequence had to be *cis, trans, cis, trans* [49]. This was subsequently confirmed for the crystal [50]. Thus, if an intramolecular H bond is present in the c-(X-Y₃) molecule (with the —NH— CO— group in the *trans* configuration and the —NR—CO— group of the central N-alkyl-α-amino acid residue also in the *trans* configuration), it can give rise only to a conformer containing an eight-membered ring. A recent X-ray diffraction study revealed that in the solid state the c-(Gly-Sar₃) and c-(DL-Ala-Sar₃) molecules indeed adopt the *trans-, cis-, trans-, cis*-amide configuration sequence [51]. However, the single N—H group of each molecule is not involved in intramolecular H bonding but is bonded to a neighbor molecule. The occurrence of intermolecular H bonding was previously also proposed for the single conformer of these molecules in a CHCl$_3$/dimethyl sulfoxide solution [52]. It should be mentioned, however, that only slight rotations of the ring torsional angles are required for turning from the observed conformation to that involving the eight-membered-ring intramolecularly H-bonded peptide structure. It is possible that such a conformation would exist in solvents of lower polarity at higher dilutions. Two intramolecular H bonds have been found in the solid state in the naturally occurring cyclotetrapeptide dihydrochlamydocin [this peptide is of the type C-(X$_3$-Y)] [47], giving rise to two 3 → 1 intramolecularly H-bonded peptide structures, involving, respectively, an α-aminoisobutyric acid (Aib) and a D-proline residue (see below). But in this compound the amide configuration sequence is *all-trans*, a new observation for a cyclic tetrapeptide.

It is also possible that the 2 → 3 intramolecularly H-bonded peptide conformation would exist in linear peptides in a solvent such as carbon tetrachloride at a very low concentration. Fully protected dipeptides ("tripeptide" models) of the type Ac-X-Y-NMe$_2$ appear to be the major candidates. Experimental investigation on such compounds are in progress in our laboratory.

THE 3 → 1 INTRAMOLECULARLY H-BONDED PEPTIDE CONFORMATIONS

The 3 → 1 intramolecularly H-bonded peptide conformations are ring structures that are folded by an H bond between the H(3) and O(1) atoms, as shown in Fig. 9. The *trans*-amide groups lie in two planes, which make an angle of 115°. When R in —NH—CHR—CO— is not a hydrogen atom, two different conformers can exist, depending on the inclination of the C$^\alpha$—R bond with respect to the intersection line of these two planes. These two forms (equatorial and axial), proposed recently in several studies [4,36,38,53, 54] as stable structures, are represented on the usual conformational map by

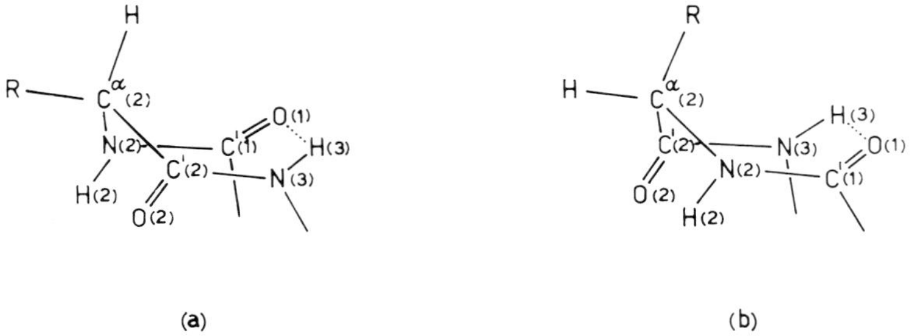

(a)   (b)

Fig. 9 The equatorial (a) and axial (b) 3 → 1 intramolecularly H-bonded peptide conformations.

two centrosymmetric points, the coordinates of which are respectively, $\phi = -75°, \psi = +50°$ (for the equatorial form) and $\phi = +75°, \psi = -50°$ (for the axial form). While the H bond is very strongly bent it has a normal $H \cdots O$ distance, and it still makes a sizable contribution to the stabilization of the folded structures. Actually some variation of the ω values ($|\Delta\omega| \simeq 10°$) is necessary for the stabilization of the $3 \rightarrow 1$ intramolecularly H-bonded peptide conformations. However, as already noted, the small energy of torsional rotation is more than compensated for by the energy of the H bond [46].

The $3 \rightarrow 1$ H bond has been demonstrated in a crystal only recently in the natural cyclotetrapeptide dihydrochlamydocin [47]. The *all-trans* structure of dihydrochlamydocin contains peptide units that are considerably less planar than have been observed in other peptides. In particular, the ω angles for the Aib and D-Pro residues involved in the two $3 \rightarrow 1$ intramolecularly H bonded forms are $+162°$ and $-156°$, respectively (the torsional angles for the Aib residue are $\phi = +72°, \psi = -65°$, respectively, and those for the D-Pro residue are $\phi = +82°, \psi = -73°$, respectively). The $N \cdots O$ distances are 2.82 and 2.94 Å, respectively. It should be noted that in the case of the seven-membered-ring structure of Aib (α,α-dimethylglycine) only one conformer can exist. In addition, the pyrrolidine ring of the Pro residue is such that the seven-membered-ring form can only be an equatorial one. In this context, it is of interest that it has been demonstrated by X-ray diffraction that Ac-L-Pro-NHMe does not adopt such a structure in the solid state [55]. An X-ray diffraction investigation of the model compound Ac-Aib-NHMe [56] should be rewarding.

The occurrence of the $3 \rightarrow 1$ intramolecularly H-bonded forms in solution has been shown by several groups using ir absorption, pmr, and dipole moment measurements [33–35,37,57–60]. In most cases, there is experimental evidence that the equatorial form is the most probable one. In the case of derivatives of trifunctional amino acid residues these folded conformations can be abnormally stabilized by additional intramolecular side-chain–main-chain interactions [34,35].

The $3 \rightarrow 1$ intramolecularly H-bonded forms also represent the main feature of the γ turn (Fig. 10), which allows complete reversal of a polypeptide chain over three amino acid residues [61]. This chain reversal can connect two strands of intramolecularly H-bonded antiparallel-chain pleated-sheet β structure like the well-recognized β turn; however, it requires one less residue for this purpose. This turn is stabilized by two H bonds. One of them ($3 \rightarrow 1$) is bent and gives rise to the seven-membered-ring form; the other H bond ($1 \rightarrow 3$) is straight and is analogous to those in antiparallel-chain pleated-sheet β structures. Conformational energy computations indicated that the γ turn is a stable structure and that it can occur in a sequence consisting of L-amino acid residues. It was also proposed that pmr coupling constants can distinguish in a unique manner between the two types (β and γ) of turns [61].

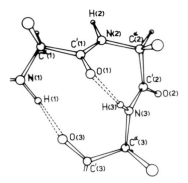

Fig. 10 Perspective drawing of a tripeptide segment folded into a γ turn.

There is only one example in the literature of γ turn in the solid state. It occurs in the sequence 25–27 (-L-Ser-L-Thr-L-Tyr-) of thermolysin [62,63] and has dihedral angles very similar to those originally proposed by Némethy and Printz [61].

In the same year (1972) a conformation containing a γ turn was proposed for the linear octapeptide hormone angiotensin II (at the L-Val-L-Tyr-L-Val-sequence) in aqueous solution [64]. Subsequently, however, it has been shown that all conformations (including the γ-turn structure) that require at least one small peptide NH—α-CH coupling constant can be eliminated as significant contributors to the conformational equilibrium of angiotensin II [65,66].

More recently, using pmr and carbon magnetic resonance (cmr) techniques, Urry and co-workers [67,68] suggested that, in the elastin model synthetic sequential polypeptides $($L-Ala-L-Pro-Gly-L-Val-Gly-L-Val$)_n$ and $($L-Val-L-Pro-Gly-L-Val-Gly$)_n$, a γ turn, involving the sequence Gly-L-Val-Gly, is present.

Finally, in the solid state a sequence of two 3 → 1 intramolecularly H-bonded peptide forms exists in horse heart ferricytochrome *c* (Ala-43-Pro-44-Gly-45-Phe-46) [69]. Other examples of such folded structures are discussed in the section dealing with intramolecularly bifurcated H-bonded peptide conformations [38,40–43,70] (see below).

THE 4 → 1 INTRAMOLECULARLY H-BONDED NONHELICAL *cis*-PEPTIDE CONFORMATION

A series of recent papers [13,38,47,71–76] has demonstrated the importance of the 4 → 1 intramolecularly H-bonded nonhelical peptide conformations [also referred to as H(4)···O(1) H-bonded forms, β loops, β turns, β bends,

β twists, C_{10} ring forms, hairpin bends, U folds] in polypeptides and proteins as key factors in their three-dimensional structure.

Investigations in this field have been performed on suitable model compounds, and the first study is due to Venkatachalam [13], who employed a simple yes-or-no criterion on a "tripeptide model" (see Fig. 3). This first essay, essentially qualitative, indicated only that the existence of the $4 \rightarrow 1$ H-bonded nonhelical peptide conformations was favored by satisfactory van der Waals contacts. More sophisticated model studies have been performed through potential-energy evaluations by the teams of Popov [71], Scheraga [72], and Ramachandran [73] and through a quantum mechanical approach by Maigret and Pullman [38]. These chain reversals have been analyzed in the native structures of globular proteins, as found by X-ray diffraction methods, by Scheraga and co-workers [72] and by Crawford *et al.* [74].

Three different kinds of $4 \rightarrow 1$ intramolecularly H-bonded nonhelical peptide forms have been found in crystalline peptides namely, *trans-I*, *trans-II*, and *cis* [47]. It should be noted that each of these conformations can have a mirror image if L-peptide units are replaced with D-peptide units and vice versa. The effect of such a replacement is to change the signs of the values of the torsional angles. The numerical values for these angles are quite similar for all the substances, virtually independent of the side groups on $C^\alpha(2)$ and $C^\alpha(3)$.

In this section we discuss only the most unusual type of $4 \rightarrow 1$ H-bonded nonhelical form, i.e., that having the central amide group in the *cis* configuration (Fig. 11). It is obvious that in this structure, as in the *trans-I* and *trans-II* structures, the terminal amide groups are in the *trans* configuration. Model building indicates that, because of the dimensions of the ring generated by the $4 \rightarrow 1$ intramolecular H bond, only one type of *cis*-peptide ring structure can exist, i.e., that with both C=O and N=H bonds of the central amide group pointing outward.

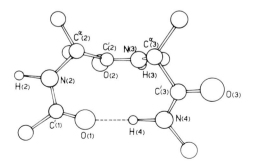

Fig. 11   The $4 \rightarrow 1$ intramolecularly H-bonded nonhelical *cis*-peptide conformation.

Studies in which the $4 \rightarrow 1$ intramolecularly H-bonded nonhelical *cis*-peptide conformation is discussed are extremely rare [72,76,77]. No conformational energy calculations have been carried out for bends of this type. Chain reversals with X-Pro bonds in the *cis* configuration have been found in a number of globular proteins [72,76]. However, in none of them, perhaps with the single exception of the -Gly-Tyr-Pro-Gly- (166–169) segment of subtilisin BPN', is an intramolecular H bond present [76].

The only case where $4 \rightarrow 1$ H-bonded *cis*-peptide structures have been definitely demonstrated to occur in the solid state is represented by the cyclic heptapeptide ilamycin B_1, as shown by a recent X-ray diffraction analysis [77]. The amide bonds connecting residues 2 and 3, and 6 and 7 are both *cis*, as seen from the values of ω angles ($\omega_2 = -11°$ and $\omega_6 = +1°$, respectively) and serve to fold the peptide chain to form a cyclic structure. It should be noted that both residues 3 and 7 are N-methyl-L-leucine, an N-substituted amino acid. Two transannular intramolecular H bonds [$N(4) \cdots O(1)$ of 2.98 Å and $N(1) \cdots O(5)$ of 2.80 Å] stabilize the conformation. There is also one shorter transannular close contact [$N(5) \cdots O(1)$ of 3.32 Å] for which a weak H bond might be considered. The torsional angles are $\phi_2 = -61°$, $\psi_2 = +126°$; $\phi_3 = -121°$, $\psi_3 = +38°$; $\phi_6 = -86°$, $\psi_6 = +117°$; $\phi_7 = -128°$, $\psi_7 = +99°$. Interestingly, the ϕ,ψ values of residue 4 ($\phi_4 = -123°$, $\psi_4 = +2°$) lie in the forbidden region, at the borderline between the α and β conformations (Fig. 2).

A 300 MHz pmr study of this molecule, on the basis of the H $\rightleftarrows$ D exchange rate of amide hydrogen atoms at room temperature, indicated that the exchanges of H(1), H(4), and H(5) took several days, whereas those at H(2) and H(6) took only a few hours [78]. As is clear from the solid-state structure [77], only H(1), H(4), and H(5) turn to the inside of the molecule, which causes a strong interaction with the carbonyl oxygen atoms, and the exchange rate of these hydrogen atoms should consequently be reduced. It appears that this conformation is only slightly solvent dependent.

Thus, the observation of the $4 \rightarrow 1$ intramolecularly H-bonded *cis*-peptide structure in ilamycin B_1 [77] reveals that this also could be an important structural element in polypeptide molecules.

THE OXY ANALOGS OF THE $3 \rightarrow 1$ AND $4 \rightarrow 1$ INTRAMOLECULARLY H-BONDED PEPTIDE CONFORMATIONS

In the oxy analogs of the $3 \rightarrow 1$ intramolecularly H-bonded peptide conformations, which could represent an essential feature of polypeptide chains near their C-terminal end, the C-terminal —OH group and the

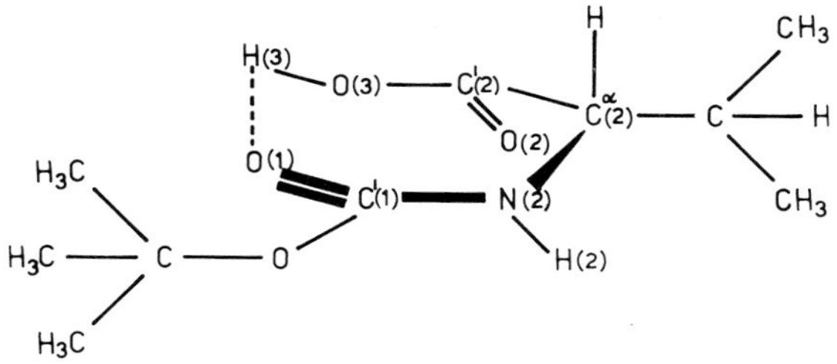

Fig. 12 The equatorial oxy analog of the $3 \rightarrow 1$ intramolecularly H-bonded peptide conformation for t-BOC-D-Val-OH.

carbonyl group nearest the C-terminal are involved in a seven-membered ring containing a somewhat bent H bond. As in the case of the $3 \rightarrow 1$ intramolecularly H-bonded peptide forms, (1) such an H bond is possible only in the *trans*-amide isomer, and (2) when R in —NH—CHR—CO— is not a hydrogen atom, two different conformers (equatorial and axial) can exist. An example of the equatorial oxy analog of the $3 \rightarrow 1$ intramolecularly H-bonded peptide conformation is illustrated in Fig. 12.

The ir absorption spectra of the t-BOC derivatives of all N-alkylamino acids so far examined in the solid state indicate that the urethane (carbamate) C=O is H bonded (obviously with the —OH group) [78,79]. Among the t-BOC derivatives of the various aliphatic amino acids, t-BOC-D-Val-OH shows strong evidence of an H bond involving the urethane carbonyl [78]. To ascertain *inter alia* if the urethane carbonyl of these compounds takes part to an intramolecular H bond (giving rise to the oxy analog of a $3 \rightarrow 1$ H-bonded peptide conformation) or to an intermolecular one, X-ray diffraction analyses of t-BOC-L-Pro-OH [80], t-BOC-L-Aze-OH (azetidine-2-carboxylic acid) [81] and t-BOC-D-Val-OH [78] were carried out by our group and others. The molecular structure of t-BOC-D-Val-OH is shown in Fig. 13. The prerequisite for the formation of the oxy analogs of the $3 \rightarrow 1$ H-bonded peptide forms, i.e., *trans*-amide configuration, is not met by the three protected amino acids. An intermolecular H bond (2.63–2.65 Å) linking the —OH and urethane carbonyl is apparent in all cases. In the t-BOC-D-Val-OH molecule, which contains a secondary urethane, an additional intermolecular H bond is present between the N—H and the C=O group of the carboxylic acid moiety (2.92 Å). A preliminary X-ray diffraction investigation indicated the absence of such folded forms also in the case of Z-Gly-OH (Z denotes benzyloxy-carbonyl), which also possesses a urethane N-protecting group [82].

N-Acetylamino acids as model compounds of the C-terminal regions of polypeptide chains are superior to N-*tert*-butyloxycarbonylamino acids. The higher basicity of the amide carbonyl with respect to that of the urethane carbonyl and the weaker van der Waals interactions of N-acetyl derivatives if

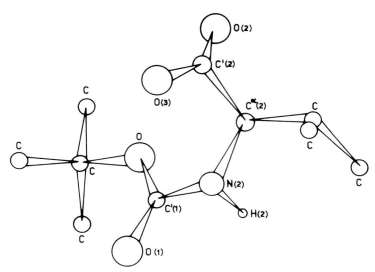

Fig. 13 Molecular structure of *t*-BOC-D-Val-OH.

compared with those of *N-tert*-butyloxycarbonyl derivatives could lead to different types of structures. In this context it is encouraging that Ac-L-Nva-OH, although not forming a seven-membered intramolecularly H-bonded structure in the solid state, has the —CONH— group in the *trans* configuration and the amide carbonyl and the C—O bond of the carboxylic acid group on the same side of the plane passing through the two carbonyl carbon atoms [83]. However, it should be admitted that Ac-Gly-OH, although having the amide group in the *trans* configuration, does not adopt such a conformation [84–86]. Both *N*-acetylamino acids possess two types of intermolecular H bonds. The former links the —OH to the amide C=O (2.55–2.56 Å), while the latter links the —NH to the C=O of the carboxylic acid group (2.95–3.04 Å). A hydrogen bond between the hydroxyl group and the amide carbonyl has been recently described in the solid state also in the case of Ac-L-Leu-OH [87].

The presence of the oxy analogs of the various types of ten-membered $4 \rightarrow 1$ intramolecularly H-bonded peptide conformations has been recently proposed by Deber [88] when the sequences Gly-L-Pro, L-Pro-Gly, and L-Pro-D-Pro occur in the two residues at the C-terminus of the polypeptide chain. In *t*-BOC-dipeptides experimental evidence for this folding was obtained from the observation in the ir absorption spectra of a $\simeq 30$ cm^{-1} shift to lower frequency of the urethane carbonyl band, due to H bond formation. The structure proposed by Deber [88] for *t*-BOC-Gly-L-Pro-OH is illustrated in Fig. 14. In the molecular structure of *t*-BOC-Gly-L-Pro-OH,

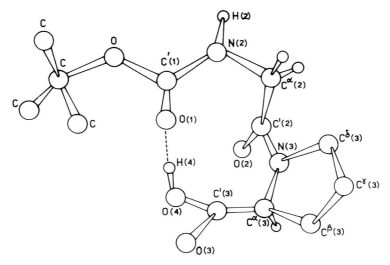

Fig. 14 The oxy analog of the type II′ 4 → 1 intramolecularly H-bonded nonhelical peptide conformation proposed by Deber [88] for *t*-BOC-Gly-L-Pro-OH in the solid state.

as determined by our group using the X-ray diffraction technique, two different types of H bond occur [39] (Fig. 7). The first is an intramolecular H bond between the urethane N—H and the amide carbonyl (2.59 Å). As discussed above, this H bond gives rise to the formation of a five-membered ring in the molecule. The second type of H bond is intermolecular between the O—H and the urethane C=O (2.64 Å). Thus, the hypothesis of the folding of the *t*-BOC-Gly-L-Pro-OH molecule in the solid state, put forward on the basis of the ir absorption results [88], has to be rejected. An extended conformation was also suggested in the solid state for Z-Gly-L-Pro-OH from a preliminary X-ray diffraction study [82].

   In summary, the X-ray diffraction studies discussed in this section have demonstrated the absence of the oxy analogs of the 3 → 1 and 4 → 1 intramolecularly H-bonded peptide forms for the various N-protected amino acids and peptides in the solid state. Intermolecular H bonds due to crystal packing effects prevail. Crystallographic data on other N-protected amino acids and peptides are required before any permanent conclusion can be drawn about the occurrence of the oxy analogs of the 3 → 1 and 4 → 1 H-bonded peptide conformations in the solid state. It is possible that specific crystallization solvents may also play a role in favoring these folded forms. In addition, as Deber correctly pointed out [88], "X-ray crystallography remains the method of choice for substantiation of the postulated intramolecularly H-bonded structure." In fact, merely on the basis of the ir absorption data, it is impossible to rule out unequivocally intermolecular H-bonding effects in the crystals of these acids as the source of the shifted carbonyl frequencies.

By means of pmr and cmr, ir absorption, optical rotatory dispersion, and CD spectroscopies it has been demonstrated rather conclusively that the oxy analogs of the $3 \to 1$ and $4 \to 1$ intramolecularly H-bonded peptide conformations are present in solution. The extent of their population in the conformational equilibrium mixture is dependent on structure, temperature, and solvent [39,78,88–93].

THE INTRAMOLECULARLY BIFURCATED H-BONDED PEPTIDE CONFORMATIONS

The first example of a bifurcated H bond, i.e., one hydrogen atom shared by two H bond acceptors, was reported by Albrecht and Corey in crystals of the α form of glycine [94]. This situation was pointed out by Pauling [95] to represent an exception to the condition that "the coordination of hydrogen does not exceed two." The position of the hydrogen atoms was later verified by Marsh in a detailed X-ray diffraction study [96]. Some structural details of such an unusual H bond are presented in Fig. 15, in which it can be seen that, although the nitrogen is closer to one oxygen, the hydrogen is closer to the other because, as suggested by Marsh [96], of the more favorable angle of attack.

In the last 25 years the structures of a number of molecules containing bifurcated H bonds have been published [42,97]. However, to the best of my knowledge, intramolecular bifurcated H bonds in peptides without artificially

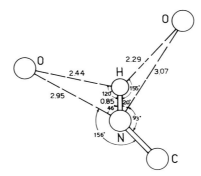

Fig. 15 Details of the intermolecularly bifurcated H bond in α-glycine.

introduced heteroatoms (e.g., the chlorine atom of the monochloroacetyl group) have never been reported as occurring in the solid state.

Two recent papers by Néel and colleagues clearly demonstrated the occurrence of intramolecularly bifurcated H-bonded peptide conformations in carbon tetrachloride solution [41,42]. Infrared absorption and pmr were employed for such an investigation. If the general formula of the N- and C-protected linear dipeptides examined is considered to be R-CO-L-Pro-X-Y, it appears that the stability of the structure containing the intramolecular bifurcated H bond depends on the nature of R, X, and Y. The Nancy group found that Ac-L-Pro-D-Ala-OMe showed the highest content of this conformation in the inert solvent. The structure proposed for this peptide is illustrated in Fig. 16. The steric repulsion between O(2) and the methyl group in the side chain of the D-alanine residue dictates a minimal distance $O(1) \cdots O(3)$ of 3 Å, thus permitting the formation of a rather strong intramolecular bifurcated H bond, $O(1) \cdots H(3) \cdots O(3)$. This type of H bond stabilizes a $3 \to 1$, $3 \to 3$ conformation (also called the $C_7 C_5$ form), where an equatorial $3 \to 1$ form involving the L-proline residue ($\phi = -75°, \psi = +50°$) is juxtaposed in part to an extended $3 \to 3$ form involving the D-alanine residue ($\phi = +158°$, $\psi = +180°$). The two amide bonds are in the *trans* configuration. The perturbation experienced by the H(3) atom results in a large bathochromic frequency shift for the corresponding $\nu(N-H)$ absorption maximum, which is observed at 3277 cm^{-1} in the ir absorption spectrum, and in a relevant frequency shift to lower fields for the corresponding proton signal in the pmr spectrum.

An interesting example of intramolecularly bifurcated H-bonded conformation has been proposed in solvents of low polarity by Madison for cyclo(-L-Pro-Gly)$_3$ on the basis of a thorough investigation using pmr and cmr, theoretical and experimental CD, and potential-energy calculations [43].

Recently, the $3 \to 1$, $3 \to 3$ peptide form has been taken into consideration in two theoretical papers devoted to the prediction of the most probable conformations in cyclic and linear peptide molecules [38,40].

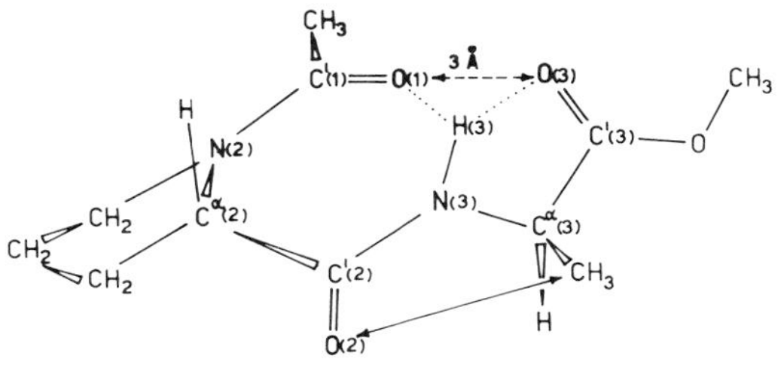

Fig. 16 The proposed intramolecularly bifurcated H-bonded peptide conformation of Ac-L-Pro-D-Ala-OMe.

It can be concluded that intramolecularly bifurcated H-bonded peptide conformations are rare but certainly not extinct. More examples will doubtless be discovered in the future.

cis-AMIDE GROUPS IN PEPTIDE STRUCTURES

In the solid state the peptide group usually adopts the *trans* configuration [46] and is essentially planar. The ω angle rarely deviates by more than 15° from 180°. As already mentioned, nonplanar deformations of the amide unit with $|\Delta\omega|$ up to 15° are quite probable, since the energy increase for such deviations from planarity are only of the order of 0.5 kcal/mole [46].

In cyclic dipeptides, because of the steric requirements of the six-membered ring, the amide groups are constrained to be *cis* ($\omega \simeq 0°$). However, the *cis* state also occurs in larger rings [4,47,48,51,77,98–107]. In Table 1 the homodetic cyclic peptides, in which *cis*-peptide bonds have been found by X-ray diffraction are listed. Only for cyclic nonapeptides is no example known. It is possible that the X-ray diffraction study of cyclolinopeptide A quoted by Naider *et al.* [108] will fill this gap. In fact, this peptide contains the Pro-Pro sequence, already found in the *cis* configuration in the structurally related cyclic decapeptide antamanide [47]. With a single exception (dihydrotentoxin), the peptide groups adopting the *cis* state in cyclic tripeptides or higher peptides are those with a substitution on the nitrogen atom, i.e., Pro, Hyp, Sar, MeLeu, MeIle. The occurrence of an N-substituted residue in a cyclic polypeptide does not guarantee a *cis* configuration, but the occurrence of a *cis* configuration has involved an N-substituted peptide unit in all but one of the known structures. Most unexpectedly in dihydrotentoxin are the secondary amide bonds involving the Gly and L-Leu residues, not the tertiary bonds involving the L-MeAla and D-MePhe residues, in the *cis* configuration. Rather surprisingly, however, no explanation was given by the authors for such an unusual result [103]. In all the cases listed in Table 1 the *cis*-amides have both C=O and N—R bonds pointing outward with respect to the cyclic structure.

Crystal structure results on *linear* oligopeptides indicate that the uncommon *cis* configuration is present only in *t*-BOC-L-Pro-L-Pro-L-Pro-L-Pro-OBzl [109] and in an equimolar mixture of Z-Gly-L-Pro-L-Leu-OH and its enantiomer [110]. It should be noted that in the former case it is the tertiary amide of the N-terminal urethane group which takes the *cis* state ($\omega = -1.1°$), while in the latter case it is the tertiary amide of the peptide Gly-L-Pro group ($\omega = -3.8°$). Interestingly, the presence of the *cis* tertiary amide of the N-terminal urethane group in *t*-BOC-L-Pro-L-Pro-L-Pro-L-Pro-OBzl was suggested by Deber *et al.* [111] and Tonelli [112] before the X-ray diffraction study.

TABLE 1

Occurrence of *cis*-Amide Bonds in Homodetic Cyclic Peptides

Substance	No. of residues in ring	No. of atoms in ring	Conforma-tion[a]	*cis*-Peptide bonds
c-Dipeptides	2	6	cc	All (forced)
c(L-Pro)$_3$	3	9	ccc	Pro, Pro, Pro
c-L-Pro-L-Pro-L-Hyp	3	9	ccc	Pro, Pro, Hyp
c-D-HyIv-L-MeIle | L-MeLeu-D-HyIv[b]	4	12	tctc	MeIle, MeLeu
c-Sar$_4$	4	12	tctc	Sar, Sar
c-Gly-Sar$_3$	4	12	tctc	Sar, Sar
c-L-Ala-Sar$_3$	4	12	tctc	Sar, Sar
c-Gly-L-Ala-Gly-L-Ala	4	12	tctc	Sar, Sar
Dihydrotentoxin	4	12	tctc	Gly, Leu
c-Sar$_5$	5	15	ccctt	Sar, Sar, Sar
c-L-Ala-Sar$_4$	5	15	ccctt	Sar, Sar, Sar
c-Sar$_7$	7	21	tttcccc	Sar, Sar, Sar, Sar
Ilamycin B$_1$	7	21	tctttct	MeLeu, MeLeu
c-Sar$_8$	8	24	ccttcctt	Sar, Sar, Sar, Sar
Li$^+$-Antamanide	10	30	tcttttcttt	Pro, Pro
Na$^+$[Phe^4, Val^6] Antamanide	10	30	tcttttcttt	Pro, Pro

[a] c = *cis*-amide bond; t = *trans*-amide bond.

[b] HyIv = α-hydroxy-iso-valeric acid.

Among the various N-protected amino acid derivatives so far examined by X-ray diffraction only *t*-BOC-L-Pro-OH [80], *t*-BOC-L-Aze-OH [81], and *t*-BOC-D-Val-OH (Fig. 13) [78] show the *cis* configuration (about the amide bond of the urethane moiety). In all three cases the deviation from planarity of the amide group is small. As far as the *t*-BOC-D-Val-OH molecule is concerned, this is the first time such a configuration has been found in the solid state for linear secondary amides (urethanes). Also, in our laboratory we are able to demonstrate that the structure assumed by *t*-BOC-D-Val-OH in the solid state does not depend on the polarity of the solvent mixture from which the crystals were grown. This finding probably indicates that the molecular structure of this compound, and in particular the *cis* configuration at the urethane–CONH linkage, is fixed by crystal packing forces, which in turn are governed by H bonds and the unique type of van der Waals contacts present in it.

The *cis–trans* isomerism at secondary and tertiary amide bonds of N-protected amino acid derivatives and linear oligopeptides in solution has been extensively investigated by a number of physicochemical techniques. Branik and Kessler [93] and other authors have recently published [113–119] review articles and papers that authoritatively cover this area.

A search through the protein crystallographic literature suggests that the *trans* configuration is adopted by most proline in native proteins. However, nine X-Pro bonds in the *cis* configuration have been shown to exist in globular proteins, including bovine ribonuclease S, staphylococcal nuclease, subtilisin BPN′, carbonic anhydrase, thermolysin, erythrocruorin, and Bence Jones protein Rei [72,76]. Overall, for all globular proteins for which high-resolution data exist, only 5–10% of the total prolines have been suggested to be *cis*. This low estimate could be misleading, however, since Huber [76] stated that, in view of the uncertainty of protein models obtained from Fourier maps calculated with isomorphous phases, a reexamination of segments around proline residues in other protein structures should be considered. In this context it has been suggested that the slow step in protein denaturation reactions might be due to *cis* ⇌ *trans* isomerism of proline residues [120]. It may be that these serve a very important function in the overall folding of a protein chain; thus, the folded protein might have no stability with these critical prolines in the *trans* state. Recently, a new model for the triple-helical conformation of the collagenlike polymer poly(Gly-L-Pro-L-Hyp) has been developed in which the Gly-L-Pro bond in the tripeptide monomeric unit is in the *cis* configuration, thus allowing the formation of an interchain H bond between the hydroxyl group of the hydroxyprolyl residue and the carbonyl group of the glycyl residue in an adjacent chain [121]. Bansal *et al.* [122] have confirmed this structure by model building using computer techniques, and the helical parameters obtained are close to the experimentally observed values. The model is also comparable in stability with other models from energy considerations.

Finally, it should be recalled that the well-known form I of poly-L-proline is characterized by a right-handed helical arrangement of *cis*-amide groups [123–125].

It can be concluded that the current interest in this area will certainly cause the number of *cis*-amide bonds discovered in peptide structures to increase rapidly.

ADDENDUM

The $2 \to 3$ intramolecularly H-bonded peptide conformation has been proposed by Wütrich [126] in the case of -X-Pro- peptides in aqueous solution. A further example of $3 \to 1$ H bond in a crystal has been demonstrated by Petcher *et al.* [127] in the case of the cyclic undecapeptide cyclosporin A. The amino acid residue involved is Ala 7 ($\phi = -92°$, $\psi = +64°$, $\omega = -166°$); its side chain is in the equatorial position. Interestingly, cyclosporin A contains a *cis* amide bond linking the MeLeu-9-MeLeu-10 residues. Demel and Kessler [128] have proposed the existence in solution of a conformation containing a γ turn for the cyclopentapeptide c(L-Pro-L-Phe-Gly-L-Phe-Gly).

The crystal and molecular structure of *t*-BOC-L-Thz-OH (Thz denotes thiazolidine-4-carboxylic acid) has been reported [129]. No intramolecular H bonds are present. An intermolecular H bond (2.653 Å) linking the —OH and urethane carbonyl is apparent. The tertiary amide group is in the *cis* configuration ($\omega = -6°$).

No evidence of an oxy analog of a $4 \rightarrow 1$ intramolecularly H-bonded peptide conformation was found in the solid state in the case of *t*-BOC-L-Pro-Gly-OH on the basis of an X-ray diffraction study [130].

Patel and Tonelli [131] discussed conformations for gramicidin-S analogs which incorporate *cis*-peptide bonds at L-Orn-2-L-MeLeu-3.

REFERENCES

1. IUPAC-IUB Commission on Biochemical Nomenclature, *J. Mol. Biol.* **52**, 1 (1970), and references therein.
2. G. N. Ramachandran, C. Ramakrishnan, and V. Sasisekharan, *J. Mol. Biol.* **7**, 95 (1963).
3. G. D. Fasman, ed., "Poly-α-Amino Acids." Dekker, New York, 1967.
4. G. N. Ramachandran and V. Sasisekharan, *Adv. Protein Chem.* **23**, 283 (1968).
5. C. M. Venkatachalam and G. N. Ramachandran, *Annu. Rev. Biochem.* **38**, 45 (1969).
6. M. Goodman, A. S. Verdini, N. S. Choi, and Y. Masuda, *Top. Stereochem.* **2**, 69 (1970).
7. H. A. Scheraga, *Chem. Rev.* **71**, 195 (1971).
8. N. Lotan, A. Berger, and E. Katchalski, *Annu. Rev. Biochem.* **41**, 869 (1972).
9. A. G. Walton and J. Blackwell, "Biopolymers." Academic Press, New York, 1973.
10. A. J. Hopfinger, "Conformational Properties of Macromolecules," Vol. 1. Academic Press, New York, 1973.
11. R. D. B. Fraser and T. P. MacRae, eds., "Conformation in Fibrous Proteins." Academic Press, New York, 1973.
12. B. Pullman and A. Pullman, *Adv. Protein Chem.* **28**, 348 (1974).
13. C. M. Venkatachalam, *Biopolymers* **6**, 1425 (1968).
14. K. H. Meyer and H. Mark, *Ber. Dsch. Chem. Ges.* **61**, 1932 (1928).
15. W. T. Astbury and A. Street, *Philos. Trans. R. Soc. London, Ser. A* **230**, 75 (1931).
16. L. Pauling and R. B. Corey, *Proc. Natl. Acad. Sci. U.S.A.* **39**, 253 (1951), and previous papers in the series.
17. *Cold Spring Harbor Symp. Quant. Biol.* **36**, (1972), and papers therein.
18. M. Palumbo, S. Da Rin, G. M. Bonora, and C. Toniolo, *Makromol. Chem.* **177**, 1477 (1976).
19. C. Toniolo and M. Palumbo, *Biopolymers* **16**, 219 (1977).
20. B. B. Doyle, E. G. Bendit, and E. R. Blout, *Biopolymers* **14**, 937 (1975).
21. T. Miyazawa and E. R. Blout, *J. Am. Chem. Soc.* **83**, 712 (1961).
22. Yu. N. Chirgadze and N. A. Nevskaya, *Biopolymers* **15**, 607 and 627 (1976).

23. C. Toniolo, unpublished results.
24. T. J. Yu and W. L. Peticolas, in "Peptides, Polypeptides and Proteins" (E. R. Blout et al., eds.), p. 370. Wiley, New York, 1974.
25. J. S. Balcerski, E. S. Pysh, G. M. Bonora, and C. Toniolo, J. Am. Chem. Soc. 98, 3470 (1976).
26. E. S. Pysh, J. Chem. Phys. 52, 4723 (1970).
27. K. Rosenheck and B. Sommer, J. Chem. Phys. 46, 532 (1967).
28. R. Woody, Biopolymers 8, 669 (1969).
29. V. Madison and J. Schellman, Biopolymers 11, 1041 (1972).
30. A. Del Pra and C. Toniolo, in preparation.
31. C. Toniolo and G. M. Bonora, in "Peptides: Chemistry, Structure and Biology" (R. Walter and J. Meienhofer, eds.), p. 145. Ann Arbor Sci. Publ., Ann Arbor, Michigan, 1975.
32. P. Y. Chou and G. D. Fasman, Biochemistry 13, 211 (1974).
33. M. Tsuboi, T. Shimanouchi, and S. Mizushima, J. Am. Chem. Soc. 81, 1406 (1959), and previous papers in the series.
34. J. Néel, in "Macromolecular Microsymposium" (B. Sedláček, ed.), p. 201, and references therein. Butterworth, London, 1971.
35. M. T. Cung, M. Marraud, and J. Néel, in "Conformation of Biological Molecules and Polymers" (E. D. Bergmann and B. Pullman, eds.), p. 69. Academic Press, New York, 1973.
36. B. Pullman and B. Maigret, in "Conformation of Biological Molecules and Polymers" (E. D. Bergmann and B. Pullman, eds.), p. 13. Academic Press, New York, 1973.
37. A. W. Burgess and H. A. Scheraga, Biopolymers 12, 2177 (1973).
38. B. Maigret and B. Pullman, Theor. Chim. Acta 35, 113 (1974).
39. E. Benedetti, M. Palumbo, G. M. Bonora, and C. Toniolo, Macromolecules 9, 417 (1976).
40. N. Go and H. A. Scheraga, Macromolecules 6, 525 (1973).
41. G. Boussard, M. Marraud, and J. Néel, J. Chim. Phys.-Chim. Biol. 71, 1081 (1974).
42. G. Boussard, M. T. Cung, M. Marraud, and J. Néel, J. Chim. Phys.-Chim. Biol. 71, 1159 (1974), and references therein.
43. V. Madison, in "Peptides, Polypeptides and Proteins" (E. R. Blout et al., eds.), p. 89. Wiley, New York, 1974.
44. D. W. Urry, W. D. Cunningham, and T. Ohnishi, Biochemistry 13, 609 (1974).
45. S. Zimmerman and H. A. Scheraga, personal communication.
46. A. S. Kolaskar, A. V. Lakshminarayanan, K. P. Sarathy, and V. Sasisekharan, Biopolymers 14, 1081 (1975).
47. I. L. Karle, in "Peptides: Chemistry, Structure and Biology" (R. Walter and J. Meienhofer, eds.), p. 61, and references therein. Ann Arbor Sci. Publ., Ann Arbor, Michigan, 1975.
48. J. D. Dunitz and F. K. Winkler, Acta Crystallogr., Sect. B 31, 251 (1975), and references therein.
49. J. Dale and K. Titlestad, Chem. Commun. p. 656 (1969).
50. P. Groth, Acta Chem. Scand. 24, 780 (1970).
51. J. P. Declercq, G. Germain, M. van Meerssche, T. Debaerdmaeker, J. Dale, and K. Titlestadt, Bull. Soc. Chim. Belg. 84, 275 (1975).
52. J. Dale and K. Titlestadt, Chem. Commun. p. 1043 (1970).
53. G. M. Crippen and H. A. Scheraga, Proc. Natl. Acad. Sci. U.S.A. 64, 42 (1969).
54. B. Maigret, B. Pullman, and D. Perahia, J. Theor. Biol. 31, 269 (1971).

55. T. Matsuzaki and Y. Iitaka, *Acta Crystallogr., Sect. B* **27**, 507 (1971).
56. M. Marraud and J. Néel, *C. R. Hebd. Seances Acad. Sci., Ser. C* **278**, 1015 (1974).
57. V. F. Bystrov, S. L. Portnova, V. I. Tsetlin, V. T. Ivanov, and Yu. A. Ovchinnikov, *Tetrahedron* **25**, 493 (1969).
58. J. L. Dimicoli and M. Ptak, *Tetrahedron Lett.* p. 2013 (1970).
59. M. Avignon and J. Lascombe, *in* "Conformation of Biological Molecules and Polymers" (E. D. Bergmann and B. Pullman, eds.), p. 97. Academic Press, New York, 1973.
60. V. T. Ivanov, M. P. Filatova, Z. Reissman, T. O. Reutova, E. S. Efremov, V. S. Pashkov, S. G. Galaktionov, G. L. Grigoryan, and Yu. A. Ovchinnikov, *in* "Peptides: Chemistry, Structure and Biology" (R. Walter and J. Meienhofer, eds.), p. 151. Ann Arbor Sci. Publ., Ann Arbor, Michigan, 1975.
61. G. Némethy and M. Printz, *Macromolecules* **5**, 755 (1972).
62. B. W. Matthews, *Macromolecules* **5**, 818 (1972).
63. P. M. Colman, J. N. Jansonius, and B. W. Matthews, *J. Mol. Biol.* **70**, 701 (1972).
64. M. P. Printz, G. Némethy, and H. Bleich, *Nature (London), New Biol.* **237**, 135 (1972).
65. G. R. Marshall, H. E. Bosshard, W. H. Vine, and J. D. Glickson, *Nature (London), New Biol.* **245**, 125 (1973).
66. J. G. Glickson, W. D. Cunningham, and G. R. Marshall, *Biochemistry* **12**, 3684 (1973).
67. D. W. Urry, L. W. Mitchell, T. Ohnishi, and M. M. Long, *J. Mol. Biol.* **96**, 101 (1975).
68. D. W. Urry, T. Ohnishi, M. M. Long, and L. W. Mitchell, *Int. J. Pept. Protein Res.* **7**, 367 (1975).
69. C. B. Anfinsen and H. A. Scheraga, *Adv. Protein Chem.* **29**, 205 (1975), and references therein.
70. V. Madison, N. Atreyi, C. M. Deber, and E. R. Blout, *J. Am. Chem. Soc.* **96**, 6725 (1974).
71. G. M. Lipkind, S. F. Arkipova, and E. M. Popov, *Mol. Biol. (Moscow)* **4**, 509 (1970).
72. P. N. Lewis, F. A. Momany, and H. A. Scheraga, *Biochim. Biophys. Acta* **303**, 211 (1973).
73. R. Chandrasekaran, A. V. Lakshminarayanan, U. V. Pandya, and G. N. Ramachandran. *Biochim. Biophys. Acta* **303**, 14 (1973).
74. J. L. Crawford, W. N. Lipscomb, and C. G. Schellman, *Proc. Natl. Acad. Sci. U.S.A.* **70**, 538 (1973).
75. A. J. Geddes, K. D. Parker, E. D. T. Atkins, and E. Beighton, *J. Mol. Biol.* **32**, 343 (1968).
76. R. Huber and W. Steigemann, *FEBS Lett.* **48**, 235 (1974), and references therein.
77. Y. Iitaka, H. Nakamura, K. Takada, and T. Takita, *Acta Crystallogr., Sect. B* **30**, 2817 (1974).
78. C. Toniolo, M. Palumbo, and E. Benedetti, *Macromolecules* **9**, 420 (1976).
79. R. K. Olsen, *J. Org. Chem.* **35**, 1912 (1970).
80. E. Benedetti, M. R. Ciajolo, and A. Maisto, *Acta Crystallogr., Sect. B* **30**, 1783 (1974).
81. M. Cesari, L. D'Ilario, E. Giglio, and G. Perego, *Acta Crystallogr., Sect. B* **31**, 49 (1975).
82. Y. Sasada, K. Tanaka, Y. Ogawa, and M. Kakudo, *Acta Crystallogr.*, **14**, 326 (1961).
83. Gy. Lovas, A. Kalman, and Gy. Argay, *Acta Crystallogr., Sect. B* **30**, 2882 (1974).
84. G. B. Carpenter and J. Donohue, *J. Am. Chem. Soc.* **72**, 2315 (1950).

85. J. Donohue and R. E. Marsh, *Acta Crystallogr.* **15**, 941 (1962).
86. M. F. Mackay, *Cryst. Struct. Commun.* **4**, 225 (1975).
87. A. Waskowska, K. Lukaszewicz, L. G. Kuzmina, and Yu. T. Struchkov, *Bull. Acad. Pol. Sci., Ser. Sci. Chim.* **23**, 149 (1975); *Chem. Abstr.* **83**, 89101t (1975).
88. C. M. Deber, *Macromolecules* **7**, 47 (1974).
89. V. Madison and J. Schellman, *Biopolymers* **9**, 569 (1970), and preceding paper.
90. J. T. Gerig, *Biopolymers* **10**, 2435 (1971).
91. M. Christl and J. D. Roberts, *J. Am. Chem. Soc.* **94**, 4565 (1972).
92. C. A. Evans and D. L. Rabenstein, *J. Am. Chem. Soc.* **96**, 7312 (1974).
93. M. Branik and H. Kessler, *Chem. Ber.* **108**, 2176 (1975), and previous papers in the series.
94. G. Albrecht and R. B. Corey, *J. Am. Chem. Soc.* **61**, 1087 (1939).
95. L. Pauling, "The Nature of the Chemical Bond," p. 266. Cornell Univ. Press, Ithaca, New York, 1939.
96. R. E. Marsh, *Acta Crystallogr.* **11**, 654 (1958).
97. J. Donohue, *in* "Structural Chemistry and Biology" (A. Rich and N. Davidson, eds.), p. 433, and references therein. Freeman, San Francisco, California, 1968.
98. I. L. Karle, *in* "Chemistry and Biology of Peptides" (J. Meienhofer ed.), p. 117. Ann Arbor Sci. Publ., Ann Arbor, Michigan, 1972.
99. F. A. Bovey, *in* "Peptides, Polypeptides, and Proteins" (E. R. Blout *et al.*, eds.), p. 248. Wiley, New York, 1974.
100. E. R. Blout, C. M. Deber, and L. G. Pease, *in* "Peptides, Polypeptides, and Proteins" (E. R. Blout *et al.*, eds.), p. 266. Wiley, New York, 1974.
101. Yu. A. Ovchinnikov and V. T. Ivanov, *Tetrahedron* **31**, 2177 (1975).
102. I. L. Karle, H. C. J. Ottenheym, and B. Witkop, *J. Am. Chem. Soc.* **96**, 539 (1974).
103. W. L. Meyer, L. F. Kuyper, D. W. Phelps, and A. W. Cordes, *Chem. Commun.* p. 339 (1974).
104. P. Groth, *Acta Chem. Scand., Sect. A* **29**, 38 (1975).
105. J. Dale and K. Titlestadt, *Acta Chem. Scand. Ser., B* **29**, 353 (1975).
106. K. Titlestadt, *Acta Chem. Scand., Ser. B* **29**, 153 (1975).
107. G. Kartha and G. Ambady, *Acta Crystallogr., Sect. B* **31**, 2035 (1975).
108. F. Naider, E. Benedetti, and M. Goodman, *Proc. Natl. Acad. Sci. U.S.A.* **68**, 1195 (1971).
109. T. Matsuzaki, *Acta Crystallogr., Sect. B* **30**, 1029 (1974).
110. T. Yamane, T. Ashida, M. Shimonishi, M. Kakudo, and Y. Sasada, *Acta Crystallogr., Sect. B* **32**, 2071 (1976).
111. E. M. Deber, F. A. Bovey, J. P. Carver, and E. R. Blout, *J. Am. Chem. Soc.* **92**, 6191 (1970).
112. A. E. Tonelli, *J. Am. Chem. Soc.* **92**, 6187 (1970).
113. F. A. Bovey, *in* "Chemistry and Biology of Peptides" (J. Meienhofer, ed.), p. 3. Ann Arbor Sci. Publ., Ann Arbor, Michigan, 1972.
114. F. A. Bovey, *J. Polym. Sci., Macromol. Rev.* **9**, 1 (1974), and references therein.
115. O. Oster, E. Breitmaier, and W. Voelter, *in* NMR Spectroscopy of Nuclei Other than Protons" (T. Axenrod and G. A. Webb, eds.), p. 233. Wiley, New York 1974.
116. K. Wütrich, C. Grathwohl, and R. Schwyzer, *in* "Peptides, Polypeptides, and Proteins" (E. R. Blout *et al.*, eds.), p. 300. Wiley, New York, 1974.
117. P. E. Young and C. M. Deber, *Biopolymers* **14**, 1547 (1975).
118. I. Z. Siemion, T. Wieland, and K.-H. Pook, *Angew. Chem., Int. Ed. Engl.* **14**, 702 (1975).

119. S. Fermandjan, S. Tran-Dinh, J. Savrda, E. Sala, R. Mermet-Bouvier, E. Bricas, and P. Fromageot, *Biochim. Biophys. Acta* **399**, 313 (1975).
120. J. F. Brandts, H. R. Halvorson, and M. Brennan, *Biochemistry* **14**, 4953 (1975).
121. R. A. Berg, Y. Kishida, Y. Kobayashi, K. Inouye, A. E. Tonelli, S. Sakakibara, and D. J. Prockop, *Biochim. Biophys. Acta* **328**, 553 (1973).
122. M. Bansal, C. Ramakrishnan, and G. N. Ramachandran, *Biopolymers* **14**, 2457 (1975).
123. J. Kurtz, A. Berger, and E. Katchalski, *Nature (London)* **178**, 1066 (1956).
124. A. Elliott, E. M. Bradbury, A. R. Downie, and W. E. Hanby, *in* "Polyamino Acids, Polypeptides, and Proteins" (M. A. Stahmann, ed.), p. 255, and references therein. Univ. of Wisconsin Press, Madison, 1962.
125. J. Arnott and S. D. Dover, *Acta Crystallogr., Sect. B* **24**, 599 (1968).
126. K. Wütrich, ed., "NMR in Biological Research: Peptides and Proteins," p. 157. North Holland, Amsterdam, 1976.
127. T. J. Petcher, H. P. Weber, and A. Rüegger, *Helv. Chim. Acta* **59**, 1480 (1976).
128. D. Demel and H. Kessler, *Tetrahedron Lett.* p. 2801 (1976).
129. F. Robert, *Acta Crystallogr., Sect. B* **32**, 2376 (1976).
130. E. Benedetti, V. Pavone, C. Toniolo, G. M. Bonora, and M. Palumbo, *Macromolecules* (in press).
131. D. Patel and A. E. Tonelli, *Biopolymers* **15**, 1623 (1976).

Index

A

Actinomycins, as antitumor antibiotics, 99–100
Acyl derivative, carboxyl-terminal, 74–75, 81
O-Acylhydroxamic acid, C terminal, synthesis of, 81–82
Adenine
    deamination, 249–251, 253–254, 257
    nucleosides, syntheses of, 236–237, 240
    reaction with α-ketoaldehydes, 261
Adenosine
    and bisulfite deamination, 254
    nucleosides, syntheses of, 231–234, 236, 239
    reaction with nitrous acid, 250
Adenosine deaminase
    in conversion of deoxyinosine to deoxyadenosine, 223–224
    in nucleoside syntheses, 232, 240
Aggregation number, affecting properties of micelles, 139–140, 142–143
Alanine
    conformation studies of, in peptides, 179–182, 186, 188–190, 192, 196–197
    in proteinoid production, 124–133
    tRNA, defining on DNA duplex, 211
Alkylating agents, antitumor, 95–96, 98, 109–119

Alloisoleucine, in proteinoids, 124, 129
Amide groups, configuration in peptides, 272–274, 277–281, 283–287
Amines, configuration in micelles, 148–149
Amino acids, *see also* names of specific amino acids
    aromatic, in peptides reacting with DNA, 51–56, 59
    in C-terminal degradations, 75–79, 85–87, 91
    as precursors to protocells, 24–25, 28–30
    in proteinoids, 123–135
    residues
        in hormone conformation studies, 178, 190–195, 198–199
        in peptides, conformation of, 270–271, 273, 275, 279–281, 285–286
    sequences
        in lac repressor, 63–64
        in polypeptide genes, 215–218
        in proteins, 210–211, 219
α-Aminobutyric acid, in proteinoids, 124
α-Aminoisobutyric acid
    in meteorites, 124
    residues, configurations of, 274–275
Aminoquinone moiety, in antitumor antibiotics. 99–118
Amino terminus, degradation of polypeptides from, 72–76, 85

Angiotensin
    conformation of, 276
    DNA duplex coding in, 217
Anhydride procedure, in synthesis of gluta-
    mate oligomers, 179–180
2′,3′-Anhydroadenosine, synthesis of, 233–
    234
Anhydronucleosides, sugar transformations
    in synthesis of, 222
Anisotropy, in DNA–peptide complexes,
    56–57
Antibiotics, see also individual ones by name
    containing aminoquinones, 99–118
    reaction with DNA, 95–99, 246
    synthetic transformations of, 221–241
Antineoplastic activity, chemical features of,
    95–96, 118
Antiparallel β structures, of peptides, 268–
    270, 275
Antitumor antibiotics, 95–96
    from chemical modification of nucleic
        acids, 246, 261–262
    containing aminoquinones, 99–118
    inhibitors of nucleic acid syntheses, 96–97
Arginine, in hydrolyzates of proteinoids,
    129–130, 132
Aspartate, in peptides, synthesis of, 179, 182,
    184, 192, 193, 197
Aspartic acid
    in C-terminal peptide degradations, 74–75,
        78, 87–89
    in meteorites, 124
    in production of protenoids, 125–130
    in thermal polymers, 123, 131–134
Auratins, as antitumor antibiotics, 99–100
Axial peptide conformations, 274–275, 279
Azide
    C-terminal, synthesis of, 80–81
    method of peptide degradation, 89–91
    procedures in synthesis of glutamate
        oligomers, 179
Aziridine ring, in mitosanes, 106–106, 112,
    115
Aziridinoquinones, as antitumor antibiotics,
    99–100
α, α′-Azobisisobutyronitrile, in purine nu-
    cleoside synthesis, 221, 233–234, 239

B

Bacteria
    attack on foreign DNA, 206–208
    genetic information in, 204, 218

Bacteriophages
    DNA attacked in invading, 206
    sequencing of nucleotides in, 209–212, 218
Base–sugar coupling, in syntheses of nu-
    cleosides, 221–223, 230–231
Benz[a]pyrene method of binding reactive
    intermediates, 157–164
Benzyloxycarbonyl, amine-protecting group
    in oligopeptides, 177, 180–181, 183–
    187, 191, 193, 195–197
Bifurcated H bond, in peptide con-
    formations, 282–284
Binary fission, reproduction in micro-
    spheres, 28, 31
Binding, see also Hydrogen bonds
    β-chain DNA, 58–59
    of DNA, to small molecules, 33–50
    of peptides, to DNA, 51–67
Biology, relation to chemistry in origins of
    life, 22–24, 30–31
Bisulfite modification of nucleic acids, 246,
    252–57
Budding, reproduction in microspheres,
    28–29, 31
Butyloxycarbonyl
    amine-protecting group in oligopeptides,
        177, 189–190, 193–198
    compounds, conformation studies of,
        271–272, 279–281, 284–285, 287

C

Cancer control, see Antitumors
Carbodiimides, in peptide C-terminal residue
    determination, 74, 82–88
Carbohydrate transformations, development
    of reactions for, 221–241
β-Carbolines, from cataractous lens pro-
    teins, 164–166
Carbon magnetic resonance
    measurements of gonyautoxins, 170–171
    in peptide configuration studies, 276,
        282–283
C-terminal
    amino acid residues, in proteinoids, 123,
        131–134
    sequencing of proteins, 71–92
Carboxymethyl cellulose chromatography,
    studies of hemoglobin, 168–169
Carboxypeptidase, use in C-terminal
    sequencing of proteins, 73
Carcinogenesis, from chemical reactions of
    DNA, 247, 261, 262

Catalase, role in streptonigrin reactions, 103, 115

Cataract, isolation of β-carbolines from proteins of, 164–166

Catenane, structures in DNA bindings, 66–67

Cell, emergence of first, 21–31

Cetyltrimethylammonium bromide catalyses, as micellar enhancements, 146–148

Chain arrangement
β, in peptide-DNA binding, 52, 58–61, 65–66
in β-forming peptides, 267–270, 277

Chain reversals, in peptide structures, 275, 278

Chemistry, in origins of life, 22–24, 30–31

6-Chloropurine deoxyriboside, synthesis of, 222–224

Cholestanol, molecular bonding in, 9–10

Cholesterol, function in membranes, 3–5, 7–16, 18

Chromatin
and protein–DNA binding specificity, 33, 47, 62
structure, 66–67

Chromosomes, as targets for cross-linking, 256

Circular dichroism spectra,
of adducts of DMBA-5,6-epoxide and poly(G), 157–160, 163
in conformation of oligopeptides, 178, 183–190, 192, 195–198
in DNA binding, 34–37, 39, 51–52
of gonyautoxins, 170
in intercalation process, 42–46
of L-valine heptamer, 270, 282–283

Cis-amide groups, see Amide groups, configuration of

Codon redundancy, affecting nucleotide sequence determination, 215–217

Collagen, conformation studies of, 194–195

Cordycepin, synthesis of, 233–236

Counterions, around micelles, 138–144, 147, 149

Critical micelle concentration, in formation of micelles, 137, 139–143

Cross-linking, of DNA strands, 96–98, 101, 104
in reaction with mitamycin C, 106–107, 111–118
by S$_1$-endonuclease, 107–108

Curtius rearrangement, in peptide C-terminal residue determination, 80–81, 90–91

Cyclonucleosides, transformations of, 222, 233–23, 236, 239

Cyclosporin A, stereochemistry of, 286–287

Cystine, in proteinoid production, 126

Cytidine
deamination, 250–252, 254
in nucleic acid structure studies, 248
in nucleoside transformations, 227, 229
reaction with α-ketoaldehydes, 261

Cytosine
deamination, 249–256
nucleosides, synthesis of, 223–229
reaction with ninhydrin, 261–262

D

Deamination
of micelles, stereochemistry of, 148–149
of nucleic acid components, 249–256

Deazaguanosine, see Q-nucleosides

2'-Deoxyadenosine
conversion from deoxyinosine, 223–224
synthesis of, 233, 235–236

2'-Deoxyinosine, conversion to deoxyadenosine, 223–224

Deoxynucleosides, transformations of, 223–224, 233–236, 241

DNA
bacterial attack on, 206–208
chemical modification of, 245, 262
covalently closed circular fluorescense assays of, 98, 101–104, 106, 114–115
lac repressor binding in, 204–206
peptide complexes, 34, 47, 51–61
protein recognition, 33, 45–47, 50–56, 58–59, 61–67
reactions with
antitumor agents, 96–119
mitomycin, 106–113
streptonigrin, 104–106
reporter complexes, 34–50
sequences
elucidation if, 208, 210
expressed in polypeptide structure, 214–219
synthesis
of duplex, 211–212
inhibited, 96–97, 99–101

Deoxyuridine
derivatives, hydrolysis of, 258–259
synthesis of, 253

Deuterochloroform, in oligopeptide conformation studies, 177, 192, 193

Dichroism, *see also* Circular dichroism
flow, in DNA peptide complexes, 51–52, 57
in intercalation, 42–43, 49
reduced, by electrostatic forces, 53–54, 57
Dimethylformamide, 177, 183
in purine transformations, 221–223
in pyrimidine transformations, 230, 240–241
4,4-Dimethyl-3-pivaloxypent-2-enoyl, in nucleoside syntheses, 221, 234–236, 238–239

E

*Eco*RI, *see* Endonuclease, restriction
Edman reagent, for degradation of polypeptide chains, 72, 74, 76, 80, 91
Elastin, conformations in, 272, 276
Electrostatic forces, in DNA binding, 33–34, 42–45, 52–54, 58, 61, 64–66
Endonucleases
restriction, recognition of "minimal DNA," 206–208, 218
S₁, to confirm cross-linking in DNA, 107–109, 118
Endonucleolysis
avoided by bacteria, 206
in synthesis of tRNA, 213–214
Enzymes, *see also* specific ones by name
as catalysts in synthesis of nucleotides, 203–204
as control mechanisms in nucleosides, 221, 223–224, 240
in sequencing nucleic acids, 247–248
in joining synthesized nucleotide fragments, 212, 216
Equatorial peptide conformations, 274–275, 279
Equilibrium dialysis
studies in DNA–peptide complexes, 51–52, 55
studies in DNA–reporter complexes, 42–43, 45, 47
Erythrocyte membranes, elements in, 3–4, 7
Escherichia coli
antibacterial activity against, 117
DNA
cross-linking in, 108–109, 111
fluorescence in, 110
lactose-metabolizing, 204

*Eco*RI restriction endonuclease in, 206
polymerase I, catalyst in "repair" synthesis, 210–212
treated with streptonigrin, 103
Ethanolamine, in membranes, 3, 16–17
Ethidium fluorescence assays, 97–99, 101–104, 108–111, 114, 119
Ethyl ester, carboxyl protecting group in oligopeptides, 177, 180–181, 193, 195–197
Evolution, of cells, 22–24, 28–30, 124

F

Fagarol, antisickling agent, 166–167
Fatty acids, unsaturated, in hydrophobic bonding, 3, 6–9, 11, 13–14, 16
Fluorescence
of β-carbolines from cataractous proteins, 164–166
of gonyautoxins, 172
in intercalative dyes, in nucleic acid reactions, 96–99, 101–104, 107, 114, 118–119
5-Fluorocytidine nucleosides, synthesis of, 227–229
5-Fluorocytosine nucleosides, synthesis of, 223–229
5-Fluoropyrimidines, synthesis of, 223–229
5-Fluorouracil nucleosides, synthesis of, 223–229
5-Fluorouridine nucleosides, synthesis of, 225–229
Formycin, in nucleoside transformations, 230–231, 233, 237

G

Galactose, in membranes, 3, 16–17
β-Galactosides, synthesis blocked, 257
Galactosides, effect on *lac* repressors, 204–205
Genes
G, relationship to "spike" protein, 210
for polypeptides, 214–218
RNA, attempts to synthesize, 210–214
structural, for lactose-metabolizing enzymes, 204
Glutamate, in oligopeptide conformation studies, 179, 182–186, 191, 193, 197–198

Glutamic acid
γ-linkage, determination of, 78, 87
in production of proteinoids, 125–131
in terminal composition of thermal polymers, 123, 131–134
Glycerolipids, function in membranes, 2–3
Glycine
bifurcated H bond in, 282
in bisulfite transaminations, 256
in C-terminal peptide sequencing, 86
residues in peptides, conformation studies of, 198–199
Glycolipids, function in membranes, 3, 7, 18
Glycosyl bond cleavage, in transformation of nucleosides, 222–223, 236
Glycyl derivatives, H-bonded peptide structure in, 271, 286
Glyoxal, reaction with guanine nucleotides, 259–261
Glysine, in proteinoid production, 124, 126–132
Gonyaulax tamarensis, shellfish poison, 168–170
Gonyautoxins, new toxins from *Gonyaulax tamarensis,* 168–174
Guanine
alkylation of in DNA, 111–112
deamination, 249–254
reaction with α-ketoaldehydes, 259–261
Guanosine
adducts of DMBA, 158–163
reactions with α-ketoaldehydes, 260–261
reaction with nitrous acid, 250

H

Helix, DNA
bending, in partial intercalation, 47–48, 55, 57, 65–66
in binding to reporter molecules, 33, 37, 41
coil transition, interaction with peptides, 51, 67
dissymmetric recognition of, 45–47
dynamic structure of, 49–50
in peptide β-chain binding, 58–60
Hemoglobin, normal and sickle cell, 167–169
Histidine, in polymerization process, 128–130
Histone
in DNA complex, 61–62, 66–67
proteins, effect on DNA in chromatin, 47, 61

Hydrodynamic studies, of DNA–reporter complexes, 37–39
Hydrogen belt, in membranes, 3–4, 10–12, 14–16, 18
Hydrogen bonds
of DNA, to small molecules, 33, 49, 52
in DNA–protein interactions, 61–62, 64
in membrane structures, 4, 10–18
in peptide structures, 265–267, 270–286
Hydrolysis, acidic, of nucleosides, 257–259
Hydrolytic deamination, of cytosine and cytidine, 250–252, 255, 257
Hydrophobic
amino acids, in peptide–DNA binding, 60–61
binding of DNA, to small molecules, 33–34, 44, 52
binding of membrane lipids, 1, 3–4, 6–7, 10, 12, 15–16, 18
core, in membrane lipids, 5–12, 15, 18
interactions, in DNA–protein complexes, 61–62, 67
Hydroxycerebroside, molecular model of, 5, 13–14
Hypochromism, in DNA–reporter complexes, 34–36, 42, 44–46

I

Infrared absorption, of homooligopeptides, 178, 269–71, 275, 279, 282
Inhibition, of nucleic acid syntheses, 96–97, 99–103, 115, 118
Intercalation
ethidium, into duplex nucleic acids, 96–97, 110
full vs. partial, 33–34, 39, 42, 47–48, 53–57, 62, 64–67
mode in DNA binding, 44–50, 52
model, in absorption studies of DNA, 36–37
steric requirements for, 42–43
Ionization, of Stern layers around micelles, 143–144
Isoleucine, in polymerization process, 126–131
L-Isoleucine, structure of peptides derived from, 268–270

K

Kethoxal, reaction with guanine nucleotides, 259–261

α-Ketoaldehydes, reaction with nucleic
    acids, 248, 259–261
Kinetics
    of DNA–reporter binding process, 49–50
    of micellar reactions, 146–148

L

Lac operator–lac repressor model, 61–67
Lac repressor binding, in DNA, 204–206, 218
Laurate micelles, effect on ester hydrolysis,
    141, 145–146
Leucine, in polymer fractions, 127–132
Leucine residues in peptides, conformation
    studies of, 182, 189, 196
Leukemia, chemotherapy for, 95, 101, 106,
    117–118
Life, origins of
    criteria of validity and plausibility for,
        22–25
    flowsheet, 21–22
    proteinoids in, 123–124
Ligase
    in resealing endonucleolytic cleavage, 207
    in synthesis of 21-mer duplex of DNA, 205
Linkages, *see also* Cross-linking
    of amino acids in proteinoids, 132–133
Lipids, membrane, molecular designs of,
    1–18
Lossen rearrangement, use in peptide
        C-terminal residue determination,
        80–89
Lysine
    in proteinoid production, 126, 128–130
    residues in polymer terminals, 132–134
    in thermal condensation polymers, 123
Lysinoalanine, toxic factor in foods, 124–125
Lysolipids, in membrane fusion, 6, 14

M

Membrane
    lipids, molecular designs of, 1–18
    qualities in microspheres, 28, 30
Meteorites, amino acids in, 124
Methionine
    in oligopeptides, conformation studies of,
        182
    in polymerization process, 128, 130, 134
    residues in peptides, configuration of, 182,
        189, 192, 197–199
2-Methoxy[2-ethoxy(2-ethoxy)]acetyl,

amine protecting group, 177, 180,
    188–189
Methylamine, and bisulfite procedure, 254–
    256
Methylase, modification, recognition of
    "minimal DNA," 206–208, 218
Methylation, to avoid endonucleolysis, 206,
    207
Methyl ester, carboxyl-protecting group in
    oligopeptides, 177, 181, 193, 196–198
3-Methyluridine, synthesis of, 230–231
Micelles
    kinetics of reactions, 145–147
    stereochemistry of reactions, 148–149
    structure of, 137–139
Microspheres, proteinoid, *see* Proteinoids
Mitomycin C, interaction with DNA, 96,
    99–101, 105–115, 118
Mitosanes, antitumor activity of, 105–118
Molar rotation, in oligopeptide studies,
    180–183, 186, 190
Mutagens
    for deamination studies, 246, 249–252, 254
    point of attachment to proteins, 96
Mutations, chemical, types of, 246–247, 256
Myelin membrane, hydrogen bond donors in,
    6, 13, 16–17

N

Neoplastic diseases, control of, 95, 118
Ninhydrin reaction with cytosine deriva-
    tives, 261–262
Nitroaniline, labeled binding with DNA,
    34–35, 42–45
N-terminal, amino acid residues at, in pro-
    teinoids, 123, 131–134
p-Nitrophenyl esters, hydrolysis inhibited by
    micelles, 145, 146
Nitrous acid deamination, of nucleic acids,
    246, 248–250, 253–254
Nuclear magnetic resonance
    in DNA–small molecule studies, 34, 39–49
    in DNA–peptide complex studies, 51–56,
        59–61
    in oligopeptide conformations, 183, 190–
        195, 197–198
    in micellar complexes, 143, 144, 146, 149
Nucleic acids, *see also* DNA, RNA
    acidic hydrolysis in, 257–259
    bisulfite modification, 252, 257
    hydrolytic deamination of, 250–252

nitrous acid deamination of, 249–250
  in organic chemistry, 245–246
  in origins of life, 123
  reaction with α-ketoaldehydes, 259–262
  structure and function studies, 247–249
  synthesis, inhibitions of, 96–97, 99–100
Nucleosides
  acidic hydrolysis of, 257–259
  in organic chemistry, 245–246
  structure and function studies, 247–248,
    250
  synthetic transformations of, 221–241
Nucleotide sequence
  affected by codon redundancy, 214, 219
  of DNA, recent advances in, 208–211
  of tRNA, 212–214
Nucleotides, in organic chemistry, 245–246,
    250

O

Oligonucleotides
  syntheses of, 203–219
  transamination by bisulfite, 255–256
Oligopeptides, see also specific ones by
    name
  binding with DNA, 34, 41, 47, 50–54,
    59–61, 65–66
  conformational studies of, 177–199
  parallel β structure of, 269, 285
Optical rotary dispersion
  in conformation studies of oligopeptides,
    178, 184, 196–197
  to determine critical micelle concen-
    trations, 141
Oxy analog, to peptide ring structures, 278–
    282, 287

P

Pancreatic ribonuclease, cleavage of RNA,
    247–248, 256
Parallel-chain β structures, in peptides,
    266–270, 275
Paramagnetic resonance
  of gonyautoxins, 170, 172–173
  of Q-nucleosides, 155–156
  of guanosine adducts, 158–159, 161–164
Partially relaxed Fourier transforms
  of gonyautoxins, 170
  of BP–guanosine adduct, 161–163

Parturition, reproduction in microspheres,
    28, 31
Peptides, see also individual peptides
  carboxyl-terminal sequencing and end-
    group determination, 71–92
  DNA reactions, and binding with, 51–62
  hormone angiotensin II, 217–218
  ring structures of, 270–273, 277–279, 281,
    284–286
  S, genetic instruction of, synthesized,
    215–216
  stereochemistry, 266–287
Phenanthridine trypanocide, see Ethidium
Phenylalanine
  in polymerization process, 128, 130, 132
  residues in peptides, conformation of, 182,
    189–190, 197
L-Phenylalanine, structure of peptides de-
    rived from, 268–270
Phosphatidylcholine, molecular model of, 5,
    16
Phosphatidylethanolamine, in membrane
    stabilization, 16–17
Phosphatidylinositol, molecular model of, 5,
    8–9
Phosphatidylserine, in membrane stabiliza-
    tion, 16–18
Phosphodiester bonds, cleaved by EcoRI
    endonuclease, 207–208
Phosphoglycerides, in hydrogen belt of
    membrane lipids, 11–17
Phospholipids, function in membranes, 3–18
Phosphorycholine, in membranes, 3–4,
    16–17
Photoactivity, primordial, on proteinoids,
    29–30
O-Pivaloylhydroxylamine, in C-terminal de-
    gradation, 82–89
Plasma desorption mass spectrograph
  of gonyautoxins, 170
  of Q-nucleosides, 153–155
Plasmalogens, molecular, 3, 5, 9–10, 12–13,
    15
Plasmids, affecting endonuclease activity,
    206
Pleated-sheet, structure, of peptides, 267–
    270, 275
Polyguanylic acid, reaction with DMBA-
    5,6-epoxide, 157–160
Polymerase
  catalyst in "repair" synthesis, 210–211
  RNA, in synthesis of tRNA, 213

Polynucleotides, syntheses of, 203–219
Polyuridylic acid, affected by bisulfite, 257
Porfiromycin, antitumor antibiotic, 99–100
Proline
  configuration in proteins, 286
  in meteorites, 124
  in polymerization process, 126–133
  residues, in peptide conformations, 194
Proteins
  amino acids in, 124, 126, 132–134
  carboxyl-terminal sequencing of, 71–92
  in cataract development, 164–166
  conformation studies of, 178, 181, 189,
    194, 199
  and DNA recognition, 33, 47, 50–56, 58–67
  fibrous, stereochemistry of, 268
  globular, stereochemistry of, 270, 277–
    278, 286
  interaction with drugs, 95, 96
  in membranes, 1–9, 15–18
  molecular weights and mobility of, 129
  mutations effect on, 246–247
  nucleic acid interaction, 204–206, 219, 248,
    250, 256–257
  precursor of angiotensin II, 217
  role in origins of life, 123
  "spike," relationship to gene G, 210–211
  synthesis, inhibited, 101, 118
Proteinoids
  amino acid composition of, 123–134
  branching of, 123, 133–135
  microspheres, 25–31
  molecular size of, 123–124, 130–131
Protocells, origin of, 21–31
Proton magnetic resonance, in peptide
    configuration studies, 273, 275–276,
    278, 282–283
Purine nucleosides
  mechanisms of, 258–259
  syntheses of, 222, 229–230, 233–235
Pyrimidine nucleosides
  hydrolytic deamination of, 257–259
  synthesis of, 222, 228–230, 236–241
Pyroglutamic acid, in thermal polymers, 131,
    133

Q

Q-nucleosides, structure of, 153–157
5,8-Quinolinequinone moiety, in streptonig-
    rin, 104–106

R

Racemization, in synthesis of oligopeptides,
    179–180, 183
Reporter, interactions with DNA, 34–50
Repressor binding, requirements for, 204–
    206
Reproduction, in microspheres, 28, 31
RNA
  alkylation of, 111
  chemical modification of, 245–262
  configuration of, 34, 36, 59, 209
  nucleosides isolated from, 221, 229–230,
    232
  in polypeptide genes, 210–215
  reactions with antitumor agents, 96–97
  synthesis, inhibited, 96–101, 115, 118
tRNA
  biogenesis, precursor molecule, 211–214,
    218
  source of Q-nucleosides, 153–157
Ribosomes
  formed by protein–nuclaic acid interac-
    tion, 248, 257
  protectors against nuclease degradation,
    209
Rifamycin, as antitumor antibiotic, 99–100

S

Saxitoxin, compared to gonyautoxins, 169–
    170, 172, 174
Scission, of DNA by antibiotics, 96, 98, 101,
    115
Serine
  in C-terminal peptide sequencing, 85
  in hydrogen belts of membranes, 3, 15, 17
  in proteinoids, 130
Sickle-cell disease, antisickling agents,
    166–168
Sodium bisulfite, see also Bisulfite modifica-
    tion of nucleic acids
  deamination by, 252, 255
Sodium borohydride, in DNA reactions with
    mitomycin C, 110, 112–113, 115
Sodium dodecyl sulfate, effect on ionization
    of micelles, 143, 150
Solid-state conformations, of peptide struc-
    tures, 265, 269–285
Sphingolipids, function in membranes, 2–5,
    11–15

Sphingomyelin, molecular model of, 5, 16
Sphingosine, molecular bonding in, 9–10, 15
"Spike" protein, N-terminus, amino acid sequence of, 210
Spin lattice, relaxation times of micelles, 144, 149
Sporulation, reproduction in microspheres, 28, 31
Stannous chloride, use in nucleoside syntheses, 229–231
Stereochemistry
    of micellar reactions, 148–151
    of peptides, 266–287
Stern layer, around micelles, 138, 142–149
Streptonigrin, as antitumor antibiotic, 99–106, 114
Sugar
    derivatives in synthesis of nucleosides, 221–223, 227, 229
    residues in tRNA base, 157
    transformation in nucleosides, 233–238, 241
Sulfur dioxide, air pollutant, 252, 256–257
Superoxide dismutase, role in streptonigrin reactions, 103, 115
Surfactants, in formation of micelles, 137–151
Symmetry
    DNA, in repressor protein recognition, 204–205
    in EcoRI restriction endonuclease, 206
    in tyrosine tRNA, 211, 214
Synthesis
    of a cell, as constructionism, 21–22, 30-31
    of DNA fragments, 203–219
    of modified nucleosides, 221–241

T

Tetrahydrofuran, in pyrimidine nucleoside transformation, 221, 241
Thermal condensation, of amino acid mixtures to polymers, 123–135
Thermal peptides
    molecular weight and mobility of, 129
    terminal group composition of, 131–134
Thin-layer chromatography, in nucleoside synthese, 221, 229
Tobacco mosaic virus, protein–nucleic acid interactions in, 250, 254

Torsion
    angles of polypeptide structures, 265–266, 271, 273, 278
    rotation, 274–275
Transamination, of nucleic acid components, 253, 255–256
Transformations, synthetic, of nucleosides, 221–241
Trenimon, as antitumor antibiotic, 99–100
Triazine, procedure in terminal labeling, 78–79
Trifluoroacetic acid, in oligopeptide conformation studies, 177, 181, 192–193, 196–197
Trifluoroethanol, in oligopeptide conformation studies, 178, 183–197
Trifluoromethyl hypofluorite, in synthesis of 5-fluoropyrimidines, 224–225
$N^6,N^6,O^{2'}$-Trimethyladenosine, synthesis of, 232, 233
Tritiation, Matsuo, in C-terminal identification, 77–79
Tryptophan, in C-terminal peptide sequencing, 85
    in oligopeptide conformation, 190
    in proteins from cataractous lenses, 165–166
Tubercidin, transformation of, 229–230, 233, 237–241
Tyrosine
    in C-terminal peptide sequencing, 85
    in oligopeptide conformation, 190
    in polymerization process, 128, 130
    tRNA gene, synthesis of, 211–214

U

Ultraviolet, in oligopeptide conformations, 178, 183–186, 190, 192, 196–197
Unsaturation, in membranes, 7–11
Uracil
    conversion by nitrous acid, 249–250
    nucleosides
        acidic hydrolysis on, 258–259
        synthesis of, 236–237
        reaction with bisulfite, 256–257
Urethane carbonyl, in H bonding of peptides, 279–281, 287
Urethanes
    carcinostatic group, in mitosanes, 106
    cis configuration in, 284–285

V

Valine
  in meteorites, 124
  residues in peptides, conformation of, 182,
      189, 197
  role in polymerization process, 126–130
L-Valine, structure of peptides derived from,
      268–270
Viruses
  formed by protein–nucleic acid interac-
      tions, 248
  tobacco mosaic, 250, 254
Viscosity studies
  in DNA–peptide complexes, 51–57

  of DNA–reporter complexes, 37–39
  in intercalation, 42, 44, 47–49

W

Water holes, in membranes, 8–9, 16

X

Xanthine, conversion by nitrous acid, 249
Xanthoxylol, antisickling agent, 166–167
X-ray diffraction, for determining peptide
      conformations, 269–271, 273, 275,
      277–285